Praise for
Kevin Dockery's Navy SEALs Series

Navy SEALs: A History of the Early Years
Navy SEALs: A History: The Vietnam Years
Navy SEALs: A History: Post-Vietnam to the Present

"A collection of oral histories knitted together with carefully researched narrative . . . traces the sea, air, and land force's evolution from its WWII precursors to its current highly trained, high-tech incarnations."

—*Publishers Weekly*

"[An] oral history of the teams' harrowing combat duties in Vietnam. Loosely organized around the chronology of the conflict in Southeast Asia, the book is, at its heart, a tribute to the collective spirit exhibited here. A commitment to teamwork and excellence echoes throughout."

—*Kirkus Reviews*

"Dockery blends oral history and conventional narrative with consummate skill."

—*Booklist*

STALKERS AND SHOOTERS

A HISTORY OF SNIPERS

KEVIN DOCKERY

BERKLEY CALIBER, NEW YORK

THE BERKLEY PUBLISHING GROUP
Published by the Penguin Group
Penguin Group (USA) Inc.
375 Hudson Street, New York, New York 10014, USA
Penguin Group (Canada), 90 Eglinton Avenue East, Suite 700, Toronto, Ontario M4P 2Y3, Canada
(a division of Pearson Penguin Canada Inc.)
Penguin Books Ltd., 80 Strand, London WC2R 0RL, England
Penguin Group Ireland, 25 St. Stephen's Green, Dublin 2, Ireland (a division of Penguin Books Ltd.)
Penguin Group (Australia), 250 Camberwell Road, Camberwell, Victoria 3124, Australia
(a division of Pearson Australia Group Pty. Ltd.)
Penguin Books India Pvt. Ltd., 11 Community Centre, Panchsheel Park, New Delhi—110 017, India
Penguin Group (NZ), 67 Apollo Drive, Mairangi Bay, Auckland 1311, New Zealand
(a division of Pearson New Zealand Ltd.)
Penguin Books (South Africa) (Pty.) Ltd., 24 Sturdee Avenue, Rosebank, Johannesburg 2196,
South Africa

Penguin Books Ltd., Registered Offices: 80 Strand, London WC2R 0RL, England

Cover design by Steven Ferlauto
Cover photograph courtesy of Leif Skoogfors/Corbis
Book design by Tiffany Estreicher

PRINTING HISTORY
Berkley Caliber hardcover edition / December 2006
Berkley Caliber trade paperback edition / July 2007

Berkeley Caliber trade paperback ISBN: 978-0-425-21542-5

The Library of Congress has catalogued the Berkley Caliber hardcover edition as follows:

Dockery, Kevin.
Stalkers and shooters: a history of snipers / by Kevin Dockery
p. cm.
Includes bibliographical references and index.
ISBN 0-425-21002-2 (alk. paper)
1. Sniping—History. I. Title.

UD330.D63 2006
356'.162—dc22 2006043034

PRINTED IN THE UNITED STATES OF AMERICA

10 9 8 7 6 5 4 3 2 1

To the personnel of the military and law enforcement sniper community—those few who lie behind the sights and shoot straight—this book is respectfully dedicated.

Particular thanks are given to the organizations that support our snipers, especially americansnipers.org, who act to help those on the front lines of the war against terrorism.

CONTENTS

PROLOGUE TO A MISSION

BIRTH OF THE SNIPER

THE WAR ON TERROR OPENS

LAW ENFORCEMENT

THE SHOOTERS—LAW ENFORCEMENT

THE SHOOTERS—MILITARY

PROLOGUE TO A MISSION

Sniper! One shot, one kill. *USAF*

1

TO BEGIN—THE STORY OF A WORD

There is a quick little bird, well known in England, which is a hard challenge to hunters. The difficulty is not because the bird is dangerous in any way, but because it is so very hard to find in its native habitat. Slightly larger than a common sparrow, this bird will lie hidden on the ground cover of a marshy area, completely confident in its natural camouflage. Once it is located and goaded into flight, the sudden, erratic burst of whirring feathers and wings makes the bird hard to hit, even with today's best shotguns and ammunition.

A few centuries ago, hunting this agile little game bird with a flintlock was a tremendous challenge. The hunter who tried to hit this mottled brown target in its darting flight had to account for the delay in time between when he pulled the trigger and when the spark and powder flash actually fired his weapon. He had to not only aim along the flight path of the bird, but also decide where it would be when the shot finally reached it.

It took a master hunter to bag this elusive game. The popular hunting of this bird throughout Great Britain soon caused its name to become a term for only the most accomplished of shooters. The bird was of the family *Scolopacidae*, of which there are ninety species worldwide. These shorebirds, as they are known in North America, include sandpipers, woodcock, curlews, and snipe. In the United States, one of the most common game birds of this family is the woodcock. In England, the prime game bird of this group is the snipe. By the late 1700s, to hunt the snipe was to go out sniping, reduced at around the middle of the century from the term "snipe shooting." And the successful hunter of this minuscule feathered missile would be proud to be referred to as a sniper by his fellows.

The term quickly took on an application among the military, many of the officers of which were sportsmen and hunters themselves. In *The History of the*

British Army, a monumental work in twenty volumes by the Honorable J. W. Fortesque, he quotes one of the earliest military uses of the term. In a "Letter from India," contained in Volume III, the sniper term is used in a mocking manner: "[The soldiers put their hats on the parapet for the enemy to shoot at] and humorously called it sniping."

In less than fifty years, the humor had completely gone from the application of the term. In 1824, a British military report from East India said in all seriousness: "Several sepoys were killed and wounded by the enemy snipers." But for military shooters, the terms "snipers" and "sniping" were not used. In England and elsewhere, the term "riflemen" indicated a group of soldiers noted for their marksmanship. For the individual, the common terms also used were "marksman" or "sharpshooter." By the end of the 1800s, "sniper" was back in somewhat common use in British newspapers reporting on the situation in South Africa and the 1899–1902 Boer War. Boer settlers were farmers who had spent their adult lives with a weapon in their hands. Using a rifle came as second nature to them, and the British Army paid dearly when they failed to take that into account. The long-distance marksmanship of the Boer snipers helped account for many of the 7,582 battle casualties of that African conflict. And 10 percent of those British casualties were officers. Taking out the leadership of the other side is a common mission today for military snipers.

"Marksman" was a term in common use for much of the world by the 1800s. It had been used as a descriptive term since before the gunpowder age. Crossbow shooters and archers could be called marksmen, as they could hit their mark. The term simply means a shooter who is skilled at striking the target when he aims at it. The word is still in use today primarily as a classification of military shooter. Marksman is the lowest rating a soldier can hold and still be considered qualified. "Marksmanship" is a derived word used to explain the art or skill of firing at a target. Marksmanship requires practice to maintain a dependable level of skill that makes a soldier more valuable on the battlefield.

During the mid-1800s, a civilian marksman, an excellent shooter who was skilled well above average, was referred to as a "crack shot." In the military world, the term for this kind of rifleman was "sharpshooter." Derived from the translation of the German word *scharfschütze*, the term had the particular meaning "a soldier or hunter who is very accurate in the placement of his shots." It was in the mid-1800s, during the beginnings of the American Civil War, that "sharpshooter" became an inextricable part of the military lexicon. With Hiram Berdan's establishment of the U.S. Sharpshooters in 1861, the skills and abilities that would become the modern sniper were added to the U.S. Army. After that war, and when the public had seen what sharpshooters could do in battle, highly skilled marksmen both in the military and elsewhere were referred to as sharpshooters. Today, it is the second highest rating for marksmanship that can be held by a soldier in the U.S. military.

In the second decade of the 1900s, a new type of war broke out in Europe, a wide-ranging conflict that brought most of the world into the engagement. In the first year of what would become World War I, the fighting soon stalemated into trench warfare. In 1914, the British and allied forces faced the German Army over yards of churned-up dirt, mud, shell holes, and barbed wire. The introduction of the machine gun quickly put an end to the traditional charge of one force of soldiers against the other. In the wet muck of the trenches, the Germans brought back their tradition of the *scharfschütze*; the Jägers (hunters) were skilled soldiers with a specific mission to perform—to hunt men.

The British newspapers during World War I brought back an old term, one remembered by writers who had covered and British officers who had served during the Boer War. Men in the trenches were facing "snipers," and the term has stayed in the military lexicon ever since. In the peacetime lull that followed World War I, the sniper and his employment were forgotten by many of the world's militaries, only to be remembered and the mission resurrected as new conflicts arose.

Since the Second World War, there has always been a sniper contingent within the U.S. military. Sometimes this contingent was little more than a very small group of competitive marksmen who worked to maintain the shooting skills that made a sniper so deadly. During times of conflict, formal schools of sniper training were formed. Missions were created for the graduates of these schools. Sometimes, the snipers were badly utilized by commanders who didn't know the best way to employ such specialists. Other times, the snipers were given their lead to go out and make their presence known to the enemy. In the latter half of the twentieth century, the sniper has become an integral, particularly valuable part of the U.S. military.

In the present age of warfare, there is a constant demand for new weapons, precision weapons, and smart weapons. The desire for such hardware comes from the need to minimize, if not eliminate, civilian casualties. The term "collateral damage" is considered unacceptable to the public at large. To simply blow down a building to eliminate a single enemy combatant is an efficient military tactic; but not a politically correct one. Even in the civilian world, the use of force is something to be minimized. Lethal force is the action of last resort, only to be applied when there is an imminent risk to innocent lives.

Increased precision is desired for a multitude of reasons, even as simple a one as preventing waste. For the military, this desire resulted in the weapons popularly termed "smart" being developed. Smart weapons are self-guided to their targets, aiming themselves and correcting their flight path along the way. Ordnance such as this, normally in the form of missiles or bombs, can be launched from hundreds of yards, or miles, away from the final target. Or they can be dropped from aircraft thousands of feet in the air. The smart aspect of these almost exclusively explosive devices is that they can target a particular building, even a specific opening, prior to impact.

Such accuracy in the application of violence has become so commonplace that it is expected, even required, by the public in today's conflicts. Using hundreds of aircraft dropping thousands of bombs to eliminate a single priority target, destroying all of the surrounding countryside in the process, is not even considered in present conflicts. This is a new way of conducting war, but there has been one consistently successful precision weapon on the battlefield for hundreds of years. There is a particular weapon system that stands orders of magnitude above the efficiency of the cutting-edge designs of today. That weapon system is the sniper and his rifle, a trained, experienced, expert shooter armed with a select, close-tolerance firearm with an advanced optical sighting system. He is capable of selecting only the most high-value targets in a crowded battle zone, and destroying them.

The sniper is a master of his craft; he has to be in order to survive on the battlefield. One of the greatest fears of the average soldier in combat is the sniper. Troops will go to great lengths to obliterate an area where a sniper is known or even suspected. If a sniper is spotted, he can expect the full firepower of the enemy to come in on him. If he lives and is captured, a sniper can expect little mercy at best from the hands of the soldiers so in fear of him. The greatest enemy of the sniper is another sniper, someone on the other side with the same skills of observation, marksmanship, and camouflage. Over the years of warfare, there have been legends told about master snipers stalking each other across the battlefield. The normal outcome of such stories is that only one man lives.

But the sniper is far more than just a precision instrument for destruction. He is accomplished at observation. Skilled in communications and concealment, a sniper can be the eyes of a commander well forward of the lines of conflict. Overlooking a meeting, a sniper can call in fire when the targets of highest value have gathered. He can direct artillery and aircraft strikes with as much precision as he can fire his weapon. Or he can withdraw unseen to report and prevent the action an enemy might have been planning. Above all, a sniper can pick out that one target that could mean the most in a conflict, eliminate that leader, destroy that piece of equipment, and change the outcome of a major event.

For over a hundred years, the term "sniper" has meant a highly trained soldier, a skilled military shooter who can pick off single enemy combatants from a place of concealment. From the intense dislike a regular soldier has for an enemy sniper, that shot from the dark that can take out even the most prepared trooper, the modern media has corrupted the term "sniper" into something that it is not and was never intended to be.

Since the 1960s, the term "sniper" has been used for anything but its original intent of "an expert marksman—military or civilian." This has changed the meaning of the term, making it have only the most negative connotations for a misinformed public.

The random, inaccurate shots of a rifleman toward a crowd of townspeople in a conflict zone have been called "sniping." If the shooter is someone who actually has some marksmanship skills, the terror he spreads through an area as he slaughters usually unarmed civilians is even worse. The actions of criminals, just killers armed with a long gun, have been mislabeled by the media as "sniper attacks." These are not snipers and never have been, but the term has been spread out to include them and blur the line between the professional shooter and the fringe psychopath. Individuals employing the skills or marksmanship of a trained sniper have been thankfully rare in the criminal ranks. But there is a true sniper in the civilian world, an operator with every reason to lay claim to the title. That shooter is the law enforcement precision marksman.

The negative connotation that has been added to the word "sniper" has increasingly forced law enforcement shooters to use other terms to describe their jobs. "Tactical marksman," "counter-sniper," "sharpshooter," "precision shooter," and "precision marksman" are all less emotionally loaded terms that are used to describe a police sniper. No matter what the name, the assignment is a difficult and demanding one.

Though rarely facing the same level of physical danger that a military sniper could expect—few criminals can call in artillery strikes or concentrated fire—the law enforcement precision marksman faces his own unique threats. If a military sniper misses his shot, he may be able to follow up with another round. Or he may have to use all of his considerable skills in stalking and concealment to extract from a combat zone, slip away, and return another time.

The law enforcement precision marksman does not have those options. If he misses a shot, his fellow officers may pay the price of his error with their lives. The nightmare possibility of a missed shot could also mean the death of a hostage, perhaps even a child. These are some of the reasons a police precision marksman practices regularly and intensely. Normally, the selection process for such men only takes the most stable, steady, experienced officers with years on the job. And there are more dangers than just the ones that could come about from a missed shot; there's also danger in hitting the target.

Tremendous legal pressure is brought to bear on the police marksman every time he is called out to action. Liability follows the path of the bullet; either a hit or a miss can affect a career. Even if the shooting is an unquestionably justified one, the marksman faces the very real possibility of a lawsuit. An uneducated public, brought up on the information from popular television shows and movies, can demand to know why an officer could not have just shot to wound, or just shot the weapon out of a perpetrator's control. Such shots are rarely even possible, and almost never advisable. A speeding projectile striking the hardened steel of a criminal's firearm could careen off into the distance, doing unknown damage downrange. A wounded criminal can still pull a trigger, can still shove a knife, and can still fire a bomb.

The ultimate stress of the police marksman's job is that he can expect to be put in the position of being the final arbiter of the law. If the situation is such that lives are endangered—law enforcement personnel, hostages, or just bystanders—the police marksman may have to make that split-second decision to fire. And he must place his bullet to end the situation with finality. Even if given the green light by higher authority to use lethal force in a situation, the final decision to pull the trigger and send the bullet on its way belongs to the marksman and to no one else.

The true sniper is an absolute professional, committed to his job. He is chosen from volunteers among the best in the ranks, and even then he must pass grueling testing, undergo demanding training, and absorb detailed knowledge. In the military, volunteers are qualified as expert marksmen, the highest ranking of that skill, before they even report for training. Once they have completed their course, they are masters of their weapons, far above expert in their marksmanship. In spite of only superior troops being accepted for such training, many fail to complete it. It is only after he has completed all of this work successfully that an individual can proudly accept the term "sniper" applied to him.

And the path of this hard-won label, a dedicated job description, can be traced back to a little brown bird.

2

TECHNOLOGY OF THE ERA PREHISTORY TO 1200

The mission of the sniper has always been dependent on the development and application of the technology of the era. The nature of the job has required the primary tool of the sniper, the sharpshooter, or the marksman to be a projectile weapon. This class of weapon sends an object out against a target normally much farther away from the operator than could be reached by a hand-thrown missile. This is a very elementary fact often overlooked in the history of what has become the modern sniper.

Probably the oldest universally recognizable projectile weapon is the bow and its missile, the arrow. The bow stave is the body of the weapon; bent and held in place with a string, it stores the energy put into it by the archer's arm and, by straightening, drives the arrow forward and downrange. Being made of organic materials—wood, horn, ivory, sinew—bow staves have tended not to survive as intact artifacts over the centuries. The part of the bow and arrow that has most survived over time is the arrowhead. Specimens of arrowheads have been unearthed in archeological digs in Bir-El-Ater in Tunisia and other North African sites dating from fifty thousand years ago. In the area of what was to become Hamburg, Germany, fragments of bows have been unearthed dating from twenty thousand to seventy-five hundred years in the past.

As a military weapon in Europe, the bow reached its zenith of development and application in the form of the longbow in the hands of the English archer. Bows had been used in England for centuries, but it was under the direction of King Edward I (1272–1307) that they were officially adopted for military use.

Because the English had suffered under the accurate and long-range fire of Welsh archers, it was an obvious choice to bring in the longbow and incorporate it fully into the tactics and training of the English military.

Though the Welsh were the archers who "introduced" the English to the effectiveness of the longbow in warfare, the weapon probably didn't originate there. The Saxon, Welsh, and later English longbows all share common characteristics. They are long, from five-and-a-half to six feet—the length of the archer's outstretched arms; D-shaped in cross-section; and made of wood, generally yew. These bows launched an arrow roughly twenty-seven inches long, about a third the length of the bow, commonly called a "cloth-yard" shaft. The length and thickness of the bow, combined with the material it was made of and its basic design, gave the English longbow a pull (draw weight) of between 100 and 175 pounds. Such power stored in the wood of the bow would launch an arrow to an effective range of 200 to 350 yards. The larger bows, those with a pull of 160 to 175 pounds, had a maximum effective range of 320 to 350 yards, but it would take an exceptional archer to utilize such a weapon.

The English system of developing users of the longbow helped create exceptional archers. Traditionally, young men were taken at the age of seven and instructed in the use of the bow. Training was mandatory and practice extensive. The English were the only European military to really give the weapon and its user the high degree of training required for dependable, effective use on the field of battle. To be an English longbow man was a dedicated way of life. Local tournaments helped determine the best archers, and these were in the forefront of the English military units. The average bowman in the English military could launch ten to fifteen arrow flights a minute. The accuracy of the archers could put all of those flights into a man-sized target at two hundred yards. The power of the longbow and it's "cloth-yard" shaft could drive a steel point through an inch of oak at two hundred yards. At close range, the arrow could penetrate nearly four inches of oak. That meant that even the armor plate of a knight did not protect against an English archer at less than a hundred yards.

The medieval longbow of England gave its army an extremely powerful and flexible weapon to command the battlefield. Heavy flights of arrows coming in at a rapid rate could literally pin an enemy force to the ground. But the bow had its drawbacks for use as a sniper's weapon. Even though the English archers had a long history of effective hunting with the longbow, it was extremely difficult to use while concealed. Holding the bow back for a length of time at full draw, whether to the chin or under the ear, was a strain for even a strong man. And the bow had to be used while the archer was standing; the length of the longbow prevented anything else.

The alternative to the longbow throughout much of medieval Europe was the crossbow, also known as the arbalest. Familiar to the population at large as a hunting weapon, the crossbow had been in use for over four hundred years before really coming into its own with the military. By the 1100s, most European armies had adopted some version of the crossbow, bringing sights and mechanical triggers into widespread military use. The weapon was so popular that it retarded the development and further adoption of the longbow in many countries. The weapon itself is simple; a bow (the prod) is attached across the end of a stock (the tiller). The bowstring is pulled back (spanned) and held by a catch (the nut) which releases the bowstring when a lever on the bottom of the tiller is pressed in. The ammunition (bolt) lies in a groove cut into the top of the tiller, this groove holding the bolt straight and guiding it when the nut is released. The basic design had been around for centuries, having been used by the Romans as a form of hand ballista, and by the Chinese hundreds of years before that.

The design of the crossbow calls for a very heavy pull prod, much heavier than a normal bow would be. This is because the energy of the straightening prod can only act on the bolt for a short distance. In a standard bow, the arrow is driven by the bowstring for more than double the distance found on the longest handheld crossbows. Close in, the crossbow was a devastating weapon that could send a bolt smashing through all but the heaviest armor a mounted knight could wear.

The greatest advantage of the crossbow lay in its simplicity of use; any commoner could use the weapon after relatively little instruction. Men skilled in its use could use the crossbow to dominate a battlefield. By the middle ages, the crossbow had become the leading handheld projectile weapon in the European military. Mercenary groups of trained crossbowmen, notably Italians from Genoa, hired out to many European leaders, bringing with them a recognized capability for destruction.

The shortened arrow launched by the crossbow is the quarrel or bolt. It is a wooden shaft roughly between one-half and three-quarters of an inch thick and between ten and twelve inches long. At the butt end of the bolt are usually three vanes making up the fletching. The stiff vanes are made of metal (brass), wood, or even parchment rather than the more common feathers as used on an arrow. A war bolt usually has a pointed heavy iron tip, square, pyramidal, or conical in cross-section. The strong metal tip mounting a stout shaft gives a war bolt an armor-piercing capability. By the twelfth century, such a bolt loosed from a powerful crossbow could penetrate a shield with more than enough velocity left to continue on through chain mail of armor and into the body of a target. The development and more common use of plate armor simply forced the creation and use of even more powerful crossbows which continued to penetrate the armor of mounted knights.

The crossbow was the first weapon to see widespread use that put a foot soldier at an advantage to an armored knight. As the antitank weapon of its day, the crossbow allowed a man-at-arms or, considered even worse, a common soldier, to take an armored knight down from his horse. The range of the crossbow over that of the handheld weapons of the knight meant that the armored man would never have a chance even to reach the crossbowman who had killed him.

In 1139, under the direction of Pope Innocent II, the Second Lateran Council stated in their twenty-ninth decree that the crossbow was an anathema, a detestable weapon unfit for combat among Christians. The papal edict condemned the weapon and forbade its use in warfare. It was the first record of an extensive attempt at weapons control in Europe and the beginning of the bias against a sharpshooter in the field. The Catholic Church wanted to ban the ability to accurately place a single shot—a shot that could possibly have a battlefield effect completely out of proportion to the relative importance of the shooter. A low-ranking conscript armed with a crossbow had the possibility of intentionally eliminating a noble, or even a king. The Vatican was not going to allow anyone as unimportant as a low-born soldier or vassal to possess the means to be an effective one-on-one threat to the nobility, who also happened to be the patrons of the Church.

The decree had little practical effect in removing the crossbow from use by Europe's militaries. Even the renewal of the papal decree by Pope Innocent III (1198–1216) was mostly ignored. By the thirteenth century, the crossbow and men trained in its use were an important part of much of the general military planning of the day. The weapon continued in its development and evolution in spite of the papal anathema. More sophisticated and complex prod materials gave increasing pull weights. The bowlike prods were initially made of wood and could be easily spanned (the bowstring drawn back to the nut and held, cocked for loading and use). More complex prods were structures of wood, bone, sinew or tendon, and horn, all bound together with glues. Finally, steel prods were developed to a dependable stage. Pull weights of the prods increased from a few hundred pounds at most to nearly a thousand pounds near the end of the era.

With the increasing power of the crossbow came more and more complicated ways of spanning the heavy bowstring. At first, crossbows could be spanned by the crossbowman simply placing the butt end of the tiller against his stomach or chest and drawing the bowstring back to the nut with only the strength of his arms. A foot-stirrup in front of the prod allowed the crossbow to be spanned with one or both feet bracing the weapon against the ground. When pull weights had advanced to the point that hand strength alone was no longer enough to span the bowstring, a hook was attached to a waist belt. The belt-claw arrangement worked by having the ends of the claw placed over the bowstring and the operator push the crossbow away with his foot in the stirrup in front of the prod. Alternately, the crossbowman could bend at the waist, hook onto the bowstring, hold the stirrup down, and just straighten up, using the back and leg muscles to span the weapon.

Finally, military crossbows became so powerful that they had to be spanned with a complicated assembly of cords, pulleys, hooks, and handles known as a windlass. With such arrangements, crossbows with draw weights of seven to eight hundred pounds and more could be easily spanned by one man. The rate of fire for such powerful crossbows was low, dropping from the four or more bolts per minute for hand- or foot-stirrup crossbows to one bolt a minute or less for weapons requiring a windlass to prepare them for use.

The crossbow became an important weapon for sieges and ground combat. The crossbowman could remain under cover for an extended length of time with his weapon cocked and ready to fire the moment a worthwhile target presented itself. That gave him a tremendous advantage over the bowman, as the crossbow could remain spanned with a bolt in position for launching for far longer than even the strongest man could hold a longbow at full draw. The crossbow could be used from cover, whether by a hunter stalking game or a soldier waiting in ambush.

3

ANATOMY OF A SHOT 1066 TO 1307

1066

The first century of the second millennium saw major changes in how warfare was to be conducted in the Middle Ages. Centuries of conflict between the Saxons in England and the Normans in northern France were to come to a conclusion in a final epic battle that would decide the fate of a country. Both the Normans and the Saxons were descendants of the Danes who had raided up and down the coastlines of Britain and Europe for hundreds of years. The Normans had settled in northern France, where they gradually absorbed much of the culture. The Saxons had spread throughout much of England and established their own line of nobility there.

In January 1066, Edward the Confessor, the Anglo-Saxon king of England, died without leaving a direct blood successor. Harold Godwinson, an Anglo-Saxon nobleman, had been recommended by Edward to be his successor, and he was crowned the king of England on January 6, 1066. The English Army at that time consisted primarily of infantry units who fought with handheld weapons from behind shields, interlocked together to form a shield wall. The weapons of the soldiers included swords, spears, great two-handed battle-axes, and slings that threw lethal projectiles over reasonable distances. These tactics and weaponry had worked well for combat since the times of the Romans and served Harold II well. During the summer and into the fall of 1066, his army fought against invading ships and men from Norway, led by King Harold III Hardraade, who also considered himself qualified to hold the throne of England. The Norse foes were defeated by Harold and his men on September 25.

Duke William of Normandy considered himself to be the proper successor to the English throne and at the head of a Norman army he landed in England on September 28, 1066. The Norman Army was one that would be recognized and long connected with medieval England. The main attacking force William had available to him was made up of armored knights on horseback, armed with shields, swords, and lances. The Norman knights had trained for battle in tournaments across France and were more than capable with their weapons and mounts. In addition, William's army numbered among its ranks both crossbowmen and archers, something the English forces were completely lacking.

The cavalry of the Normans were not the heavy full-plate armored knights of later years in England. Norman knights rode unarmored horses while the men themselves wore long coats of chain mail. In addition, the knights wore helmets and held long, kite-shaped shields. Their primary weapon was the lance, or long spear, but as well they carried maces and swords. The tactic of the knights was to charge their opponents directly, standing in their saddle stirrups and leaning forward with their lances held tucked in under their arms.

On October 14, 1066, the final contact between the Norman forces under William and the English gathered under the banner of King Harold II took place near Hastings in Sussex. The story of the Norman invasion and the final battle is recorded in a unique form, stitched into scenes on a cloth known as the Bayeux Tapestry. On the twenty-inch-wide, 230-foot-long cloth are images of the chain-mail-armored knights along with their shields and long spears fighting the troops of King Harold II.

The battle raged throughout that fall day until late in the afternoon when a Norman arrow struck King Harold in the eye. The stories of the battle, and the information shown on the Bayeux Tapestry, are lacking in detail of the exact circumstances of King Harold's injury. The Normans under William were supposedly running low on arrows for their bows. Because the English did not use such weapons, the Normans were unable to pick up arrows from the battlefield to help maintain their stocks. Running low on ammunition tends to force an archer to pick and choose his targets more carefully. It may have been a marksman's aimed shot that struck King Harold; the angle of the arrow held by the Harold figure on the Bayeux Tapestry reinforces this supposition. The final result of the battle was that the English forces fell and King William I of Normandy was crowned the king of England on Christmas Day, 1066. Now Norman rule would extend across England.

The Battle of Hastings was the last time that a foreign army successfully invaded and conquered England. The single shot of an unknown archer significantly changed the course of English history from that moment forward.

1199

The Battle of Hastings was not the only time that a marksman's shot had an effect that echoed down through history. And it was most likely a chance viewing of a high-value target by someone able to take advantage of the situation, rather than any form of planned event. There is also the very real possibility that it was simply a chance event of the battlefield, a situation of someone being in the wrong place at the wrong time in a lethal environment. But this latter situation, though common enough in combat, was certainly not always the case.

As a historical figure, King Richard I was a larger-than-life individual who strode across the story of England and the Crusades. A great deal of myth and legend has grown around the life of King Richard I, also known as Coeur-de-Lion—Richard the Lionheart. Even by today's standards King Richard would have stood an imposing figure; he was physically fit from his extensive military activities, blond-haired, blue-eyed, and an estimated six feet four inches (1.93 meters) tall.

Though not as politically astute as other leaders of his age, King Richard I was an accomplished fighting man and an educated military commander. He had demonstrated his military abilities from an early age and was noted for both his personal courage and chivalry. Fighting against rebellious nobles on his lands as a teenager, Richard quickly developed a taste for warfare and combat. Raised in France, Richard considered that country home for him and spent relatively little time in England during his life, in spite of being crowned king of England on July 6, 1189. During his ten-year reign, Richard spent only six months of them on English soil.

The weight of the crown changed Richard into someone more concerned with spiritual needs, as well as an individual who wanted adventure and fame. He had sworn an oath to renounce his past ways of lust and violence. Instead he would raise an army to liberate the Holy Land for the Christian Church.

Squeezing every coin that he could from the English treasury and nobility, Richard I left for the Holy Land, striking out on the Third Crusade as called for by Pope Gregory VIII and his successor, Clement III. It had been decided that the leaders of the Third Crusade would be Philip II Augustus of France, the German king and holy Roman emperor Frederick I Barbarossa, and King Richard I of England.

It is interesting that the popes would consider Richard I as a leader for the Third Crusade. He had ignored the papal injunction of Pope Innocent II and incorporated crossbows as an intrinsic part of his military forces. Having hunted with the crossbow, King Richard I was very familiar with the power and accuracy of the weapon and was not going to deprive his military of those advantages. The crossbows that King Richard I preferred for his army were the most powerful of the time, capable of defeating any portable armor of the era. Since the crossbows were going to be used against Muslim foes rather that Christians, the popes excused King Richard's ignoring of their edict.

As a more than competent military leader, King Richard I did not simply employ crossbows during the crusades, he developed innovative ways to employ the weapon. One of King Richard's methods was to field crossbowmen in pairs. At Jaffa on the shores of the Mediterranean Sea, Richard had his crossbow teams operate with one man aiming and loosing bolts while the other spanned the crossbow with a belt hook. The two men working in concert removed one of the most limiting factors of the crossbow, its slow rate of fire. In fact, the two men could put out more flights over time than even an accomplished archer would have been able to. The archer would have had his arm give out over time, while the two crossbowmen could simply exchange places and remain in operation.

The final death of King Richard the Lionheart had little to do with his military conquests and everything to do with his constant need for funds for his adventures. In March 1199 King Richard had laid siege to the castle of Chalus-Charbrol in Limosin, France. The skirmish and siege were the result of a claim King Richard had made on a treasure trove supposedly known to the viscount of Limoges. Having chased the viscount to the castle, Richard set about directing the siege that he thought would result in his receiving a share of the treasure.

One of the two knights defending the castle was Bertram de Gurdon, a young man skilled with weapons and bearing a grievance against King Richard. On March 26, while Richard was moving across the grounds near the castle, he was struck in the shoulder by a bolt loosed from a crossbow up on the wall of Chalus-Charbrol. Either due to the poor treatment he received from his own surgeons, or possibly from his own disdain of treatment for injuries, the wound became infected and gangrene soon set in. While Richard was deteriorating, the castle was taken and the siege lifted.

On his sickbed Richard asked to see the man who would be his killer, and young de Gurdon was brought in to him. The shot that had struck Richard had been fired from some concealment up on the castle wall, and had taken some significant marksmanship, unless it had simply been a fluke bit of luck, bad luck for Richard. The dying monarch asked the young knight what his grievance with Richard could have been.

"You killed my father and brother," de Gurdon is reported to have said, "and I hope I have killed you."

This was an answer that Richard could completely understand. Instead of showing any hatred or resentment against the young knight and marksman, King Richard ordered his release, and that he be awarded a sum of money for his act of marksmanship.

On April 6, 1199, King Richard I of England died of his wounds. Rather than release the young knight who had struck down the regent, Richard's followers had de Gurdon flayed alive and then hanged.

Many in the Church that had directed Richard Lionheart to take up the Third Crusade considered his death by crossbow bolt to be a fitting punishment for a man who had ignored the edict against the weapon by Pope Innocent II. The action could easily be considered the first loss of an English leader to a sharpshooter's skill. There was no question that de Gurdon intended to strike King Richard. And the monarch cut an imposing, and unmistakable, figure on the battlefield. Standing with a crossbow spanned and ready, waiting for his target to show himself, put de Gurdon in the same mission profile as a modern sniper. And the reaction of King Richard's followers also showed very much the mercy that could be expected by such marksmen.

1307

A famous act of marksmanship in the Middle Ages is well known throughout the world today, even though it probably never happened. William Tell is a fictional character—even though there was a concerted effort to try and prove him a real person—but his story is deeply part of the culture of Switzerland. It is notable that he was said to be a chamois hunter. The small mountain antelope required a rugged man to be able to stalk it across the mountains, and a steady, accurate shot to bring it down. Tell was said to be both of these—a powerful, strong man and an expert marksman.

A bailiff, a form of area governor, had been assigned by the Hapsburg emperors of Vienna, Austria, to control the area of Uri, in what was to become Switzerland. The Austrians were oppressing the people of the area while extending their control of the trans-Alpine trade routes. The bailiff, Hermann Gessler, had been sent to subjugate and control the people of Uri and he did this with a heavy hand.

Raising a pole in the center of the town of Altdorf, Gessler placed his hat on the top of it and ordered all who passed it to bow in respect. It was to be a symbol of his authority and a means of the people showing their acceptance of Austrian rule.

As the legend goes, when William Tell arrived in Altdorf, he passed by the hat and pole without bowing. Seeking to make an example of the man, Gessler had both Tell and his young son seized. Knowing of Tell's prowess as a hunter, Gessler issued the famous challenge in which Tell would have to shoot an apple off his son's head. If he refused or failed, he and his son would be killed. If he succeeded, both would go free.

The result of that first shot is the heart of the legend and it is well known that Tell struck the apple from his son's head. It is the second shot fired in the story that is not as well known. When Tell was asked why he had held a second bolt in his belt, he replied that it was intended for Gessler if he had struck his son. Immediately, Gessler had Tell arrested for the threat and told he would be taken to Gessler's castle at Küssnacht. There, Tell would be held in the dungeons for the rest of his life.

On the trip across Lake Luzerne to reach the castle, a storm rose up and Tell was able to escape when the boat approached the shore. Not only was Tell able to escape Gessler's custody, he recovered his crossbow and the single bolt he had reserved during the challenge. As a hunter and a man experienced in woodcraft, Tell struck out across country to reach Küssnacht ahead of Gessler and his party of guards.

Hohle Gasse was a dark lane that had to be traveled to reach the castle at Küssnacht. As Gessler approached his castle and the safety it offered, he was struck by a bolt fired from ambush. That was William Tell's second shot of the story. He killed the Austrian bailiff with the skill of a sharpshooter and the concealment of an experienced hunter. Escaping back into the woods, Tell safely removed himself from the area and returned back home to his son. The example Tell set of personal bravery, skill, and reaction to oppression is said to have led to the overthrow of the Austrians' hold on what became Switzerland in 1308.

Marksmanship skills, both with a crossbow and firearms, are held in high esteem in Switzerland even today. Matches are held with crossbows that are as accurate as many modern firearms. The legend of William Tell is said to be one of the reasons for this celebration of skill. It is interesting to contrast Tell effectively sniping at an oppressor from ambush and being held up as a hero, and the treatment received by the marksman who took down King Richard I.

4

TECHNOLOGY OF THE ERA 1250 TO 1430

The Beginnings of the Gun

The future conduct of warfare changed dramatically, though slowly, beginning in the thirteenth century. That was the period when gunpowder was introduced in Europe. Known in China since the middle of the ninth century, the first popular explosive compound saw some use in Asian warfare, both in limited cannon and smaller handheld guns, as well as the more common rockets, incendiary arrows, and fireworks. By the thirteenth century, the Chinese were also utilizing iron-bodied fragmentation bombs filled with gunpowder and dropped from fortification walls.

The discovery of gunpowder in Europe cannot be solidly pinned down to a single date, or even an individual. The most generally accepted history is of the Franciscan Monk Roger Bacon of England writing the composition of a crude form of gunpowder in a treatise he wrote circa 1260. The explosive properties of the powder described by Bacon were graphically described in this phrase: "The sound of thunder may be artificially produced in the air." The actual composition of the powder Bacon carefully hid in his book, coding it in the form of an anagram. This early cryptography was quite good, the meaning of the text remaining obscured until deciphered by a Royal Artillery officer in 1904.

The fact that Bacon's exact formulation remained unknown for almost 650 years shows that there were a number of other scholars of his era who knew the "secret" of gunpowder. Count Albert of Bollstadt, more universally known as Albert Magnus, was a contemporary of Roger Bacon. Magnus also described the composition of gunpowder, in his book *The Marvels of the World*. Lastly, there is the mythical figure of Berthold Schwartz, the Black Berthold of history, who is supposed to have not only discovered gunpowder, but also invented guns, in his alchemical laboratory in 1353 and 1380 respectively. Many historians consider Berthold Schwartz to be a legendary figure, existing only in stories and folk tales.

The opinion of some historians that even Bacon's coded formulation may have been a historical forgery of some kind makes the European history about the beginnings of gunpowder extremely murky at best. One of the most likely avenues of gunpowder's introduction into Europe is by way of the Arabs, who had more extensive trade and information exchange with the Far East and China. However the formulation and idea of gunpowder entered Europe, it had a tremendous influence on the development of weaponry and tactics.

The first major military application of gunpowder in Europe was primarily its use in siege engines, cannons used during an assault on a castle or other fortification. In a work prepared in 1325 and later presented to the young King Edward III, the first known illustrations of a cannon are shown. The device is pear-shaped with a bell mouth, looking much like a vase, and is lying on its side on a heavy trestle table. The projectile being launched in a belch of smoke and flame is a large, heavy arrow. The cannon is being fired by a chain-mail-wearing figure holding a red hot iron above the thickest part of the cannon.

The pear shape of the early cannons soon gave way to the more recognizable tube form. These guns could be made of cast bronze but were often shaped from staves of wrought iron, forged down around a mandrel. The rough-forged barrel would be surrounded with hoops of white-hot iron that would shrink as they cooled, binding the barrel solidly together. These ribbed-barrel guns were what would have been used by King Edward III at the Battle of Crecy on August 26, 1346. This major event of the Hundred Years War was won by the English through the skillful use of archers and dismounted infantry. It has been chronicled that guns were used at the battle, but their effect on the combat was little more than to frighten the horses of the French knights.

The small cannons of the Edward III era were soon overshadowed by monstrous siege guns built to batter down walls. The projectiles of these massive engines of war were usually stone balls, shaped by masons from natural rocks. Iron projectiles also were used, but these were much less common for the big guns. The scale of these weapons can be shown by a description of one of the biggest ones, used in the assault on Constantinople by the Moslem Turks in 1453.

The bombard was an immense cast-bronze construction made by a Hungarian engineer named Urban at the direction of Mohammed II, the sultan of Turkey and one of the world's first great artillerymen. The cannon, known as the "Mahometta" took three months to form and cast and was ready in January 1453. The length of time to create the weapon is not excessive when you take into consideration that the barrel was twenty-six feet long and weighed nineteen tons. It could throw a half-ton stone projectile over a mile when loaded with around three hundred pounds of gunpowder. About seven hundred men were assigned to operate, guard, and transport Mahometta, along with the fifty pairs of oxen that could barely move it. The huge weapon could only be fired about once every two hours, but the destruction caused by the massive projectile on the stone walls of Constantinople was startling to the occupants of the city. The rate of fire of the huge weapon was so slow that the damage could be largely repaired in the time between shots, but the thunderous discharge of the monster piece of ordnance was terrifying to those it was aimed at, and there were two of the guns, one slightly smaller than the other and firing only a six-hundred-pound stone ball.

During the same time that huge cannons were growing out of the gunpowder weapons concept, much smaller handheld versions were also being developed. These early weapons were neither truly shoulder-fired nor exactly handheld. Often, they were little more than miniaturized versions of cannons with tillers or shafts attached to them. The weapons themselves were not much more than metal tubes with one end closed off. They were made either with an open socket in the base where a shaft could be inserted, or they were strapped down into a groove cut into a wooden tiller made from a rough-shaped board. From these crude weapons are descended all of today's sophisticated firearms.

The exact development of early firearms is uncertain; they were developed at a time when records were poorly kept at best. Descriptions in inventories and manuscripts from the era are not specific enough to determine exactly what they are talking about. Even the name "gun" has a muddied background. Cannon is easy enough; it derives from the Latin *canna* meaning "a tube." Most researchers feel that the term "gun" derives from the Teutonic (German) stem word *gun* meaning "war," as in such names as "Gunhilde." Such feminine names were used for specific weapons such as the "Domina Gunhilda," the Lady Gunhilda, a siege weapon listed in the inventory of Windsor Castle in 1330–1331. Other researches come up with a different background for the term. It is thought to derive from "gyn," an abbreviation for "engyn." This was the term used in Old English to describe an engine of war, what we would today call ordnance. The archaic spelling of the term is "gonne" and the kind of weapons that became modern shoulder arms were "handgonnes."

Something that could only be described as a hand cannon was discovered near Loshult, Sweden, and dates from the early 1300s. The weapon has a barrel length of thirty centimeters with a tapered bore measuring from thirty-six to thirty-one millimeters. Cast of bronze, the gonne weighs slightly over nine kilograms and has no provision at all for the attachment of a shaft or tiller. Having a reinforced muzzle ring and pear-shaped body, the specimen looks so much like the 1325 manuscript illustration of a medieval cannon that it could be nothing more than a founder's model. That would be something a lot easier for a producer to carry around and show to prospective clients than a full-sized weapon.

The earliest handgonne that can be dated with absolute certainty is called the Tannenberg hand cannon. Found in the bottom of a well during an 1849 archeological dig at the Tannenberg castle in what is now Hessen, Germany. The date of the weapon can be set at no later than 1399, since that was the year the castle was destroyed, a fact well documented in local records.

The cast-bronze Tannenberg hand cannon predates by several years, if not decades, the destruction of the Tannenberg Castle. Relatively small, the gun is thirty-three centimeters long with an octagonal shape. Actually a sophisticated device, the gun has a 156-millimeter-long 17-millimeter smooth bore with an additional 107-millimeter-long 9-millimeter powder chamber. The end of the gun is socketed for the attachment of a shaft for holding.

The small touchhole at the rear of the powder chamber has no real provision for a quantity of priming powder to be placed in and around it. Instead, the Tannenberg weapon was probably fired by the gunner shoving a red-hot poker down the touchhole. From tests of reproductions, it was found that it could also be fired by mounding some powder on top of the touchhole and setting a slow match (a length of cord treated to burn continually and slowly with a hot coal at the tip) into a split stick.

Firing the Tannenberg involves loading the weapon with powder and a lead ball (the original was found still loaded, with a lead ball in place), priming it, and placing the shaft up underneath the arm of the operator. The opposite hand holds up the weapon and points (aims) it. Since the weapon had no sights at all, aiming it is a matter of instinct and training. The opposite hand holds the split stick with the burning slow match. Touching the match to the powder sets off the cannon with a loud boom.

During testing, an inexperienced man had no real trouble keeping four out of five rounds fired on a man-sized target at about twenty yards. With a fully loaded powder chamber and a lead ball wrapped in a cloth patch, the Tannenberg had no trouble punching a hole in two-millimeter steel plate at two meters distance. This is a lot closer than an operator would normally allow an armored knight to approach him in combat, but knights also didn't wear armor that was two millimeters thick—it would have been too heavy. So the effective range of even this early handgonne against armor would have been at least ten yards or more.

Technical improvements continued on handgonnes as experience was gained with them, both by the operators who fired them and the military commanders who employed the operators. The very basic ingredient of the weapon, gunpowder, also underwent an evolution in terms of production and quality. A simple mixture of saltpeter (potassium nitrate), charcoal (carbon), and sulfur—basically an oxidizer and fuels—the formula ratio of these ingredients settled on roughly 71 percent, 16 percent, and 13 percent respectively. Modern chemistry gives the ideal formula as 74.64 percent, 13.51 percent, and 11.85 percent respectively, given pure ingredients. The purity of the basic ingredients, especially the saltpeter, improved as the demand for powder grew.

As a propellant and explosive, gunpowder had the field to itself for nearly seven hundred years in Europe. This worked well for the material as gunpowder is not a very good propellant. Given other options, it would have been abandoned early on except that there weren't any other options for the better part of a millennia. When ignited, only about 45 percent of the original mass of material in a charge of gunpowder converts to a rapidly expanding hot gas. The balance of the charge turns into a sticky, thick black residue. The residue inside a barrel quickly builds up on itself as firing is continued. Finally, it can clog a bore so badly that even ill-fitting projectiles can't be rammed down it for firing. Then the residue must be removed with water and scrubbing before the weapon can be effectively used again.

To create the mixture, the three components of gunpowder were first separately ground very fine and then mixed together. The resulting material was referred to as "serpentine" powder and had the consistency of flour. This flour-like serpentine powder had the characteristic of separating into its component parts if carried and jostled for a length of time, such as being in kegs and carried on carts. Transporting containers of serpentine powder was particularly dangerous as the material could put a very explosive dust into the air. A local spark or open flame could obliterate the cart, contents, and driver in a blinding flash of smoke and flame.

The safest way to transport serpentine powder was as prepared separate components that were mixed on-site by a master gunner shortly before use. This procedure made each batch of powder unique, with slightly different burning rates and power. Accuracy suffered greatly as a result. Using the material also carried its own difficulties. If serpentine powder was packed too tightly in the breech of a cannon, it might not fire at all. Packed too loosely and the power offered by the propellant was lost. It would burn only slightly and the projectile would pop out of the muzzle to land only a few yards away. Master gunners earned their pay.

In 1429, the process called "corning" was first described in writing. This procedure involved mixing the ground components completely before wetting down the mixture and forming it into a flat cake. Urine was considered one of the best wetting down agents for the corning process, the urine of wine drinkers being preferred over that of beer drinkers. Considered the best agent of all was the urine of wine-drinking bishops. By supplying the desired ingredient, the Church was able to bestow its blessing on powder manufactured by the faithful.

The cake resulting from the wetting process was allowed to dry before being broken up by rolling crushers. The resulting grains were separated by size through the use of sieves of differing mesh. Large grains, those the size of kernels of corn, were reserved for use in cannons. Smaller grains went to smaller weapons. The finest grains of all, but still larger than serpentine powder, were used for priming weapons.

Corned powder burned well and consistently. It was also too powerful for the early weapons, especially those made from bad castings, poor metallurgy, or ring-bound iron staves. Many of those weapons would burst when fired with corned powder. This resulted in serpentine powder remaining in use for cannons for another several hundred years. The smaller handgonnes used much smaller charges of powder in relation to the thickness of their barrels and were able to make good use of the regular-burning corned powder from the time of its development.

Along with improvements in the propellant, there was continuing development of the weapons that used it. Instead of being held by the tiller or shaft under the arm, above the shoulder, against the chest, against the cheek, in the hands, or snugly against the shoulder, tillers and shafts gradually resolved into the more recognizable stock shape that could be braced against the shoulder and properly aimed, or at least pointed. The longer barrels of the handgonnes in relation to their thickness put them in the class of gunpowder weapons known as "culverins," a class that also included a number of light cannons.

The ignition system for handgonnes also improved quickly for a time. The slow match soon replaced the hot poker, removing the need for the gunner to have an open heating fire readily at hand. That gave at least a somewhat safer situation for the gunner and his volatile explosive powder. Formulations and manufacturing of the burning slow match also improved, resulting in a material more commonly known as matchcord with a constant burning rate that could be conveniently measured in inches per hour.

To allow for better sighting of a handgonne, the gunner would often brace it against a wall, shield, or staff, giving him both hands to hold and aim the weapon. Many of the early handgonnes were fitted with hooks under their barrels or stocks to secure them against a wall or shield. These weapons were given the name *Hackenbüchse*, the German for "gun with a hook." In English and other languages, this name quickly became harquebus, or later, arquebus.

While the gunner was holding his harquebus or culverin, an assistant would be standing by with a burning length of matchcord. At the command from the gunner, the assistant would touch off the gun with the tip of his matchcord. The need for this third hand of an assistant was eliminated with the invention of the serpentine lock, the most elementary form of mechanical firing mechanism. The serpentine lock had a single moving part, an S- or Z-shaped arm. Attached to the side of the stock at a single point, the arm of the serpentine was free to rotate around the central pivot point. The upper arm of the serpentine was split at its end to take and hold a length of matchcord. By pulling up, or back, on the bottom arm of the serpentine, the operator caused the upper arm to swing down to where it could touch a small pile of priming powder on the barrel of the weapon with the burning end of the matchcord.

This was a revolutionary weapon, first described on the pages of a military manuscript written in 1411. The simple firing lock allowed a single man to effectively hold and aim his weapon prior to firing it when he wished. The system had major drawbacks. Any knock to the weapon or breath of wind could send the pile of priming powder flying out of position. And the priming powder holder, a simple depression in the barrel of the weapon centering on the touchhole, could quickly flood out from the least exposure to rain or spray. But the mechanism did work, and it was the beginning of aimed, purposeful firing of handheld gunpowder weapons.

5

ANATOMY OF A SHOT 1429

The Hundred Years War was one of the most extended conflicts ever seen in Europe. Running intermittently from 1337 to 1453, the war was basically fought over who would hold sovereignty over parts of France. The war continued on under a number of kings, both English and French, each of whom wanted to extend the reach of his rule. It was a war that literally spanned generations and saw a number of new technologies spill the blood of soldiers across the fields of France. Archers stood with crossbowmen and faced charging infantry. Knights were brought down off their horses with their armor pierced through by feathered shafts. And the stink and roar of cannons sounded out across the battlefield.

Among all these new means of war, an age-old one strode across the countryside in the 1340s when the Black Death, bubonic plague, swept through Western Europe. The lull in the fighting was another temporary one, as the conflict started up again in 1355. Under the banner of a new king, Henry VI, the English continued to take control of much of northern France. By 1428, English forces were poised to invade the south of France. In October of that year, the English forces laid siege to Orléans, a major city only seventy miles south-southwest of Paris.

The English siege wore on well into the next year. It was only when French relief forces arrived in Orléans in April 1429 that the siege was finally lifted and the tide of war turned hard against the English. The lack of detailed accounts and the proximity of a significant historical figure combined to conceal the contributions by lesser known individuals. At the siege of Orléans, Jean (Jehan) de Montesiler from Lorraine was called a particularly skillful culveriner in the testimony of witnesses. He was a master gunner, and his actions with firearms in helping to lift the siege established him as probably the first combat sharpshooter in history. His only difficulty in receiving the recognition that is due his feats was that his actions were overshadowed by those of his leader, Jeanne d'Arc, popularly called Joan of Arc.

Following the divine guidance she received from the voices that spoke to her, Jeanne d'Arc knew that she was destined to help lead France to victory in the war against the English. Convincing the authorities of both her divine inspiration and enthusiasm, Jeanne worked with the French military commanders in leading troops to the relief of Orléans in April 1429. The city had been under siege for seven months when Jeanne and her military entourage entered the city with supplies on April 29. Only a few days after her arrival, Jeanne was inspired to attack the English.

Leading a group of citizen militia and men-at-arms, on May 6 Jeanne helped conduct an assault against the English, who withdrew to the fortified convent of Les Augustins. There, nearly five hundred Englishmen fought against the French attacks with arrows, crossbow bolts, and cannonballs. Fought through the day, the battle raged first against the English, then against the French, and once again against the English. Attack and counterattack caused heavy losses on both sides.

In later retellings about the Battle at Orléans, several gunners fighting on the side of the French are noted. The men with their crude gunpowder weapons struck down selected targets among the English forces at ranges that would be considered long for the handgonnes of the day. One of these gunners was a Spaniard called Alfonso de Patada. The other, and most notable, master of the weapon was Jean de Montesiler.

One of the French commanders later testified that he had directed Jean to take down a "large Englishman" who was successfully defending a gate from the French attackers. With a shot from his weapon, Jean struck that warrior, opening the way for the French forces to rush in and seize Les Augustins.

The man was considered a phenomenal shot with his culverin, and that's where some confusion arises in the story. It isn't known exactly what kind of weapon Jean de Montesiler used during his actions at the lifting of the siege of Orléans. He had been conducting his own battle against the English for some time prior to Jeanne d'Arc joining the attack against Les Augustins. From a hidden position under a pier near a significant bridge, de Montesiler killed a number of the enemy, operating in what would today be recognized as a classic sniper style. At one point, he is said to have killed five English with two shots.

One of his five victims was "Richard, lord of Grey and nephew to Salisbury." So a sniper took out the equivalent of an English officer, certainly a leader and nobleman.

The reported fact that he killed more than one person with a single shot suggests that de Montesiler was using a significantly powerful weapon. It is suggested that his weapon was a small mounted culverin, rather than a handheld weapon. That would launch a ball large and heavy enough to pass through several people with a lined-up shot. But de Montesiler could not have hidden with such a weapon, as is described in the testimony. Instead, it is most likely that he used a large-bore handheld weapon, possibly leaning it against a wall, rubble, or shield for support. To aim such a weapon carefully would not only take experience, it would require the use of the then still new serpentine lock. At no time is an assistant or aid to de Montesiler mentioned.

It is also a strong possibility that de Montesiler made use of corned powder, which would give him a much higher velocity shot from a small weapon. And the weapon could stand up to such power while also delivering more accuracy—the faster the shot travels, the flatter its trajectory. The skill level with which de Montesiler handled his weapon—no matter what type it may have been—shows that the man had made a careful study of his craft. It is very likely that such a man would have made use of the very newest and best technologies available to him. This would have included such developments as a serpentine lock and corned powder. And consistent practice would have made him able to strike targets that would be considered out of range for such a weapon by people unfamiliar with the master gunner's art. These were errors of underestimating that would be appearing again as firearm warfare continued over the centuries.

The marksman's role demonstrated so ably by Jean de Montesiler was not the main reason for the adoption of small hand firearms by Europe's militaries. The massed use of such weapons by only slightly trained peasant conscripts would help disrupt attacks by infantry or mounted troops. That was the appeal of such weapons to the military leaders of the time. But the other effect of a hand firearm, the much darker one of striking down an armored nobleman or officer, was not lost on the English, who'd faced a French gunner at Orléans.

6

TECHNOLOGY OF THE ERA 1500 TO 1600

The Coming of the Rifle

The simple serpentine lock with its single moving part evolved into the snapping matchlock. In this mechanism, the arm that holds the burning matchcord would stand in an uncocked position far enough away from the flash pan and the priming charge to minimize accidental ignition. When the operator wanted to fire the weapon, he would cock back the arm holding the matchcord, and it would be mechanically held in place against light spring pressure. Slipping a cover back from the flash pan would expose the priming charge. Now a more careful aim could be taken or the weapon held while level, waiting for a command to fire. Pressing back on a "tricker" (trigger) would release the arm holding the burning match, and it would swing past its uncocked position to push the tip of the glowing matchcord into the priming charge.

This type of matchlock allowed as large a group of operators as desired to be lined up and discharge their weapons in a single, rough volley. This type of volley fire was desired by the military as it minimized individual marksmanship, removing the need for extensive training, yet would put a swarm of projectiles across the battlefield. Even if a soldier had little to fear from a single matchlock being aimed at him from eighty yards away, thirty of the guns going off together could make his time on the battlefield particularly short.

In some European armies, the use of the matchlock was on a roughly equal level with the crossbow in the early 1500s. By the middle of the century,

Through the AN/PAS-13 heavy thermal weapons sight on his M107 rifle, this sniper examines the cover far across the field in front of him. The thermal abilities of the AN/PAS-13 help expose hidden individuals by their heat signatures even when sight of them is blocked by the distant vegetation. ***Private Collection***

the matchlock arquebus had almost completely replaced the crossbow for military use throughout Europe. This was not due to any greater effectiveness of the matchlock over the crossbow. It was the much greater simplicity of the matchlock, both in use and in logistical support, that gave it an edge over the crossbow for military use. With the matchlock, there was no specialized cocking tackle to be used, and possibly broken or lost. It could take only a few days to train someone, even a peasant recruit, to operate an arquebus effectively enough to take part in volley fire. And the arquebus was finally much cheaper to make and employ than the crossbow, especially the more powerful steel crossbows.

With the development of a satisfactory matchlock weapon for the military in the form of the arquebus, some of the driving force for firearms development centered on the sporting use of the weapon. For the hunter and the target shooter, two things were demanded—accuracy and surety of fire. Hunting tended to be a rich man's sport in much of Europe and England, as the bulk of the lands were under the control of some sort of nobleman. Owning the land meant that you also owned the game that lived on it. This meant that the hunters were nobles themselves, or worked directly for them, and that they had the money to support their desires, as well as to indulge their taste for the special and unique.

Some mechanical genius had probably witnessed a hand-turned grinding wheel striking sparks from a piece of steel as it was ground. Reversing the situation put a steel wheel being turned against a stone to strike sparks, which is basically what a wheel lock does. In the jaws of what was called a dog-head, a piece of iron pyrite, a sulfite mineral, would be pressed down against a spring-driven serrated steel wheel. When the wheel was released, it would spin against the pyrite, causing a sudden shower of sparks. By directing these sparks into a flash pan full of priming powder, an operator could make a gun fire. This meant that there was no burning match, no constant adjustment of a matchcord as it was consumed, and that firing a weapon was as easy as pulling a trigger.

The wheel lock was the first completely mechanical firing lock developed for gunpowder weapons. It was also one of the most significant mechanical devices of its day, second only to a clockmaker's product in complexity and expense. The exact date of the development of the wheel lock is not known, and it cannot be pinned down positively to a specific country. It is suggested from manuscripts and a few surviving specimens that the wheel lock existed around 1500 if not earlier and was most common in that area of Europe that would later become Germany.

No less august a person than Leonardo da Vinci showed an interest in firearms and in the wheel lock in particular. In his Codex Atlantico, assembled from his handwritten notes and drawings after his death, da Vinci detailed two wheel lock mechanisms. One of the drawings is of the internal workings of a wheel lock, showing a compression coil spring operating the wheel. The other drawing is unmistakably that of a wheel lock intended for a firearm. The exact date that da Vinci made the drawings is not known, but some scholars estimate their date of creation as being sometime between 1480 and 1485. Other researchers make the date of the drawings as recent as 1505. Whatever the exact date, the drawings make a strong case for the wheel lock having been known, if not invented, in Italy.

The wheel lock made a firearm far more efficient than the matchlock. The expense and complexity of wheel lock–equipped firearms kept them from being adopted in any real numbers by the military. The only place this didn't hold true was in the cavalry. The firearm helped force the end of the armored mounted knight being the controlling force on the battlefield. The extensive use of horses with mounted cavalry remained. Wheel lock pistols allowed the cavalry to handle their mounts with one hand and still effectively aim and fire a weapon with the other. Since many of the early cavalry were members of the lesser nobility, the expense of the wheel lock handguns was not as difficult to meet as it would have been for an army-wide issue of the weapon.

In the years before the development of the wheel lock, the demand for accuracy in a firearm had been met with the invention of rifling. It had been known for thousands of years that canting the fletching (feathers) on an arrow would make it fly more accurately. The canted feathers caused the arrow to spin in flight, giving it an amount of gyroscopic stability, though that principle was unknown to the men who used it.

Barrels with straight grooves cut inside their bores were made as early as 1460. The straight grooves did not spin the bullet, so they had no effect on the accuracy of the weapon. They did, however, give the fouling produced by the burning gunpowder some place to build up without blocking the loading of the weapon. It was not a great evolution for a master gunsmith to make the grooves follow a spiral twist around the bore of a weapon. These twisted grooves, known to have existed by 1475, were the rifling that caused a ball to spin in flight.

Projectiles used in gunpowder weapons were intentionally made slightly smaller than the bore of the weapon. This was so that the ball could be loaded into even a fouled barrel. The difference between the diameter of the ball and the inside of the bore was called windage. The windage allowed the easier loading of the projectile, a very valuable characteristic for combat weapons that would be firing multiple shots during an engagement, but it did nothing for the accuracy of the gun. Unless a ball closely fitted the bore of a weapon, when fired it would just careen down the barrel, bouncing from side to side until it finally left the muzzle. Driving an oversized ball down a barrel with an iron ramrod and a mallet insured that it was a tight fit to the bore and had zero windage. But the technique also deformed the ball so badly that it couldn't fly straight.

The answer to the windage problem was to use a ball just slightly smaller than the bore of the weapon and make up the difference in size with a layer of very thin greased leather or fustian (coarse cotton cloth), what we would today call using a patched ball. This technique not only eliminated the windage problem, it allowed a ball to remain round while being driven down a barrel and gripping the rifling. The patch would peel away from the ball in flight while leaving the projectile spinning from the rifling, if the barrel had it. The technique of patching a ball was written about by Espinar in his 1644 treatise on "The Art of Shooting and Horsemanship." In spite of that, the most popular method of loading even a rifled barrel for accurate shooting involved ramming down an oversized ball, forcing it along by brute force. It was the opinion at the time that the harder it was to drive the bullet down the barrel, generally the better it would shoot.

Given a well-made barrel with a straight, parallel-sided bore, even a smoothbore matchlock arquebus was capable of fair accuracy. Making a truly cylindrical bore was not easy for the blacksmiths of the time. Barrels were not bored out but were made by welding a long strip of iron around a mandrel. Gunsmiths who could make barrels true and straight, and well-finished on the inside, were able to demand a premium price for their product. Their weapons went to the nobility and those rich enough to afford them. On a calm, dry day, a quality .60- to .70-caliber arquebus loaded with a well-fitting ball could be expected to hit a playing card at seventy yards, kill an animal the size of a wild boar at eighty yards, and bring down a deer at over one hundred yards. With a rifled barrel on the same matchlock, that playing card could be struck at one hundred yards.

Locating a well-fitting ball for any firearm during the first few hundred years of their existence could be a task for a traveler who wasn't casting his own bullets. Gunpowder was developing into a known quantity; the formulations changed slightly from location to location, but they all had to burn properly to sell in the marketplace. The measurement of a barrel diameter was quite another thing. There was no universal measurement for size in Europe. Having a .76-caliber gun would mean today that it would require a round ball that measured 0.76 inches in diameter, or 19.3 millimeters if you were going metric. The lack of any universal unit of measure meant that every barony, city, state, kingdom, or country would have its own version of an "inch."

It was in England that a means of measuring bore size was developed that was universal enough to work when applied to firearms. Going by a unit of weight, the caliber of a weapon was measured in the number of balls of pure lead it would take to weigh one pound. The scales of the day were able to determine as little as a 1/100th of an ounce weight differential, and the weight of a pound met a royal standard throughout England. A royal weight standard could be cast of brass and passed around the country, additional weights being cast as needed.

Pure lead was also an easy material to acquire, as molten lead tends to clean itself when it is melted. So having a weapon with a bore measuring 0.786 inches (a pure lead ball of that size weighing one-tenth of a pound) meant that it was a ten-to-the-pound weapon. A bore size of 0.747 inches would be a twelve-to-the-pound weapon. By 1540 in England, the caliber of a weapon was stated as the bore number of a gun. Today, this unit of measurement remains for shotguns and the term has been changed to gauge.

Originally, the arquebusier, caliveer, and later musketeer carried their ready rounds for firing in stoppered wooden vials hanging from a bandoleer draped across their chests. Each vial would hold a measured unit of powder sufficient for one shot. In a pouch at their waist, the gunners would also have lead bullets along with a flask of fine priming powder. It was an involved procedure to set down the weapon, charge it from a vial, drop the ball down the barrel, and seat it solidly with the ramrod. If the individual was not armed with a wheel lock weapon, he would have to do all of the loading procedure with a burning length of matchcord held in one hand and kept away from the various containers of gunpowder about his person.

By 1550, this cumbersome procedure had been streamlined somewhat by the invention of the paper cartridge. Inside a tube of rolled paper would be held a charge of powder sufficient for one shot along with the proper projectile. By tearing off the end of the paper tube, the gunner could pour the powder down the barrel of his weapon and then press the projectile, along with the rest of the paper cartridge, down the barrel and ram it home. The paper not only acted as a consumable container for the powder charge and bullet, it also acted as wadding for the shot, holding the bullet firmly in place as the weapon was moved about.

An average smoothbore gun of the sixteenth or seventeen century would have had a more universal sporting use, unlike a rifled weapon. This was due to the fact that a smoothbore weapon would not only be able to fire a properly fitting ball at reasonable hunting ranges; it would also be able to fire shot. Shot would be made up of a large number of very small lead pellets. The much larger quantity of shot pellets in a load would put out a swarm of projectiles rather than a single bullet. This swarm would open up into a growing circular pattern ever wider as the range from the muzzle increased. Such a swarm would be much more efficient than a single shot for taking out fast-moving small game, such as rabbits or hares, and would be of greatest use in bringing down flying targets such as ducks or other birds. Firing the same load of shot in a rifled weapon would be very inefficient, as the same spin that would make a single ball more accurate would spoil the shot pattern for moving game. The spinning load of shot would open up into a circle, spreading much faster than it would have if fired from a smoothbore weapon. The spinning shot pattern would have a huge hole in its center, the shot pellets being concentrated on the outside edge of the pattern. The open pattern of shot would easily let game be completely missed by a shot, even if the gunner had aimed perfectly at his target.

The use of firearms by the gentry for hunting or target use is exemplified in the appearance of their weapons. The plain barrel and wood stock of a military arquebus would never been seen in the hands of a proper gentleman or officer. Instead, the hunters and officers of means would have highly decorated weapons, steel barrels and locks chased and deeply engraved with hunting scenes, wooden stocks made of the finest woods, carefully polished to show the natural whorls and patterns of the grain. In addition, the wood of the stocks would be carved and have inlays of ivory plates or even silver or gold wires in intricate patterns. The landed gentry of the English countryside would be more practiced with their weapons, using the finest powder and shot available. The hunting masters and gamekeepers working for the gentry and the larger estates would be the most practiced of all. Theirs would be the plainer quality weapons, none the less efficient for the lack of decoration. These men would bring in the game necessary to feed the local manor as well as their own families.

In parts of Europe hunters primarily would gather at fairs and compete with each other, firing at difficult targets with bows, crossbows, and firearms. These competitions were particularly popular in the southern areas of Germany as well as Switzerland. Each competitor fired a total of twenty-four shots. The targets were at varying ranges from about 175 to 200 yards and were of differing sizes, the smallest being around twenty-four inches while the largest was forty inches. During a match in 1584, the shooters were specifically banned from using rifled weapons, smoothbore matchlocks being the weapon of choice. The scores show that even the smoothbore musket had the capability of shooting at fair range. Of the 133 shooters, nearly a third of them struck the targets with between twenty and twenty-four shots. These men were demonstrating skills that would quickly become an advantage in a military conflict.

A meet in Basel, Switzerland, in 1605 stands as an excellent example of what was expected from a rifled and a smoothbore weapon. During the competition, both types of weapons competed, but at different ranges and with different sized targets. For the smoothbores, the range was 570 feet. The smoothbore target was a disk with a diameter of thirty inches. For the rifled guns, the distance was increased to 805 feet, over 260 yards, with a slightly larger target, having a diameter of forty-two inches. All weapons were fired from the standing position. At the time, shoulder stocks had not yet become as popular as they would be. So most of the weapons, both smoothbore and rifles,

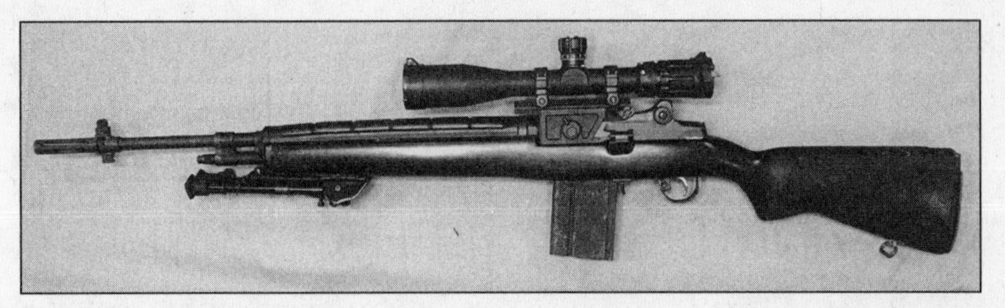

The Army Special Forces group-developed M25 sniper rifle, an improved version of the earlier M14-based M21 sniper rifle. This weapon has a nearly unbreakable McMillan synthetic stock and a very solid telescope mount base attached at two points to the receiver. A Harris folding bipod is also fitted to the underside of the stock. This weapon has been used by both the Army and Navy Special Warfare Units since it was first developed in the 1986–1988 time period. It is more accurate and dependable than the earlier M21 without requiring as much specialized maintenance as had the earlier weapon. ***Kevin Dockery***

were held in the hands, not braced against the shoulder. Some of the weapons were designed to be held in the hands and the shooter's cheek used as a leaning point. This aided aiming and did a little for stability. It was a true demonstration of strength, endurance, and marksmanship and well demonstrated the capabilities of champion shooters.

7

ANATOMY OF A SHOT 1643

Approaching the middle of the 1600s, the standard military firearm in England had settled on a design that would still be recognized as a shoulder weapon today. The term "musket" was now being used widely to describe the matchlock weapon that could be used by one man and had a formed and proportional stock shaped so that it could be solidly held against the shoulder when fired. The long barrel of the military musket was nominally made as a ten-bore weapon with a caliber of 0.786 inches. The standard issue ammunition of the day consisted of twelve-bore lead balls. Since the issue projectile had a diameter of about 0.747 inches, there was plenty of windage between the ball and the bore of the barrel.

The issue musket was not generally considered capable of hitting a discrete man-sized target at fifty yards with issue ammunition. When it was used in the military style of volley fire on command, the accuracy of the individual musket was not considered important. And in the relatively small English Army of the era, individual marksmanship was not a skill taught in the military. If it had been, some use could have been made of the fact that the military musket was capable of being aimed and hitting a man-sized target at 120 yards when fired with a tight-fitting projectile and employed by a capable shooter.

The musket in the hands of the army became much more significant in the 1640s. In October 1643, the first major battle of the English Civil War took place as fighting broke out between the Royalist forces under King Charles I and the Parliamentary forces. Combat went on across the country between men separated primarily by their political and religious beliefs. The rapidly built-up military forces of both sides were made up primarily of conscripts and volunteers, with a small handful of professional soldiers as a cadre. Among these troops, there were a handful of sportsmen and experienced hunters on both sides of the conflict. These men brought their own sporting guns and knowledge with them into the battle.

The guns the sportsmen and hunters used were not the ill-made military matchlock muskets of the time. Many of the weapons were rifles imported from Europe, where the gunsmithing art was flourishing. Besides sometimes being ornate and decorated pieces of hardware, these rifles were accurate and capable of long-range shooting at specific targets. Many of the men so armed in both the Royalist and Parliamentarian ranks were gamekeepers with years of experience firing their weapons and expecting to hit what they aimed at. Only now the gamekeepers weren't hunting deer or stag; they were stalking men.

For the first time in England, marksmen were using firearms to take down the military leadership of the other side. Officers and NCOs on both sides of the conflict were being specifically targeted.

One of the most militant leaders of the Parliamentarian forces was Lord Robert Brooke. The thirty-five-year-old adopted son of the first Lord Brooke was a popular and charismatic figure both in the halls of Parliament and later among his fighting men. Lord Brooke had known that the civil war was coming and had prepared for it. He had been stockpiling weapons and ammunition since the spring of 1641, to defend against a Royalist coup, as he well knew the value of firearms and artillery in conducting the warfare of the age.

The opinions held by Lord Brooke against the Crown were based on more than just the despotism of King Charles I. Steadfast against the established political power of the Church and the bishops who ran it, Lord Brooke had railed against them for years, even publishing several books attacking the established Church. He strongly supported the Puritan causes, which ran very much against the practices of the Church.

There were equally dedicated men on the Royalist side of the conflict. Strong in their support of the sovereign rights of the Crown and King Charles, these men were equally willing to fight for the Catholic Church against the generally Puritan but also other radical Protestant Parliamentarian forces.

The early battles of the war tended to go in favor of King Charles and the Royalists. From his militant stance, Lord Brooke was firmly against any movement toward a peace settlement early in the conflict. Using his contacts among the merchant class of the city, Lord Brooke raised funds to support the war, as well as volunteers willing to fill the ranks needed to fight it. Receiving his commission as a colonel in the Parliamentarian Army, Lord Brooke stood at the head of a regiment of a thousand men he had raised from the population of London.

In spite of the initial military setbacks against the forces of the earl of Essex, the Parliamentarian commander, Lord Brooke, and his forces drove hard against the Royalists. Combat forces on both sides consisted primarily of a mix of pikemen and musketeers making up the infantry, backed up by fast-moving cavalry troops and supported by light artillery. The pikemen were a classic holdover from the earlier days of warfare. Their sixteen-foot long wooden-shafted pikes fitted with sharp steel heads were able to hold back a cavalry charge and oncoming infantry. Their support in the ranks was necessary to give the musketeers protection to operate their weapons when the enemy had approached to close-quarters range. If the pike wall fell, the battle would quickly degenerate into a morass of close-in hand-to-hand combat, the muskets being now little more than ungainly clubs as the knife and short sword held the upper hand.

The long, pitched battles between the Royalist and Parliamentarian forces resulted in heavy losses on both sides. Units were decimated. In the fighting, the regiment under Lord Brooke's command was reduced to half its original number of combat effectives. Learning that Royalist forces were settling in to their positions in Lichfield, the earl of Essex sent Brooke and his five hundred men out against them. At the head of his now battle-hardened troops, Lord Brooke gathered another seven hundred reinforcements on his march to Lichfield.

In spite of the heavy losses to his regiment, Brooke was not operating in the field without his own successes. In spite of their weakened status, his regiment was able to secure Warwickshire for the Parliamentarian forces. Under his dynamic and increasingly experienced leadership, Brooke's men were able to drive the Royalist forces from Stratford-upon-Avon in February 1643. When the earl of Essex sent him against Lichfield, Lord Brooke was fully confident that his men could wrest it from the Royalists and bring it under Parliamentarian control.

The medieval cathedral at Lichfield well fit the description of being a natural fortress. The "close" or the immediate area of the cathedral, was surrounded by a high defensive sandstone wall as well as a ditch. There were three spires on the cathedral building, the central spire being 252 feet high while the two others on the western side of the building were 190 feet high. It would be a difficult target to breach or defeat by any means other than a long siege. The Royalist garrison had prepared its positions inside the close, with ammunition stored in the cathedral itself and cannons in place at strategic locations.

The bulk of the townspeople at Lichfield sided with the Parliamentarians during the civil war. It was the cathedral people themselves, along with a small following of the locals, who remained on the side of the king and the Royalists. This gave Lord Brooke and his troops a certain amount of support from the surrounding area as he arrived at Lichfield and laid siege to the close. It was March 2, the festival day of Saint Chad, the patron saint of Lichfield, when Brooke arrived at the town. The street fighting between the Royalists and Parliamentarians quickly descended into a chaos of hand-to-hand slashing with knives and swords among the booms and smoke of musket shots. It was edges, points, and bullets on both sides as the Parliamentarian forces soon gained the upper hand, forcing their opponents through the narrow twisting streets as they withdrew to the relative safety of the close.

The tall Gothic cathedral flying the standard of King Charles was not something Brooke was going to let stand for any longer than he had to. His Puritan convictions drove him to capture the cathedral and destroy it. He hated such edifices, considering them the haunts of the antichrist. He wanted it brought down utterly. The large artillery at Brooke's command was limited, consisting of a single demiculverin, classified more as a field piece rather than a real siege cannon. The demiculverin had a four-inch bore, weighed about thirty-four hundred pounds, and launched an eight-pound iron ball out to a practical effective range of about eight hundred yards.

Along with a number of smaller guns to support his main piece, Brooke set up his forces to begin their bombardment of the close at the south gate. Their guns were placed on the firing line that filled the city street, the demiculverin set up for firing about one hundred yards from the gate.

As the Parliamentarians were setting up their positions, the Royalist forces could see them at work. There was no question that the greatest danger to the close was the guns, and the Royalists fired down from the walls into the milling crews and infantry trying to service and protect those guns. Their best means of defense was to keep the guns from firing, and that meant killing the crews.

Two men on the side of the Royalists decided to climb up to higher ground to get a better view of the situation, and possibly a better shot. The highest ground in the area was the roof of the main spire, and that would have been a place the men were familiar with. Parts of the roof of the spires were made of lead. To increase the ammunition supply for their muskets, some of the Royalists, the two men included, had been peeling back portions of the roof lead and melting it down to be cast into balls.

The square lower section of the spire was topped with a small platform surrounded by a low wall, a battlement, about halfway up the height of the entire spire. From the battlement upward was the tapering conical roof. A winding spiral staircase led to the upper reaches of the central spire. This was the means by which the two men reached one of the higher points of the cathedral. There was cover behind the stone walls of the spire itself, the walls penetrated by windows that could easily act as gun ports. The low wall surrounding the battlement also allowed a man to lie almost unseen by anyone from ground level. Ornamented conical spires stood at each corner of the battlement, the pointed structures providing additional concealment if desired for a pair of shooters.

The men going up to the highest spire that day were a pair of local brothers, John and Richard Dyott. Probably without knowing it, they were about to conduct a classic sniper mission, one British soldiers would be conducting regularly three hundred years later during World War II. The brothers were moving to the high ground to pick off the enemy gun crews serving their weapons. One brother, John, was armed with a long-barreled smoothbore fowling piece, loaded with a close-fitting ball. Richard, the other brother, could act as a spotter, seeking out the targets they desired and allowing his brother to take the shot.

It is very likely that John Dyott was the better shot of the two brothers. Known locally as "Dumb Dyott," John had been deaf from birth. His deafness resulted in him not being able to talk, giving him the nickname "Dumb." John was thirty-seven years old when he went up that cathedral spire. He had volunteered for the local militia and was prepared to defend his town, and his king, with all of his skills. He would have had no distractions when he set up to shoot. Growing up in the country with his brother, John Dyott would have had plenty of chances to develop his marksmanship, even with as clumsy a weapon as a long-barreled fowling piece. It was his only firearm, and he would have been intimately familiar with its shooting characteristics under almost all circumstances.

The limited power of his artillery was not having the effect Lord Brooke wanted in breaching either the south gate or the walls of the close. Leaving the relative safety of his command center set up a distance from the cathedral area, Lord Brooke moved up through the cover of the streets to behind a house that was facing his gun positions. Crossing through the building, Brooke was able to see his main piece of artillery in action.

It is certain that Lord Brooke would have been wearing the distinctive purple livery of his regiment. He would not have made a figure that could have been mistaken for a Royalist, not among all of the other purple-wearing attackers besieging the close. There was other significant regalia that made up Lord Brooke's uniform, such as the steel cuirass that was a badge of his rank and position. The cuirass was a steel chest piece, almost a sleeveless waistcoat, with the seams running along the sides. It was one of the last vestiges of the plate armor worn by knights and nobility in conflicts past. It was still a badge of rank by the time of the English Civil War, as well as being a fairly good piece of body armor. Along with his steel helmet with a barred face visor, Lord Brooke would have felt adequately protected for the combat of the day. A quality cuirass, such as he would have been wearing, would have been proofed, fired at with a live musket shot to establish without doubt that it could withstand a gunshot.

Whatever protection a cuirass may have offered Brooke, it would have done him little good in the end. He would have cut a noticeable figure, even in the smoke and confusion of combat. The men of his regiment would have shown him deference and the respect due his station and rank. It would have been obvious to any observers that here was a man of importance.

Standing at the front of the house, Lord Brooke watched his men struggling with their artillery. Either his orders couldn't be heard, or he wanted to get a better view of the situation. Whatever the reason, Lord Brooke stepped out from the cover of the building.

From their positions up in the central spire, the Dyott brothers could look down at the gun positions firing at the close. The distance was around two hundred yards, a significant range for a shoulder weapon of the day. Movement in the doorway drew the attention of at least one of the brothers. Cocooned in his own silent world, John Dyott had no distractions as he lined up his shot, taking careful aim. It was not only a long range to fire, it was a downward angle shot, one of the most difficult to judge properly. He squeezed the trigger, and the big weapon bloomed smoke and bucked against him as it fired.

The huge lead projectile smashed into Lord Richard Brooke's left eye, killing him instantly as it drove him down into the ground. Though there is debate to this day that his death was due to a stray bullet shot at random, the people of Lichfield have no doubt that it was divine retribution that steadied John Dyott's hand that day. He lived through the war and prospered, in spite of the Royalist garrison surrendering the cathedral two days after Brooke was killed.

The death of Lord Brooke was a significant one to the Royalist forces. His deputy, Sir John Gell, took command of the regiment and continued with the siege. But Lord Brooke had been a rising star for the Parliamentarians. Brooke

Working from a rooftop, a sniper gives his fellow soldiers cover and security during their operation in the village of Shiaha in Iraq on January 13, 2006. *U.S. Army*

had proven to be a loyal and influential supporter of the Parliamentarian cause. He had even been suggested as a possible successor to the Earl of Essex as commander of the Parliamentarian forces. Even though his death flagged his men on to utterly defeat the Royalists in Lichfield, his ongoing life could have played a significant part in English history.

1645

There were marksmen on both sides of the English Civil War who made their presence known through the skills they had with their chosen weapons. Throughout the war, men fell to the deadly skill of these individuals. On the Parliamentarian side, the actions of these men, along with the brilliant leadership of Oliver Cromwell among others, had driven King Charles and the Royalist forces that remained to him almost to defeat in late 1645. In September of that year, King Charles moved north along the Welsh border, intending to link up with forces loyal to him, forces that he didn't know no longer existed.

Near the Royalist city of Chester, the Battle of Rowton Moor took place on September 24, and King Charles I suffered another crushing defeat, one of his last. Standing on the Phoenix Tower in Chester, King Charles witnessed the rout of his forces at the hands of the Parliamentarians. The remnants of his army were running for their lives as he watched, running to the relative safety of Chester. A siege of the city quickly took place, the Parliamentarians seeing an end to the conflict waiting for them if they could capture or kill King Charles.

From a vantage point in the bell tower of the ancient Church of Saint John the Baptist in Chester, one Parliamentarian marksman almost ended the English Civil War with a single shot. As King Charles stood on the Cathedral Tower, a shot rang out from the bell tower. The distance between the two locations made for a possible, but remarkable, shot on the part of an unknown shooter. The bullet fired narrowly missed King Charles I, striking and killing a captain who was standing in attendance nearby. This was the final straw for King Charles. He fled the city that night with the six hundred horses that were all that remained of his army. He would not be captured until June the next year.

8

TECHNOLOGY OF THE ERA 1620 TO 1820

The Flintlock, the Jaeger, the American Long Rifle, and the Ferguson Breechloader

Besides the wheel lock and matchlock, a number of other firearm ignition systems had been developed around the known world by the seventeenth century. Almost all of these systems had their positive points and their drawbacks. The Snaphaunce (circa 1550), Miquelet (circa 1580), English lock (circa 1625), and dog lock (a safety catch originally for an English lock, circa 1630) were all developed throughout Europe to replace the complex mechanism of the wheel lock and the undependable but simple mechanism of the matchlock, with something that combined the characteristics of both. None operated well enough, or cheaply enough, to replace the matchlock for general military use, though extremely fine sporting weapons were built using the systems.

The basic principle of the new locks was the age-old one of striking a spark by hitting a piece of steel with a flint. The sharp edge of the flint, a form of microcrystalline quartz, would strike the surface of a piece of steel or iron at an angle, removing tiny particles of the metal as it was scraped along the surface. The friction of the flint against the steel would heat the particles to incandescence and they would fly through the air as white hot sparks. Directing these sparks into a piece of flammable material was a way of making fire that had been used for hundreds of years, and gunpowder was one of the most enthusiastic flammable substances known at that time.

In general, the shaped piece of flint was held between the jaws of a cock, so named because the action of all of the locks resembled that of a chicken pecking at the ground. Driven forward under spring pressure when released, the cock would strike the flint across a steel (frizzen), generating sparks; the sparks would fall down into a flash pan connected to the main charge in the barrel through a touchhole. The variations of the different locks included a separate frizzen from the priming pan (flash pan) cover (Snaphaunce), the frizzen combined with the flash-pan cover, with the main parts of the triggering mechanism on the outside of the lockplate (Miquelet), the triggering mechanism and related parts on the inside of the lockplate along with a safety system built into the mechanism (English lock), and a positive external safety catch that would hold the cock in place (dog lock).

All of the above details were incorporated into a single mechanism in France around 1615. This French system of ignition was the true flintlock, a design that would dominate the world of firearms until the early nineteenth century. In a flintlock, the flint was held in a cock that faced a combination frizzen/flash-pan cover. When released, the cock would snap forward, striking sparks with the edge of the flint as it drove the frizzen back against a spring. The sparks would now fall into an opened flash pan, igniting the powder train and firing the weapon. Only the cock and the frizzen, with its tension spring, were exposed on the outside of the lock; the rest of the lock parts were protected on the inside of the lockplate. There was a half-cock position on the trigger mechanism that held the cock in place. The trigger could not mechanically release the cock until it was drawn fully back to fire. If this half-cock portion of the lock was broken, there could be an accidental discharge of the weapon—the source of the expression "going off half-cocked."

The flintlock, also known as the fire-lock, spread throughout the world of firearms. The basic action of the lock remained unchanged for nearly two centuries. Improvements in the details of the flintlock were made to the point

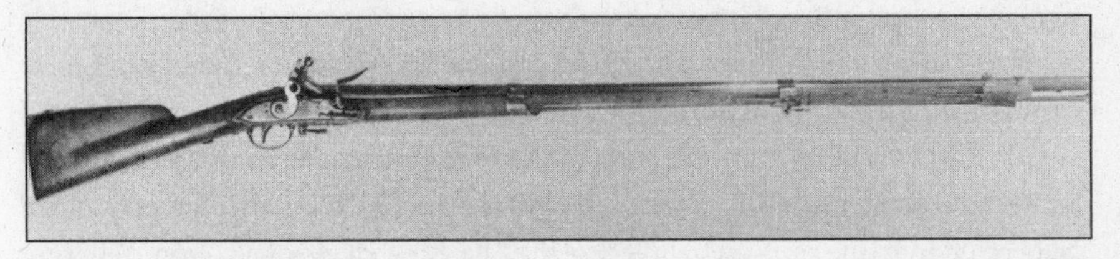

The U.S. flintlock musket, the first military weapon officially produced in numbers by the United States. It is effectively a copy of the French Charleville musket which the United States had purchased in some numbers during the Revolutionary War. ***Smithsonian Institution***

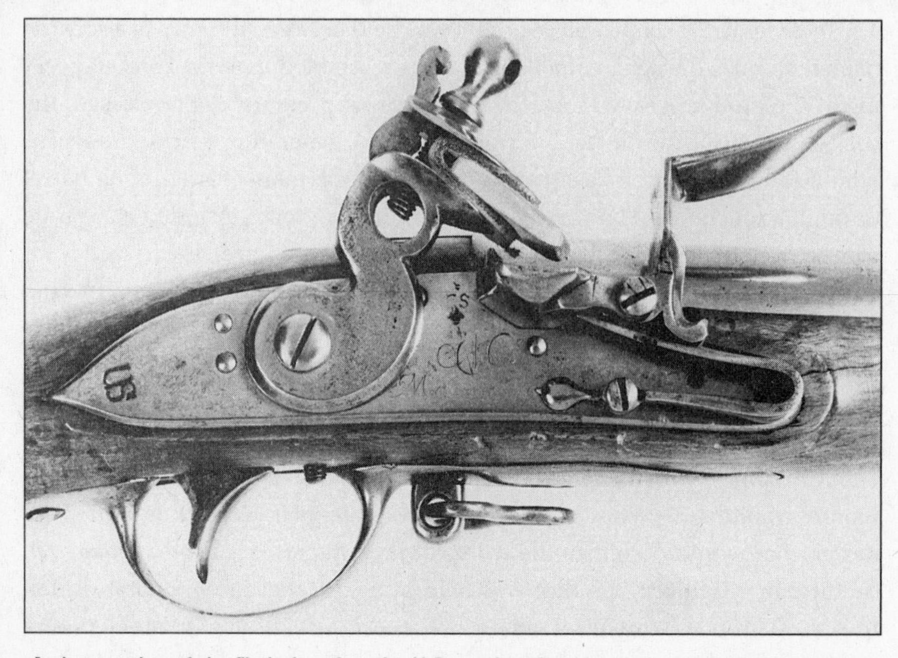

A close-up view of the flintlock action of a U.S. musket. On this specimen, the frizzen and pan cover are flipped forward as they would be when struck by the cock. The cock is also forward in the fired position. Instead of the normal piece of flint, between the jaws of the cock is a shaped chunk of wood held by a piece of leather padding, which would have also been used to secure a flint. *Smithsonian Institution*

where the loaded system could even be carried and have a fair chance of firing in the rain.

In England, the flintlock was used on a new military musket beginning shortly after 1710. In general, the new musket had a long, graceful stock, shaped with a definite "wrist" area behind the barrel and the lock where the firing hand could be wrapped around the wood for a solid grip. The barrel was forty-six inches long and nominally a ten-bore with a caliber of about 0.75 inch. Brass furniture along with steel pins and screws held the various parts of the musket onto the stock. There was no rear sight on the barrel of the musket, it not being intended for individual accurate fire. A rectangular steel stud near the muzzle of the barrel was not intended for use as a front sight. The stud was used to help hold in place a seventeen-inch-long triangular bayonet. The musket became its own pike. The long lethal blade was used against the enemy far more often than the musket ball. It was the "flashing steel" of a bayonet charge.

The new English firearm was named the first Long Land Pattern musket. In about 1750, a shorter version was developed with a forty-two-inch barrel.

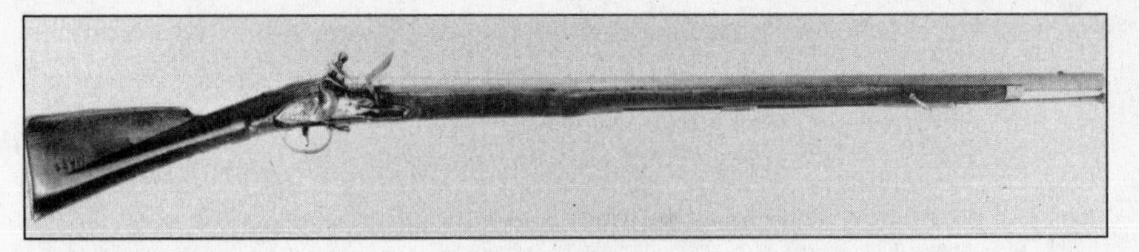

The British Brown Bess musket, the weapon most faced by the Continental Army Forces during the American Revolutionary War. *Smithsonian Institution*

This became the Short Land Pattern musket, and the official British infantry weapon, in 1768. The two muskets weighed between ten and eleven pounds, were reasonably dependable, and had a sling by which they could be carried over the shoulder. Among the troops who used it, the musket had developed the nickname Brown Bess, which has stuck with the pattern over the ages. Both patterns of musket, along with other slight variations, were used in the French and Indian Wars that took place in the colonies of North America from roughly 1740 to 1763. This was the time in which a number of colonial officers developed their initial combat experience. Many of the Brown Bess muskets used during this conflict went to the homes of the colonists, for use in hunting as well as forming militias for the common defense.

JAEGER RIFLE

As the flintlock was evolving, moving along with it was a Germanic weapon that would become very popular when used with the new lock. In general, the gun was a short-barreled rifled-bore shoulder weapon of heavy caliber. Intended for hunting in the forests and mountains of Germany, Switzerland, Austria, north central Europe, and later, Scandinavia, the style of rifle was called the jaeger, from *Jäger*, the German term for hunter.

The jaeger became a perfected style of hunting rifle, well suited for the environment and game of the countries where it was used. Its influence lasted for over 150 years, spreading out as far as the colonies in North America and spawning a new style of rifle there.

Within a few decades of its development, the flintlock jaeger became the most popular hunting weapon in central Europe, by 1700 being the standard precision weapon that others were measured by. All of this was packaged in a firearm style that measured on the average between forty to forty-five inches in overall length. Barrels were correspondingly short and stout, twenty-eight to thirty-three inches long on the average, usually with an octagonal cross section and a .60- to .70-caliber bore.

The barrel was very much the heart of the jaeger's success when combined with the other characteristics of the weapon. The bore of the jaeger was rifled, usually with from seven to ten grooves with a twist rate of around one turn in forty-eight inches. The grooves were usually fairly shallow semicircular cuts in the steel of the bore. The exact number and style of grooves in the bore varied from maker to maker, each weapon being effectively handmade and one of a kind. But one thing was shared by most of the jaegers, especially those made in Germany. They were all known for having quality bores. The barrels were straight, the bores cylindrical, and the rifling cut in a uniform manner. Finally, the bores were polished by hand-lapping.

The quality barrel of a jaeger was supported by a full-length stock that extended to the muzzle of the weapon. Underneath the barrel was held the ramrod, usually made of wood with a horn, bone, or brass tip to match the furniture on the front of the stock. The light ramrods help to prove that the jaegers were intended to be fired with patched balls. The projectiles would have been molded lead balls wrapped with a thin piece of leather or cloth, greased for additional lubrication. This patched ball would have slid rather easily into the bore, not requiring a heavy ramrod to hammer it down the bore. The patch would have fit the grooves of the rifling tightly, transferring their twist rate to the ball as it sped down the barrel when fired.

The original wheel-lock guns that initiated the jaeger style had short, thick buttstocks, often combined with enlarged, elaborate trigger guards with deep, formed trigger grooves. This style of stock and trigger guard was not intended to be used to hold the butt up against the shoulder of the firer. Instead, the weapon was to be braced against the cheek of the shooter for firing, the stock firmly held by the firing hand, the fingers slipped into the grooves of the trigger guard.

The flintlock jaeger developed a different kind of stock, one very recognizable today. The rear of the weapon had a thick, shaped butt that could be comfortably braced against the shoulder when fired. A raised comb on the stock allowed the shooter to lean his head against the stock, bracing it against his cheek for a careful aim. The thick design of the stock would have helped hold it in place even when the shooter was wearing heavy leather or woolen clothing. Another feature of the jaeger stock was that it usually had a hollowed out box in the butt, sealed with a sliding wood cover. The box would be used to hold accessories used with the weapon such as attachments for the ramrod to help clean the bore, spare flints, and small tools.

An additional characteristic of the jaeger stock and its furnishings was that the weapon almost always had a pair of sling swivels as normal accoutrements. The swivels were on the bottom of the buttstock and back a bit from the muzzle, both parts being substantial in their build. The swivels were intended for a stout sling, one capable of rough use. With the location of the sling swivels close to the opposite ends of the weapon, the jaeger-style rifle could be easily slung across the back from shoulder to hip, remaining out of the way and secure during rough travel or a hard climb.

The choice of calibers for the jaeger was significant. In general, they were large-bore rifles, smaller in caliber than the issue muskets of the era. Outside of jaegers built as wall-pieces (weapons intended to be fired from the walls of fortifications), the average hunting jaeger was of .58 caliber. Depending on how thick a patch the hunter wished to use, a .58-caliber jaeger would fire a ball weighing about three-quarters of an ounce. Pushed by 75 to 125 grains by weight of quality corned gunpowder, the jaeger ball would have a muzzle velocity of over one thousand feet per second. That would give the weapon more than enough power to take down dangerous game in Europe.

There was no question that dangerous game lived in the forests of Europe. The wild boar of the woods could stand more than three feet tall at the shoulder and weigh nearly five hundred pounds. Armored with a coat of coarse, bristly hairs over a thick skin and heavy muscle, these awesome animals were armed with a set of four curved tusks, sharp and jagged from being broken off by rooting in the ground. Driven by a mean temper and the heavy muscles of the neck and shoulders, the tusks were capable of inflicting terrible, life-threatening wounds.

One of the reasons for the short, handy length of the jaeger rifle was so that it could be quickly swung around and brought to bear on a boar deciding to take out the hunter. The power of the heavy ball launched by a jaeger could penetrate deeply enough and deliver enough energy to stop a boar's charge, even when it was close in. Another reason for the power of the jaeger was so that it could deliver a killing shot for much less dangerous game during hunting in the mountains. Mountain goats, sheep, and chamois would live among the crags and cliffs of the alpine ranges of Europe. Even though a clean shot could be over only one hundred or two hundred yards, the distance was deceptive. It could take the better part of half a day to climb down one crag and up the other in order to cross that one hundred-yard shot, and not fall in the thousand-foot ravine that lay between the shooter and his target.

The power of the jaeger for relatively light game meant that the hunter could put down his target with a single shot. This was very important given the extreme difficulty in chasing down a wounded animal over such terrain. It was also one of the reasons that accuracy was demanded of the jaeger rifle. A European hunter, especially one who climbed in the Alps, learned how to estimate ranges very well from his earliest days on the mountain. These hunters could make good use of an accurate weapon and demanded one that was dependable as well. The weight of ammunition was not of major concern since such a hunter would only expect to fire a few rounds at most during a day. But power, range, and accuracy in a compact package were considered of great importance.

In modern benchrest firing tests of a .62-caliber jaeger, the weapon was found capable of three-quarter-inch groups of shots at fifty yards, and two-and-three-quarter-inch groups at one hundred yards. This put it in the same class of accuracy as a modern hunting rifle of the time (mid-1950s). The sights of a jaeger were complex enough to take some advantage of the accuracy of the weapon. A blade or bead front sight was installed in a dovetail base, allowing it to be adjusted to the impact point of the rifle. The rear sight was usually a flat bar with a V notch in it for sighting. For differing ranges, there would be one or two flip-up leaves that were adjusted for greater distances. All of this on a weapon with an average weight of less than eight pounds.

The design proved so accurate, powerful, and useful that it had been adopted for military service in several countries by 1700. In Scandinavia, the military first bought jaeger rifles as early as 1711. By 1794, an entire Danish-Norwegian ski troop was armed with M/1785 jaeger rifles. These weapons were capable of hitting a man-sized target with individual fire at three hundred yards. Military muskets of the era, even when used in volley fire, were not considered capable of even reaching a target that far away. During the wars of 1807–1814, the jaegers proved their worth, firing several rounds a minute in combat against Swedish troops.

The jaeger rifle was the first rifle to be used extensively by any military in warfare. It also dominated the European world of hunting for well over a century. Lastly, it was unquestionably the forefather of a uniquely American weapon, the long rifle.

THE AMERICAN LONG RIFLE

Known by a variety of names, the long rifle, the American rifle, the Pennsylvania rifle, and most commonly, the Kentucky rifle, this unique American weapon was developed directly from the jaeger rifle, even as it reigned as the supreme hunting weapon of central Europe. With its long, graceful lines, medium cal-

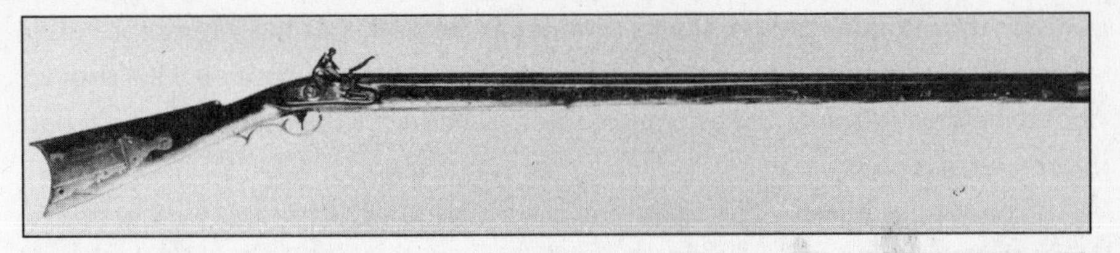

An example of the type of long rifle that would have been used by riflemen during the Revolutionary War. This specimen has a dark stock with a relatively simple brass-covered patch box at the end of the butt. *Smithsonian Institution*

iber, and flintlock action, the long rifle holds a special place in the history of the United States.

European settlers brought their skills with them when they came to the New World in the early 1700s. Among those people were German and Swiss gunsmiths who carried with them their jaeger rifles, and the ability to make more. By 1710, the jaegers had begun to evolve into the long rifle. The calibers necessary to stop a charging boar or drop a sheep in its tracks were not needed in the forests of eastern North America. Economics also helped force a change in the basic weapon of the frontiersman. The rifle had to be accurate to minimize lost shots and waste. But the caliber of the weapon could also be reduced. Gunpowder and lead were expensive, and the quantities a hunter had on hand had to be used in the most efficient way possible. This dictated a smaller-caliber rifle. The most dangerous game in the eastern forests was the larger black and brown bears, along with the armed local Indians. A .50-caliber ball was found to be effective in stopping both.

Reducing the caliber of the weapon was a simple lesson in economics. Instead of having a pound of lead cast twenty-four or twenty-five .58-caliber balls, a .50-caliber gun could have thirty-eight or thirty-nine balls cast from the same amount of materials. Economy in gunpowder was just as obvious when the propelling charge was able to be reduced from the 75 to 125 grains of the jaeger to the 30 to 60 grains of the long-rifle charge. The smaller projectiles left the barrel with a greater muzzle velocity than the heavy balls launched by the jaegers, giving them a flatter trajectory. This meant that the hunter did not have to judge distance as closely as his European counterpart did, to account for the high ballistic arc of the heavy jaeger bullet.

Along with a bore reduction from around .60-caliber to .50-caliber, the gunsmiths lengthened the barrels of the new rifles out to an average of forty-two to forty-six inches. The intent of the longer barrel was to make greater use of the burning gunpowder. It was thought at the time that the longer the barrel was, the harder and longer the powder could push on the bullet. This proved not to be the truth, but the science of internal ballistics that would prove it had yet to be invented.

The longer barrels did make for a very long sight radius, which helped give the weapon a more accurate aim. The muzzle-heavy aspect of the barrel also made the rifle steadier to hold when fired from an off-hand, standing position where the shooter is holding the rifle up to his shoulder with no other support than his hands.

Stocks were made along the same lines as the jaegers; they extended all the way to the muzzle of the rifle, but that's where the resemblance ended. The long rifle stocks were more slender and much lighter than those on the jaegers. The material the stocks were made of changed from the walnut of Europe to the striped curly maple of America. The hard, dense maple made an excellent stock wood, and the dark stripes of the grain have become synonymous with the long rifle in American lore. Careful treatment of the wood and metal of the long rifle gave it a deep mellow finish on the wood and a brown patina on the steel barrel and lock parts.

These features were not just for appearances. The lack of bright steel meant that the gun would not reflect light at an inopportune moment, frightening off game or giving the shooter's position away prematurely. The browned metal also took longer to rust, aiding the frontiersman in maintaining his weapon. The hard grain of the curly maple gave strength where it was needed in the long stock.

The results of all the development of the long rifle resulted in a standard, distinctive style of weapon by about 1740. It was a graceful-appearing weapon, long and slender with a gentle curve to the buttstock. The end of the stock was usually capped with a brass butt plate for strength, curved to give a deep pocket to quickly and naturally seat against the shoulder. The long rifle style was missing a significant accessory found on the jaegers. There were no provisions for sling swivels on the weapon. The length of the long rifle prevented it from being easily carried across the back—the extended muzzle would have caught on every low branch and bit of brush the shooter passed under. So the sling was left off the weapon, resulting in the shooter always having to carry the long rifle in one hand.

The spread of gunsmithing across the frontier prevented the European craftsmanship from being maintained by the normal apprentice system of the Old World. Quality work was demanded by the users; on the frontier a man's

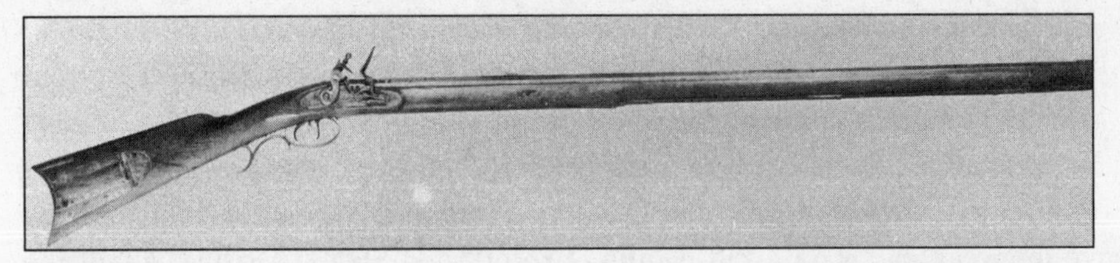

Another specimen of the American long rifle, commonly but erroneously called the Kentucky rifle. This weapon has a light-colored stock surrounding the traditional octagonal barrel. A simple hinged brass cover seals off the patch box in the butt of the weapon. There are two triggers under the weapon, with a small screw just visible sticking out between them. These are set triggers, the weight of trigger pull when "set" being adjustable by moving the small screw in or out. Pulling the front trigger in the normal fashion give a standard several-pound trigger pull that will fire the weapon. Pulling back the rear trigger until it catches "sets" the trigger. In the set condition, a bare touch on the front trigger will fire the weapon. This light trigger pull allows for maximum accuracy when the shooter is in a solid, braced position. ***Smithsonian Institution***

life easily depended on his weapons. But the fine details and polish of Old World craftsmanship made way for production and ease of repair. The results were rough and crude from a metalworker's standpoint, but they did work and tended to work well.

A good man with a long rifle could get off two aimed shots a minute, three if he was well practiced at fast reloading. The rate of fire with a long rifle was slow in comparison to the four or five rounds a minute of a standard military musket of the day. But the shots from the long rifle were very accurate, and at a much longer range than the musket could reach. The sights of the long rifle were plain and not adjustable to speak of. Usually the weapon had a blade front sight, often nickel silver, in a dovetail that allowed it some adjustment to the point of impact of the barrel. The rear sight was a simple raised V notch, adjustable for elevation only if a file was used on it.

In spite of the relative crudeness of the sights, the weapons and the men who used them were capable of excellent accuracy. Various shooting competitions were common, often held in the small towns and trading centers on the frontier. One popular sport was the turkey shoot. For this style of shoot, a live turkey was tied down behind some cover, usually a log. At a range of usually around 250 feet, the frontiersmen would line up to take their turns at trying not only to hit the turkey, but first to get the bird to raise its head up from behind the cover that hid it. This made for a small target, roughly the size of a tennis ball, which might pop up at any time, and disappear just as quickly. Firing was done from the off-hand standing position. It took a fast, accurate shot to hit the turkey in the head and take the prize—the rest of the bird—home for dinner.

During the War of 1812, frontiersmen armed with the long rifle fought off the British during the Battle of New Orleans. The men were generally from Kentucky and Tennessee, and they defeated the disciplined British solders with accurate long-range fire from fortified positions backed by artillery. The defeat of the British corresponded with the end of the war and quickly became a popular topic among the people of the United States. A ballad about the battle made the rounds and quickly became a favorite. In the tune, for the sake of rhyme and meter, the composer used the terms "Hunters of Kentucky" as well as "Kentucky Rifles." The American long rifle has been generally known as the Kentucky rifle ever since.

THE FERGUSON BREECHLOADER

The slowness of firing both the American long rifle and the Brown Bess musket was the result of the weapons being muzzle-loaded. To take advantage of the accuracy offered by the rifling, the long rifle had to be fired with a snug-fitting patched ball, slowing its rate of fire to less than half that of the drilled soldiers with muskets. The muskets could fire at a greater rate, but only if the shooters had no expectations of individual accuracy. If the muskets were rifled, they would have had greater range and accuracy, but would have been just as slow to load as the long rifle. And both weapons were best loaded while the shooter was standing up, so that the powder and shot went down the barrel and not spill out on the ground.

Captain Patrick Ferguson of His Majesty's 70th Regiment of Foot was a career officer in the British military. Of Scottish birth, Ferguson had long held a strong interest in marksmanship and firearms technology. Working from the 1721 patent of Frenchman Isaac de la Chaumette, Ferguson developed a working breech-loading flintlock rifle that he thought would be satisfactory for

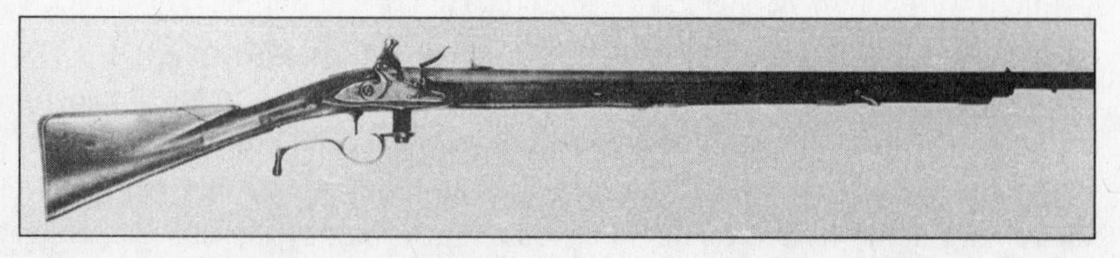

The Ferguson Breechloader with the trigger guard turned and the breech lowered for loading. A particular rarity for British military weapons prior to the 1800s, this rifle is fitted with sights on top of the barrel. The normal musket had no sights whatsoever beyond a stud near the muzzle intended for securing a bayonet. ***Smithsonian Institution***

British military use. Obtaining his own patent in 1776, Ferguson had several weapons made to his specifications for demonstrations.

The basic idea of the Chaumette action was a fast-thread screw plug at the back of the barrel, sealing off the breech. The plug was attached to the trigger guard, and by rotating the guard through one full revolution, a shooter could unscrew the plug, opening the breech for loading. After insertion of ball and powder and a quick twist of the trigger guard, the weapon was loaded.

Ferguson's modifications involved making the system less liable to jam up from powder fouling and generally making it a bit more soldier-proof. Having had two weapons made to his design by the London gunmaker Durs Egg, Ferguson arranged a demonstration of his weapon for Lord Townshend and several officers from the British Board of Ordnance. The demonstration went well enough that Lord Townshend set up a more formal demonstration of the new weapon's abilities with a number of high-ranking and influential officers observing. This was a significant time for Ferguson to demonstrate his weapon, as the British military had just issued purchase orders for rifles to be made. These weapons would have been of the standard muzzle-loading design but with rifled bores for accuracy.

On a rainy, miserable first of June, 1776, Captain Ferguson proceeded with his demonstration. As it turned out in more recent research, the Ferguson screwed-breech plug design actually benefits from a lot of moisture and humidity in the air. The moisture keeps the powder fouling soft and helps prevent it from jamming the action. Since the design of the Ferguson is a breechloader, it could be easily and rapidly loaded with a slightly oversized ball to tightly fit the bore. This meant that the design was fully able to take advantage of a rifled barrel without the slowing of the loading process. And the Ferguson already had a much faster loading process than any other weapon in the British military.

For nearly five minutes during the demonstration, Ferguson kept up a steady four-rounds-a-minute rate of fire. Then he rapidly fired six rounds in one minute. Going back to firing at a four-rounds-a-minute rate, Ferguson kept shooting and reloading while marching forward at a steady infantry rate. After pouring a bottle of water over the weapon with the breech plug open, Ferguson recharged the gun and fired within half a minute without having to remove the ball. Lastly, he shot at the target while lying on his back, still being able to reload. The range Ferguson was firing at was two hundred yards; he missed the target a total of three times during the entire demonstration.

The officers and the ordnance board personnel were satisfactorily impressed.

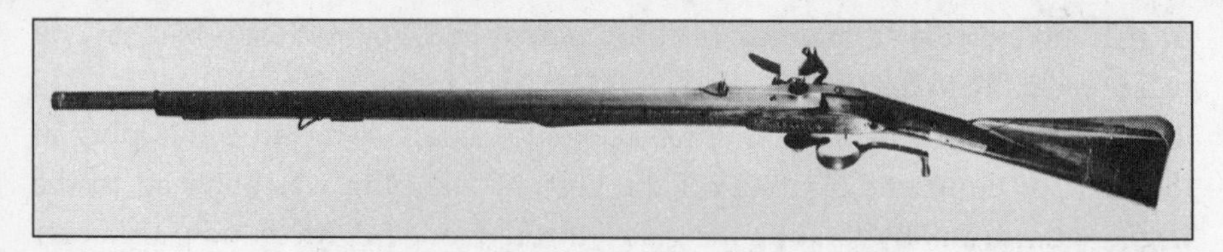

A view from the top of a Ferguson military rifle, the breech open for loading. The ball followed by the powder would be placed down the hole in the action, just next to the hammer in this picture.
Smithsonian Institution

Orders were issued for the manufacture of one hundred Ferguson rifles and corresponding bayonets for field trials. Captain Ferguson himself was put in charge of a detachment of men to be raised and trained by him for combat duty in the American colonies. He assembled the men out of volunteers from the 6th and 14th Regiments and put them through an intensive course of training. With his small Corps of Riflemen, Ferguson reported for duty in America in May 1777.

The basic Ferguson military rifle held a strong resemblance to the jaeger hunting rifle of the Continent. A flintlock weapon, the rifle was fifty inches long overall with a thirty-four-inch barrel and weighed seven and a half pounds, much more compact and lighter than the issue Brown Bess musket. The Ferguson rifle also was issued with a socket bayonet that could be firmly attached over the muzzle of the weapon. It would not be equaled until well into the 1800s.

9

WAR 1775 TO 1783

The American Revolution

The most significant conflict the United States has ever been involved in is easily the Revolutionary War which resulted in the birth of the country. Before the time of the revolution, the area that would be the United States was populated primarily with British colonies. Additional outposts along the frontier were filled with immigrants from throughout Europe, many of whom had a historical animosity toward British rule. Along the east coast of North America, the communities were well established and generally British in origin.

These cities and towns were the centers of wealth and industry in the colonies, with large areas of cleared agricultural lands surrounding them. The Indian Wars were largely over in these areas, and the military forces of the regions centered on citizen militias made up of armed volunteers. These militias drilled in the traditional sense of the British and European militaries. They fought with open-field tactics, making up ranks of the men and firing muskets in volleys at close range before making use of the bayonet. The loose ball loading and relative inaccuracy of the traditional smoothbore military musket were not considered a disadvantage as the rate of fire was thought to be a reasonable tactical compromise. Previously, many of the weapons of these militias were British Brown Bess muskets kept in private hands after the end of the French and Indian Wars.

For the frontier areas of the country, the situation was a very different one. The frontier settlements were deep in the wilderness, where life was not passive. Hunting was a constant means of obtaining food, and the frontiersmen lived with their rifles either in hand or nearby. The weapon of choice was not the inaccurate military musket but the long rifle of smaller caliber but greater range.

Fighting against an active enemy was also well known on the frontier. Many settlements were still actively engaged with an occasional Indian raid. Individual homesteads were also the target of attacks. The population situation forced the frontiersmen to fight as individuals or in small groups at most. Their method of combat was based very much on their hunting style; after tracking a foe, they would open fire at selected targets, usually at range. Supremely fit and hardened by their chosen lifestyle, these men were able to cover long distances on foot as a simple matter of necessity. As a hunter, the frontiersman could easily live off the land with no more support than the weapons and materials he carried with him.

THE FIRST BATTLES

Thomas Gage was the British general who was appointed governor of Massachusetts in 1774. His duties included the carrying out of Crown orders that helped drive the colonists into open warfare. The political unrest had been

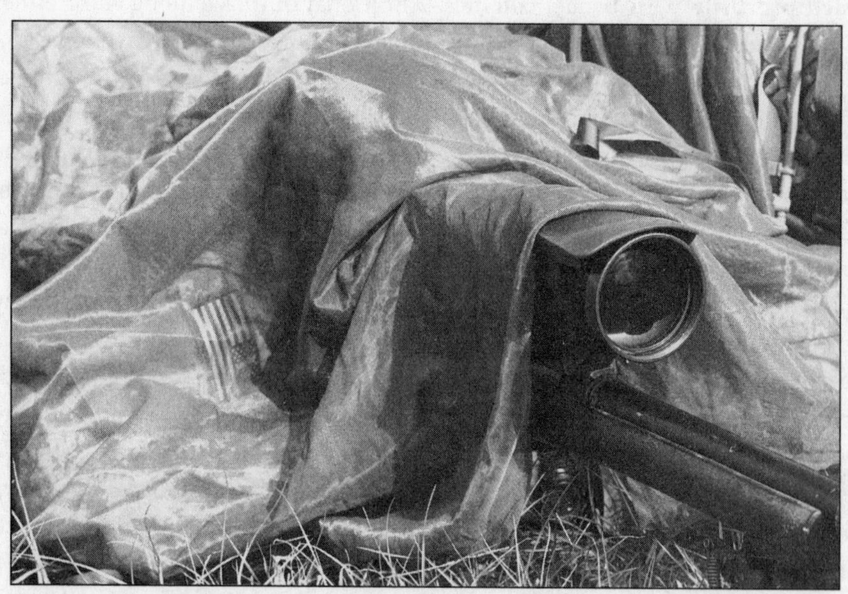

An Army PFC conceals his body and part of his M24 sniper rifle during training. The large objective lens of his AN/PVS-10 Sniper Night Sight is visible under his cloth veil. The sight was developed to give the sniper a lightweight sight that could be used both during the day and at night, with an effective range of eight hundred meters during the day and six hundred meters at night when used with the M24 rifle as seen here. *U.S. Army*

brewing for years, and Gage was not responsible for the reasons behind the rebellion itself. But it was his following of orders that did result in the first rounds being fired, and those shots signaled the beginning of the war.

Governor Gage was ordered by the Crown to suppress rebellion among the colonists by all means he had at hand. The triggering act for open rebellion was when Gage ordered seven hundred troops to march on Concord and seize the weapons and munitions stored there. It was to warn of the approach of these troops that Paul Revere and William Dawes conducted their famous ride across the countryside to awaken the populace and call out the militia.

On April 19, 1775, British forces consisting of seven hundred troops crossing into Lexington found a warned populace that was ready for them. The first responders of the Massachusetts militia, seventy of the minutemen, stood on Lexington Green and faced the British advanced guard. This was the situation that resulted in the "shot heard around the world" as the British fired a volley and pressed forward with a bayonet charge. Of the seventy militia members, eight were left dead and ten wounded. Regrouping, the British marched on to Concord to destroy the supplies there. It was at the North Bridge in Concord that the British ran into another group of militiamen and were driven back.

The British forces conducted a retreat for the twenty miles back to Boston under intermittent fire. They were harried from cover by farmers, militiamen, and woodsmen who fired and withdrew. It was a nightmarish retreat for the British, who suffered more than 250 casualties. Only the arrival of a relief force from Boston kept the British command from being wiped out. The news of the battle spread quickly throughout the colonies. The American Revolution was under way.

The Provincial Congress in Massachusetts ordered the mobilization of over thirteen thousand American troops within a week of the action at Lexington. Men of all kinds of experience—farmers, militiamen, hunters, and frontiersmen—from all around New England, gathered and laid siege to the British forces in Boston. The Second Continental Congress convened in Philadelphia with John Hancock as president. On June 14 the Congress authorized the raising of troops, including ten companies of riflemen from Pennsylvania, Maryland, and Virginia. They did not arrive in time to take part in the Battle of Bunker Hill on June 17, where the American rebels were finally taken by the British after running out of ammunition. During the assault, the British lost more than half of their effective force, with over a thousand casualties, to the four hundred casualties among the Americans.

A disproportionate number of casualties among the British forces at Bunker Hill were officers. Gunfire from the Americans, both those in the fortifications and those firing from the British flanks, was directed against officers. This was not simply volley fire; this was aimed fire from men with rifles. General Howe, who had taken over command of the British forces from Governor Gage, had his first taste of American marksmanship and the colonials' determination to stand their ground when necessary. Though the battle was considered a British victory, Howe was heard to later comment that after a few more such victories, the British military in the colonies would be wiped out.

It was not very long before the news of the colonialists and their deadly accuracy made the rounds of the regimental headquarters in England. The officer corps was soon rightfully terrified of the colonial riflemen. They considered the riflemen as little better than savages, not given to the rules of civilized warfare where officers were not to be selected out as targets to be killed before the rank-and-file troops. It became common knowledge that a young officer was to be certain to have his affairs in order before going to war in the colonies.

THE RIFLE COMPANIES

A veteran of the French and Indian Wars, Daniel Morgan was very much a self-made man. A successful farmer and captain of his local militia, Morgan was the local choice to raise and lead one of Virginia's companies of riflemen. At the head of ninety-six men, Morgan led the way along a six-hundred-mile trek from Winchester, Virginia, to Boston, Massachusetts, arriving on August 6. The combat effectiveness of such men was an unknown factor to the British, even though Morgan and a number of his troop had fought on the British side during the French and Indian Wars. Their marksmanship was considered extraordinary, conducted at almost impossible ranges by British standards. And the fighting spirit of the frontier riflemen was terrible to face.

There were problems combining the rifleman companies with more traditional troops. The riflemen did not stand for the traditional discipline that was necessary for the open-field combat of the day. They were far too independent for such actions. Open-field combat was also the weak point of the riflemen. The long rifle could not be loaded quickly, and this proved a detriment to open combat. At his best, a rifleman could get off four shots in a minute, starting with a loaded weapon. But that would almost immediately drop to two rounds a minute, while the musket-armed British were still able to fire rapid volleys. And when the combat closed to bayonet charge, the riflemen had to withdraw or be impaled. The long rifle did not take a bayonet, and the stock was not meant for the impact of close-in combat. The graceful and lethal long rifle would quickly become a clumsy club, and soon a broken one at that.

With regular troops to back them up, the riflemen could inflict terrible damage on assembled British units at long range, withdrawing before the combat became close. In the wooded areas, the riflemen reigned supreme. British forces could not assemble in their ranks and let go with their terrible volleys, no matter how disciplined, if there wasn't room to maneuver. In these situations, the riflemen could take out their chosen targets at will and withdraw, bleeding the British troops dry and leaving them confused and leaderless.

That was not to say that the riflemen had it all their own way against the British regulars. Even when firing from the cover of the woods into an open field or wide pathway, the riflemen often only had time for one or two shots at most before the British responded. With the iron discipline that made their military a force to be reckoned with over the years, hard-drilled British troops would quickly form into ranks and turn on their ambushers. The heavy volume of fire from the rapid volleys of the troops would make it suicide for the riflemen to raise their heads from cover in order to aim. Their only recourse would be to withdraw into the woods and choose another time and place to face the enemy.

It was dangerous enough, though, to be near a British officer when any colonial riflemen were about. Some officers were particular targets who were sought out by the colonials whenever possible. It would be hard to say a particular rifleman actually stalked a specific officer, as a modern sniper would be able to. Sources of military intelligence of that type were just unavailable to any of the fighting men during the Revolutionary War. But any rifleman who came in from the frontier had all the skills to conduct such a stalk if he desired to.

One particular British officer who could have singled himself out for such a stalk was Sir Banastre Tarleton. After several years in the fighting, Lieutenant Colonel Tarleton was authorized to raise a troop of loyalists from among the population, reinforced by regular British soldiers. The loyalists were American colonialists who remained loyal to the Crown and British authority. Made up of both light infantry and cavalry supported by a modest group of fieldpieces, Tarleton's troop was a complete fighting force, able to move quickly, even through rough terrain. The troop was known as the British Legion or Tarleton's Legion.

The actions of the legion were known to be brutal. In one chase after a retreating regiment under Colonel Buford, Tarleton's Legion caught up to them and slaughtered the colonials almost to a man. The term "Tarleton's

Quarter" was coined to reflect the mercy that would be forthcoming from some officers or units if any colonials were unlucky enough to be forced into surrendering to them. They could expect to be butchered and little else.

A close friend of Tarleton was Major George Hanger, the second-in-command of the British Legion. A longtime shooter and marksman himself, Hanger had a great respect for the abilities of the American rifleman which was not necessarily shared with his commander. What Hanger had little respect for was the quality and ability of the British Brown Bess musket, something he had written a scathing opinion of. His opinion of the long rifle was due to personal experience, something he also wrote about in his memoirs, *Life, Adventures, and Opinions*, published in 1801. Of a run-in he had with an American rifleman in 1780, he wrote:

> *To mention one instance, as proof of the most excellent skill of the American rifleman. If any man shew me an instance of better shooting, I will stand corrected.*
>
> *Colonel, now General Tarleton, and myself, were standing a few yards out of a wood, observing the situation on the part of the enemy which we intended to attack. There was a rivulet in the enemy's front, and a mill on it, to which we stood directly with our horses' heads fronting, observing their motions. It was an absolute plain field between us and the mill; not so much as a single bush on it. Our orderly-bugle stood behind us, about 3 yards, but with his horses' side to our horses' tails. A rifleman passed over the mill-dam, evidently observing two officers, and laid himself down on his belly; for, in such positions, they always lie, to take a good shot at a long distance. He took a deliberate and cool shot at my friend, at me, and the bugle-horn man. (I have passed several times over this ground, and ever observed it with the greatest attention; and I can positively assert that the distance he fired from, at us, was full four hundred yards.)*
>
> *Now, observe how well this fellow shot. It was the month of August, and not a breath of wind was stirring. Colonel Tarleton's horse and mine, I am certain, were not anything like two feet apart; for we were in close consultation, how we should attack with our troops, which laid 300 yards in the wood, and could not be perceived by the enemy. A rifle-ball passed between him and me; looking directly to the mill, I observed the flash of the powder. I said to my friend, "I think we had better move, or we shall have two or three of these gentlemen, shortly, amusing themselves at our expense. The words were hardly out of my mouth, when the bugle-horn man, behind us, and directly central,*

jumped off his horse and said, "Sir, my horse is shot." The horse staggered, fell down, and died.

Tarleton and Hanger both survived the incident and the war, going back to England after the conflict had ended. In looking at the situation, the details of the shot are interesting. They were two obvious British officers, recognizable by their uniforms. The fact that there were no troops about except for their bugler could also be seen by the rifleman, if Hanger's description of the concealment of the troop is accepted. In the military today, the obvious target would be the signalman, the bugler. He would act much as a radioman would today and would be the priority shot for a sniper. Take out the signaler, and the officers, unable to call for reinforcements, go down with the follow-up shots.

Even though this incident took place after five years of warfare, the two officers show little concern for the single rifleman. The idea of several of these shooters firing at them holds a higher level of threat, but not one man. Not all British officers were able to write down later how their opinions of the American rifleman went up after facing them in combat. Contempt for the rebel colonists, considered little more than traitors to the Crown, was common among the officer corps of the British Army. Their opinions were often changed at the cost of their lives.

A SIGNIFICANT SHOT LEFT UNFIRED

The use of marksmanship was not a one-sided affair during the Revolutionary War. The British made use of their own "local talent" in the form of loyalists who were also armed with rifles. These men were few in number, most loyalist units being armed with the Brown Bess musket and conducting drilled formation combat alongside the British regulars. There were scouts and skirmishers used by the British. These were primarily outdoorsmen with a high level of woods craft skills to draw on. They would travel ahead of the troops to observe the enemy and bring back information on their numbers, movements, and locations, among other military intelligence. Skirmishers would act as irregular troops, fighting out on the flanks and across the battlefront to draw out the enemy fire and guide them into position for the assault by the regular forces.

The British also brought in foreign troops to aid them in their conflict with the colonists. Prime among these foreign forces were German Jägers, the Hessians, supposedly crack troops armed with their own rifled weapon, the jaeger rifle. These troops were little more than mercenaries for the most part, and few of them were the actual quality men the British had expected. German princes and local rulers had competed against each other to offer the British the best deal on "seasoned" Jägers, capable of taking on the rebels with ease. Often enough, the Hessian troops turned out to be inexperienced, poorly outfitted, and barely trained conscripts used to fill out the ranks of units that existed solely to be rented out to the British. Nearly half of these soldiers never returned home to Europe after the hostilities were over; they opted to remain in the new United States and take up citizenship among their recent foes.

There was one unique group of marksmen who made up a single company of men in the British forces. This was Patrick Ferguson's Corps of Riflemen, a one-hundred-man unit led by Ferguson and armed with his breech-loading flintlock rifle. Arriving from England in May 1777, Ferguson and his company-sized unit were assigned to the British forces under the command of Lieutenant General Sir William Howe.

His relative youth and enthusiasm worked against Ferguson when he reported to his new commander. He was considerably hampered by two facts regarding his new unit. The first was that General Howe felt that he was one of the British Empire's foremost experts on light infantry tactics and their employment. Howe had little respect for a young captain who felt that he could show the general how to best employ men during a war. The other problem Ferguson faced besides the dislike of his commanding officer was one that would constantly plague the employment of specialized marksmen in the future military. General Howe had no idea how to employ the Corps of Riflemen to the best advantage of his campaign, and he had no intention of learning. Theirs was a brand-new technology for British warfare, and he could see no advantage in it over the tried-and-true "thin red line" of the British infantry firing their muskets in volleys before giving the enemy a taste of cold steel with a bayonet charge. In the end, Howe assigned Ferguson and his Corps of Riflemen duties as part of the division under the command of Hessian Lieutenant General Baron Wilhelm von Knyphausen.

That assignment worked out to be the best for Ferguson and the men armed with his new weapon. The Hessian general at least had some idea as to how to best employ riflemen among his forces. Knyphausen had some true Jäger units under his command, and he had some experience in the employment of riflemen. For the upcoming campaign, General von Knyphausen assigned Ferguson to be part of the vanguard of his division.

It was at what would be known as the Battle of Brandywine that Ferguson was given his chance to show what his men could do. At six o'clock on the morning of the eleventh of September, 1777, General von Knyphausen had his 6,800 men on the march to make contact with the colonial forces under the command of General Washington. His division would hold Washington's attention while General Howe brought up an additional 8,200 men along the right flank to attack from another direction. In the lead ahead of von Knyphausen's forces were Ferguson and his men along with the Queen's Loyal Rangers, an American loyalist unit that was descended directly from Roger's Rangers of the French and Indian wars.

The Corps and the Rangers had advanced only three miles when they ran into American forces. A fierce firefight erupted, one that ran on for miles as the colonials withdrew while keeping a running fire on the following British forces. The fight went on over a distance of five miles, the colonials stopping occasionally to fire an organized volley before returning to individual running fire. Ferguson's Corps and the Rangers followed doggedly, returning fire and maintaining discipline. The breech-loading rifles now proved their worth as the men armed with them could continue to load and fire while on the move, as Ferguson had done in his demonstration the year before. Finally, the Corps and the Rangers were ambushed from both sides of the road as they moved into a withering colonial fire.

With von Knyphausen bringing up his troops, the fight started to move against the colonials. Ferguson's riflemen were ordered to take up positions around a nearby house and to support the main body of troops. The Queen's Rangers along with additional British forces moved into the woods and drove the colonial forces out at the point of a bayonet. In the fighting, Ferguson was badly wounded in his right arm, a lead ball shattering his elbow. He was to be crippled by the wound for the remainder of his life, but his Riflemen had proved their worth at the Battle for Brandywine.

In spite of their gallant action, with the loss of their leader, the Corps of Riflemen was disbanded by General Howe; the men were absorbed into other units and their Ferguson rifles were put into storage. Having suffered through a number of severely painful operations, Ferguson would recover to return to combat against the colonials. Back in active service by 1778, he was promoted to major in 1779. Later, he was appointed as a temporary lieutenant colonel when he took command of three hundred American loyalists, officially known as the American Volunteers. The Ferguson rifle was brought back out of stores to arm part of the new unit, also called Ferguson's Sharp-Shooters.

Over his nearly year-long recovery period back in England, Ferguson learned to write with his left hand, his right being almost useless. One story he related to others about the battle of September 11 concerned what could have been the most significant shot in U.S. history—and how it was never taken.

During the scout ahead of the Hessians, Ferguson was in concealment at one point, prone on the ground and observing, as a pair of horsemen rode up. The men were within one hundred yards of Ferguson's position on the ground, off on his right flank. He noticed that one of the men was a Hussar, a member of the French cavalry, known allies of the rebels and enemies of the English. Following the Hussar was another rebel, a large man on a bay horse wearing what Ferguson thought was a dark green or maybe a blue uniform. One point he noticed about the other man's uniform was that he wore a remarkably large cocked hat.

Recognizing the pair as rebel officers, Ferguson ordered three of his men who he knew were good shots to steal up on the mounted pair and open fire on them. On reflection, what he had ordered his men to do—effectively ambush and kill two officers who were not conducting any open aggression—disgusted Ferguson and he rescinded his order.

As the two horsemen were moving, Ferguson stood up from his position and started moving toward the one wearing the large cocked hat. He called out to the mounted officer, who stopped and looked at Ferguson for a moment. Then the mounted man continued on his way in a calm manner. Once again, Ferguson called out and signaled to the other to stop, lifting his rifle as an implied threat. After glancing at Ferguson, the mounted officer kept his horse moving away at a slow, dignified walk.

In spite of knowing that he could have easily put six balls into the mounted man before he could have ridden out of range, Ferguson held his fire. He was impressed by the actions of the other officer, who had recognized the situation but refused to surrender to it. The other's cool, calm actions in spite of the danger made Ferguson feel that simply shooting the other man in the back would have been at best unpleasant. He let the mounted officer go, never knowing who he was or what his mission might be.

While he was among other wounded officers the next day, Ferguson recounted his tale of the unfazed officer and his actions while under threat. One of the surgeons who had been treating the wounded spoke up. He had been told about a significant rebel officer who had been out the previous morning with only a light infantry escort. The only man directly in attendance of the American was a Hussar officer. The description of the man fit every detail that Ferguson remembered about his remarkable encounter. Even after he learned the identity of the officer who had ridden away, Ferguson was not sorry that

A young sniper checks out a target through the telescopic sight on his M25 sniper rifle. The M25, an improved version of the M21 weapon, has a much more secure scope mount than the previous weapon. ***U.S. Army***

he had allowed so dignified and cool an individual to escape. If he had shot, the world, and most certainly the United States, would have been a much different place.

That mounted officer in the large cocked hat had been General George Washington.

THE AMERICAN RIFLEMAN

During his first major assignment at the head of his company of riflemen, Captain Daniel Morgan took part in the Battle of Quebec. The mission was intended to wrest control of Canada away from the British, effectively to turn it into the fourteenth American colony. Morgan was considered an effective enough leader that he was put in charge of three rifle companies for the operation.

The operation inside of Canada took place from November through December 1775. During the January 1 assault on Quebec, Morgan's column of men ran into a major tactical difficulty. The riflemen were not able to put either their skill at woods craft, or more importantly the range their weapons afforded them, to much use in the narrow streets of the city. The riflemen were surrounded by enemy forces, cut off from their support, and facing bayonets at close quarters. In spite of all of the advantages belonging to the enemy, Morgan refused to give quarter. He fought on until his back was literally up against a wall. With his sword in his hands, Morgan stood defying the British to come closer. He was finally talked into surrendering by his own men, who would rather have seen him as a prisoner than watch him die spitted against a wall by British bayonets.

Captured officers, even rebels, were regarded with some measure of respect by the British troops. Morgan did not go easily into captivity, but neither did he spend the remainder of the war as a prisoner. There was traditionally a prisoner exchange program set up between British command and their adversaries during a conflict, and the Revolutionary War came at least partly under that tradition. A prisoner exchange was only like-for-like, equal branch and rank for equal branch and rank. It was nine months before there was a British infantry captain taken prisoner by the colonialists available to be exchanged for Morgan. He returned to the American forces in September 1776.

Exchanging Morgan back to the Continental Army was going to prove a costly mistake by the British. General Washington recognized the abilities Morgan had as a leader, and the esteem his men held him in. He was put back in charge of a corps of riflemen and sent out to quickly become a thorn in the side of the British forces. Along with 180 Virginia riflemen of his choosing, Morgan was given command of an additional five hundred expert marksmen to fill out a regiment. This unit would be known as Morgan's Continental Rifle Corps. During the winter of 1776 and into the spring of the next year, Morgan's riflemen chased after the British, harassing them at every opportunity. The men of Morgan's Rifle Corps took out the British foragers as they scoured the countryside for additional food and supplies. The life of a British trooper performing a rear guard action or picket duty became forfeit to a shot from the surrounding woods.

His experiences while serving with the British had taught Morgan well about their dependence on a functioning officer corps. He told his men constantly to shoot at the British troopers who were wearing epaulettes on their uniforms, since that was a unique item of wear for the British officers. And to make sure his men would hit their targets, he had them practice their marksmanship on a regular basis.

This was a time when the regular armies of the world did not practice anything like what we would call marksmanship today. For the most part, shooting for the troops was a part of drilling. They fired when, and how, they were told to. Marksmanship wasn't a part of the traditional military equation; discipline and volume of fire was. Among the riflemen, if you couldn't hit a piece of paper the size of a dollar at sixty yards with every shot, you practiced until you could.

SARATOGA—FREEMAN'S FARM

In the fall of 1777, British General John Burgoyne devised a plan of action that would militarily separate New England from the rest of the colonies by splitting the state of New York. With eight thousand British forces available to him, including six regiments of Irish troops under General Simon Fraser, combined with the three thousand Hessian troops available to him, Burgoyne intended to take possession of the Hudson Valley. Once he reached Albany, he would join his forces with General Howe's troops approaching from the opposite direction and effectively split New York.

Burgoyne was still operating with the opinion that standard British land tactics would be successful in defeating the less organized rebel forces under Washington. He completely ignored the fact that he would be traveling through enemy territory where the entire countryside was united against him and was armed. Traditional tactics were not going to work in Burgoyne's favor, not in the kind of terrain he was heading for.

The British forces under Burgoyne would be traveling through forested country for the most part when they moved through New York. There were large stands of trees that stretched for miles, facing the few roads and large paths that cut through the rural areas of upstate New York. Farmland was open pastures and fields surrounded by more forest. It was made-to-order terrain for the American frontier woodsman.

General Horatio Gates was given the task of organizing the Continental Army forces for the defense of the New York countryside. Gates was much more an administrative officer rather than a tactical genius, which was a fact he recognized. Gates did know some very good leaders, and was more than willing to put them in charge of the divisions under his overall command. He was already very familiar with the now famous Colonel Morgan and the capabilities of the riflemen under his command; Gates had been the commanding officer of the ill-fated Canadian invasion and had led the retreat of the escaping forces back to the colonies. Knowing what he did, and having a very good knowledge of what it would be like where they were going, General Gates made a special request that Colonel Morgan and his Rifle Corps be assigned to his forces. The request was granted.

After weeks of harried movements through the area, General Burgoyne was concerned about the situation ahead of him. The British troops had won a number of engagements against the colonial forces while moving through New York, but military requirements had forced Burgoyne to leave some of his troops behind to garrison the locations he had captured. By September, his supplies were running low and his forces had been depleted enough to make him uncomfortable with the situation. What Burgoyne didn't know was that General Howe had turned away from linking up with Burgoyne. Instead, Howe was following plans of his own to attack and capture Philadelphia and had headed south.

Intelligence on what lay ahead of his forces came to Burgoyne through the use of his scouts, particularly the Indian volunteers he had available to him. During his southward march, Burgoyne had sent his Indian scouts out ahead of the main column where they could look over the land but were outside the range of immediate support from the troops. In the woods, the Indian scouts were efficiently waylaid by Morgan's riflemen and forced to withdraw back to the main column or be wiped out. Without his scouts, Burgoyne was blind to what awaited him, so he sent out a reinforced advance party to scout the area ahead.

On September 19, while under the command of General Simon Fraser, the advanced party of Burgoyne's forces was attacked by Morgan's riflemen at Freeman's Farm outside of Saratoga. Before the British troops could start to form up from their marching column, Morgan's men were inflicting heavy casualties. Before Fraser's forces could reach open ground where they could maneuver and set up ranks, the riflemen charged. Their long rifles may not have had bayonets, but the men under Morgan's command knew very well how to use the knives, tomahawks, and hatchets they carried with their gear.

In spite of Morgan's men intentionally targeting and decimating the leadership of the British forces, the troops had retained some of their long-drilled discipline, along with experienced NCOs. When Morgan's men came up against the main body of the advanced column, they were facing a prepared enemy lined up and ready to volley fire. Under the threat of the massed fire of the British forces and a bayonet assault, the riflemen withdrew back into the cover of the protecting forest.

Advancing to the attack, Burgoyne brought up his forces to the area of the farm. Nearly three thousand men under the command of General Benedict Arnold came up to face the British forces, and they made a fight of the situation. After four hours of heavy combat, Arnold pulled his men out of the fray. The British had won the farm, but at a terrible cost. Out of the 800 men who faced the active combat on the farm, 350 were either dead or wounded by the end of the battle. And the worst situation was the loss of a large number of the British officer corps to the deadly fire of Morgan's riflemen. Sharpshooters had continued their habit of picking out and targeting the officers and NCOs from the ranks of troops maneuvering across the battlefield.

Morgan had proved the value of the precision fire of the riflemen under his command. The corps of riflemen withdrew through the cover of the trees with few losses to their number. The British forces settled in on their hard-won battlefield.

SARATOGA—BEMIS HEIGHTS

More than three weeks after the battle at Freeman's Farm, Burgoyne's forces were still in the Saratoga area. The situation with the British troops in regard to their supplies was worsening. Feed for the horses was in such short supply that they were subsisting on little more than grass; they had been stripping the farm area around the encampment bare and were starving. The troops had little better in the way of food; their rations had been reduced to one-third to stretch out the supplies. Foraging parties were having little luck in gathering game. Morgan's riflemen were still in control of most of the forest, and they had seen to it that the local wildlife was gone from the area. The British troops out hunting were also gone from their area when they ran into Morgan's men.

The Continental Army forces had fortified their positions on Bemis Heights. On October 7, Burgoyne set out to take Bemis Heights while his command still had the strength to function. Leaving eight hundred men back at the encampment around the Freeman Farm, he set out to assault the rebels with a three-pronged attack, planning his march to arrive on target in the early afternoon. If the battle did not go his way, he could then plan to withdraw under the cover of darkness.

Brigadier General Simon Fraser was to slip through the forest area on the left side of the Continental positions to attack on their flank. To move quickly and quietly, Fraser's forces were made up primarily of his light infantry regiments, the loyalist and Canadian militia and ranger forces, and the remaining Indian scouts.

The militia, ranger, and Indian forces did not fare well when Morgan's riflemen fell among them and wreaked havoc. Even though his men were slightly outnumbered, Morgan's forces held the British at bay, not allowing them to come up on the flank of the Continental forces.

The rest of the battle was also going against Burgoyne. His assaults were repulsed by the numerically greater colonial forces he was facing. When he put out the orders for his Hessian forces to withdraw, the officer carrying those orders was wounded and captured. The Hessians fought on without knowing they were to have abandoned their positions.

General Simon Fraser showed his abilities as a leader, rallying the flagging British forces. Riding across the front of his troops on a gray horse, Fraser was an obvious assembly point for the men. The actions of the brave British officer did not go unnoticed by the other side.

Observing General Fraser and the reaction he was having among the enemy, General Benedict Arnold spoke up to Colonel Morgan. He considered the officer riding the gray horse to be as great a danger to the Continental forces as the troops he was directing. He had to be disposed of. Seeing the bravery of the gallant British officer, Morgan reluctantly agreed with Arnold's view of the situation.

There are conflicting accounts of just who it was among Morgan's riflemen who brought down General Fraser. Given the importance of the target, it is very likely that Morgan ordered a number of his best marksmen to bring down the British officer before he could complete rallying his troops. Without any consideration for his personal safety, Fraser was riding up and down the British line, leading his men by example and demonstrating what courage in the face of the enemy really meant.

There was some reluctance in Morgan's command when he told his men to bring the gallant officer down. The distance to the target was probably in the neighborhood of three hundred yards, though this has never been determined exactly. One of the men that Morgan ordered to target the distant rider was Tim Murphy, a rifleman of particular skill and experience.

Growing up on the frontier in western Pennsylvania, Tim Murphy developed his fighting and shooting skills early on in his life. As a young man in the wilderness, Murphy faced Indian raids, home burnings, and other outrages against his family and neighbors. When the British raided Lexington and Concord, Murphy enlisted as soon as he could. By the end of June 1775, Tim Murphy was in a company of riflemen on their way to the siege of Boston.

Through the balance of 1775 and 1776, Murphy took part in a number of engagements in the New England and New York area. He was a well-seasoned fighting man when he was chosen to be a member of Daniel Morgan's Rifle Corps. His accuracy with a rifle was noted by his peers in the Corps, all of them capable marksmen in their own rights.

When the time came that Morgan had to assign men to take out the difficult target of a man riding on horseback at long range, Tim Murphy was one of the men he turned to to make the shot. Given the value of the target, Morgan would have assigned at least several men to the shot. Too many shooters firing and Fraser might take cover. Just one and Fraser might move off during the long reload time.

Three hundred yards—that was the minimum distance between Morgan's shooters and where General Fraser was riding. And the distance to the target was not a constant one. With Fraser on horseback, riding in front of his troops, he was moving along at a diagonal angle to where Morgan's shooters were setting up their shots. As Fraser rode, the distance to the target changed, so the shooters had to account for that fact as well as just deal with a long shot on a moving target. While the other riflemen took cover on the ground, shooting from their preferred prone positions, Tim Murphy climbed up a tree to get a clearer shot.

Fraser knew he was in danger even before the first bullets flew by him, but he chose to ignore the situation. It was an open battlefield; muskets and cannon were firing all around, and the occasional musket ball would snap by almost regardless of where he was. And the general felt that his personal honor demanded that he set an example for his troops.

A number of witnesses to the incident later noted what they saw on the battlefield that day. The first shot aimed specifically at Fraser missed, passing behind him. The bullet was so close that it cut through the saddle crupper, a leather strap extending out the back of Fraser's saddle, to the horse's tail. Still, the British general rode on, extorting his men to further bravery in the face of the enemy.

The second shot also missed Fraser, passing only a few feet in front of him as the lead ball grazed his horse's ears. Up in his tree, Murphy had settled his long rifle down comfortably into a convenient notch in the branches. Pulling the cock back, he would have heard the click as it snapped into the full cock position. Aiming carefully, he squeezed the trigger and the cock snapped forward. The bright flash of the sparks would have been quickly covered by the puff of smoke rising from the flash pan. The third shot boomed out of the barrel and sped toward Fraser, the muzzle flash cutting through the white smoke as a bright orange flame. There was no fourth shot.

The bullet struck Fraser low in the abdomen, cutting through his intestines and internal organs as it passed from one side of his body to the other. It was without question a mortal wound; the men who rushed to the general's side as he tumbled from his horse would have known that. With the state of the medical arts being what they were at the time, Fraser was doomed to die from either the wound or the terrible infection that would follow the path of the lethal ball.

It was mid-afternoon when Fraser was struck. He was carried to the rough British field hospital by his men, arriving there around three o'clock. There was little enough that anyone could do. Pain medication was barely in existence, alcohol serving as a poor substitute. And pouring alcohol into the stricken man would give him no relief. With several of his men at his side, Fraser suffered through the night, occasionally speaking out as he lay dying. He was dead by eight o'clock the next morning.

A British sergeant who attended to General Fraser during that long night later wrote about Fraser's last words. In 1809, Sergeant Robert Lamb's book stated that while on his deathbed, Fraser spoke of what he had seen during those last moments of his battle. He "saw the man who shot him; he was a rifle man, and he aimed from a tree."

General Simon Fraser was far from being the only British casualty at that battle. General Burgoyne lost a thousand men from his forces during all of the Battle of Saratoga, along with most of his valued leaders. He withdrew back to the positions he had held on September 16. With his forces spent, Burgoyne retreated farther, taking his men on a withdrawal to Saratoga eight miles away. With his forces surrounded by a much greater Continental Army host, General Burgoyne finally capitulated. He formally surrendered on October 17. With his word that neither he nor his men would further serve in North America during the present conflict, Burgoyne was released along with his men to return to England.

The results of the Battle of Saratoga were very significant for the cause of American independence. It was the first major victory for the young Colonial Army and was noted throughout Europe. In France, there were victory celebrations in the streets of Paris as if the battle had been won by the French themselves rather than the Americans. Benjamin Franklin was officially received by the French Royal Court, who had agreed to recognize an independent America, striving to be free of British rule. Discussions were opened, and France prepared to form an alliance with the Americans against Britain. There were years of conflict still ahead, but a major turning point in the birth of a new country had been reached.

KING'S MOUNTAIN

Patrick Ferguson's involvement with the Revolutionary War did not end with his being badly wounded at the Battle of Brandywine. With his right arm severely crippled and stiff, the young officer taught himself how to write, and handle a sword, with his left hand. He returned to duty within a year of his being wounded, distinguishing himself during battles in the New York area. Ferguson's skills and abilities impressed his superiors, who were badly in need of qualified officers.

In 1779, Ferguson was promoted to major and selected to command the American Volunteers, a three-hundred-man unit of loyalists from the New York and New Jersey areas. The new unit was sent down to the southern colonies to aid in the British war efforts there.

After several successful operations working with Lieutenant Colonel Banastre Tarleton, Ferguson was appointed to the position of inspector general of the militia. As part of his new position, Ferguson was to raise a corps of volunteer loyalists. His personality and abilities with people lent themselves well to Ferguson's new mission. At the head of eleven hundred loyalist American troops, referred to as "Tories" by the British, Ferguson set out to subdue the rebelling colonists in the Carolinas.

The depredations and actions of both the British and loyalist forces had worked to bring a number of scattered backwoods patriots from across the mountains together into a cohesive force. Organized into rough regiments of patriot militia, these men marched against Ferguson and his men. The hatred of the backwoodsmen for the loyalists was strong enough to drive these men forward. The earlier actions of Tarleton's troops in slaughtering their foes, rather than accept their surrender, stoked the fires of the backwoodsmen's determination. They crossed rivers, mountains, and forests to catch the biggest unit of what they saw as traitors to American independence, Ferguson's Corps of Volunteers.

Considered by his peers as one of the best professional soldiers in the British forces serving in North America at the time, Ferguson knew he was operating against a very hostile populace and armed foes. He disagreed with Tarleton's methods, but the man was his superior officer and Ferguson was going to follow his orders. He knew the patriot militia was hot on his trail, and he tried to avoid direct conflict with them while sending out a request for reinforcements from higher command. Ferguson could act as a draw for the patriot militias and bring them out into the open. His greater abilities as a military officer would be enough to allow him to set up the patriots for destruction. They were at best a disorganized rabble rather than a proper military organization in his opinion, and he knew their riflemen could not stand against an organized military defense.

While trying to avoid contact with the patriot militia, Ferguson moved his command to a prominent rise in the local geography. He set up a base camp on King's Mountain, about thirty miles from Charlotte, North Carolina. On October 6 at King's Mountain, Ferguson established a loose fortification to better draw out the patriots and force them into an uphill fight.

After weeks of chasing the loyalist forces, the patriot militias, under the leadership of Colonel William Campbell of Virginia, were coming within range of the enemy. Speed became the most important factor if the patriots were to reach Ferguson and his men before any reinforcements could arrive to support him. A forced march was begun on the night of October 5. Stripping down to the essentials for combat, nine hundred rifle-armed patriot troops started out on horseback to catch up to the loyalists.

It was the afternoon of October 7 when the patriot forces reached the bottom of King's Mountain. Surrounding the area, the patriot troops cut Ferguson off from any possibility of retreat. The upcoming battle would be a short and savage one of militia against militia, riflemen in the woods against musket-armed troops holding the high ground.

Almost none of the men on either side of the battle were regular soldiers. The ranks on both sides were filled out with civilian volunteers. That fact did not lessen the fighting spirit on either side. In spite of several bayonet assaults by Ferguson's troops, the patriot militia pressed on to the camp on top of the mountain.

Heavy, accurate fire poured up from the long rifles of the patriots on the flanks of King's Mountain. Instead of being in a defensible position, Ferguson and his men were trapped by the determination and ferocity of the surrounding patriots. In little more than an hour, the slaughter among the loyalist troops ended the battle.

There is no real evidence that any of Ferguson's troops were armed with his revolutionary breech-loading rifle at the Battle of King's Mountain. Because of his earlier injury, Ferguson himself would not have been able to operate the action of the weapon that bears his name. Even the accuracy and rapid firepower of his own design, however, could not have changed the fate of Patrick Ferguson on King's Mountain that October day. The man who had spared General George Washington when he had him in his sights was not offered the same courtesy by the oncoming patriot troops.

Struck down by at least eight bullets during the hour-long battle, Patrick Ferguson was the only British officer killed during the engagement. He had a great deal of company from among the ranks of the loyalist troops. Of the nearly 1,100 Loyalists on King's Mountain that day, 225 were killed, 163 wounded, and 716 captured. On the patriot militia's side, the losses were 28 killed and 62 wounded.

The battle proved a military disaster for the British war effort in the southern provinces. They had control of a number of major population centers in the south. It was the control of the rural areas, and the people who lived there, that eluded their grasp. Each loss they faced raised the hopes of the people for final victory in the war for independence.

COWPENS

The battle that decided the fate of the British in the southern provinces was going to be fought between two well-known leaders among the troops. Health and other problems had combined to force Daniel Morgan back to his farm and away from military service. By August 1780, after having received a letter from Congress, Morgan was once more in military service to the Continental Army. Once in the south, Morgan was given the command of a light corps of men made up of two companies of Maryland Continentals and 250 militiamen from Virginia. In October, he received a promotion to brigadier general. Now, Morgan was prepared to conduct a campaign of harassment of the British troops in the Carolinas.

On the British side of the southern conflict, Lieutenant Colonel Banastre Tarleton was conducting his campaign against the colonial forces, and he was doing a very successful job of it. At Camden in the middle of August, Tarleton's forces had taken part in the ruin of General Gates's forces. The defeat of Gates was so complete that he was recalled from command and the leadership of his army given over to Nathanael Greene. Three days after helping defeat Gates, Tarleton surprised General Thomas Sumter at Fishing Creek, defeating and scattering his forces. It was appearing that Tarleton could not be beaten in the south. And then Morgan and his men arrived on the scene.

In the middle of December, Greene broke the age-old military dictum of never dividing your forces when facing a superior enemy; he did just that. The countryside had been stripped by foraging armies on both sides, and there wasn't enough food and materials to support his army where it was. Giving over half of his forces to General Morgan, Greene took the remainder of his troops away from the Charlotte area.

General Morgan now had a force of nine hundred men under his command. Many of the men were experienced troops, but there were also among the units of militiamen that Morgan did not hold a lot of faith in. These militias had not proven themselves able to stand up to the British thin red line of fixed bayonets. And such a situation was sure to arise in the coming weeks.

As the people in Britain celebrated the successes of Lieutenant Colonel Tarleton, Morgan and his men were riding through the countryside of North Carolina. He quickly became a thorn in the side of the British military as he raised the morale of the locals and militiamen came in to join his forces. When Tarleton learned of Morgan's general location, he set out to bring Morgan and his men to heel. Since Tarleton was leading a unit of crack British troops, veterans of several successful campaigns, he considered it just a matter of time before he either captured or killed Morgan and his men.

When Morgan moved into South Carolina in December, he was soon joined by an additional sixty new militiamen under the command of Colonel Andrew Pickens. His operations against the British supply lines were drawing the attention of the enemy leadership. When Morgan heard that Tarleton had started to move against him, he started to head north to stay ahead of the oncoming force.

On January 12, 1781, Tarleton's advanced scouts found signs of Morgan's forces at Grindal Shoals on the Pacolet River in the backcountry of South Carolina. With solid information on the location of his prey, Tarleton started out in a hot pursuit. Rain-swollen rivers blocked both Tarleton's chase and Morgan's withdrawal. The distance between the forces closed in spite of the rough going. On the morning of January 16, Tarleton's forces came so close to Morgan's encampments that Morgan's men had to abandon their breakfasts heating in the fires to move out ahead of the British. Deciding to make a stand and fight it out, Morgan directed his men to head for the Cowpens, a crossroad and pasturing area. There, with the raging floodwaters of the Broad River six miles to his rear, Morgan stood.

The Cowpens was a known area, and Morgan had told the local militia units that wanted to join with him to meet his forces there. During the night of January 16, Morgan walked among his men, raising their morale and fighting spirit for the coming battle. He was the hero of Saratoga and an old Indian fighter. The men had faith in their leader, and he knew how to use their skills to the best advantage.

The next morning was a cold one when dawn broke. News from the incoming scouts suggested that things were going to heat up in a hurry when they announced that Tarleton's forces were approaching. It was January 17 and the British would be there within hours. Making his final preparations, Morgan sent a number of his best sharpshooters into the tree lines facing the most likely path of enemy approach. Their job was to be the one that Morgan had been giving his shooters all during the war: Pick off the enemy, aiming for the officers whenever they could be identified. Aim for the epaulettes. When the enemy drew within musket range, withdraw to the next fighting lines.

Setting up his other forces in three ranks, Morgan prepared to face his enemy with concentrated firepower. Volley fire was not something he was noted for and wouldn't be expected by the oncoming British. At worst, the British forces would expect the ranks to break after a volley or two at most. They knew that the ragtag militia troops had no stomach for the bayonet in the hands of a British soldier.

The fast-traveling Tarleton arrived and sent out his dragoons, horse-mounted cavalry known for their shock tactics of charge and no quarter. Between two fifty-man units of dragoons, Tarleton had his artillery set up. Grape shot (heavy lead balls loaded into bags or canisters) and solid shot would mow down the ranks of the rebels. Those who were still standing would be made short work of by the charging dragoons.

The first line of Morgan's troops, his sharpshooters, opened fire on Tarleton's dragoons as they came into range. The aimed shots of the riflemen took out as many as fifteen dragoons before the battle had fully begun. Reeling under the loss of their officers and men, the dragoons broke and retreated to behind the cover of the guns. Their first mission completed, Morgan's sharpshooters withdrew 150 yards back, to the second line of defense, the militia troops commanded by Andrew Pickens.

As the British troops maintained their approach, the militia and riflemen opened fire. The militia completed two volleys of fire as the British line kept approaching. After the second volley, they turned and ran, apparently broken by the stalwart British advance.

Having regrouped, Tarleton ordered his dragoons forward after the retreating rebels. As the dragoons chased the militia troops, fighting them at close quarters among the trees where they took cover, Morgan sprung a surprise. His own cavalry came out from around the flanks of the enemy, charging in among the dragoons and overwhelming them. The dragoons were already separated by entering the trees and were quickly overwhelmed by the patriot cavalry. Eighteen dragoons went down before the survivors fled the field under heavy fire from the surrounding militia troops.

Now the British troops started forward at a trot. They were shouting at the rebels as they advanced to the sound of the fifes and drums. Knowing the value of a united front, Morgan extorted his men to shout back. The new war cry of "Tarleton's quarter!" was shouted out across the battlefield among the war whoops of the riflemen.

Seeing that the battle was about to become fully engaged, Tarleton sent his reserve forces, the 71st Highlanders, into a charge against Morgan's men. The wail of bagpipes joined in the overall din of battle. Apparently in full retreat, the companies of militia turned and withdrew from the battlefield. This was what Tarleton had expected. The rebels had no stomach for a real battle and were proving to be the cowards he knew them to be.

Even Morgan was surprised at how quickly the militia withdrew from the battle. He rode up to their commander and demanded to know what was happening. He was told in no uncertain terms that broken men did not retreat in orderly fashion, as the militiaman were doing. The British ranks had broken into a wild charge after the retreating rebels. They ran over the rise that the militia had retreated over, and suddenly found they were facing organized ranks of men, men who opened fire with volleys before closing in for the kill. Now the British were facing an organized patriot bayonet charge, and they knew well what the shout of "Tarleton's quarter" meant.

The patriot cavalry returned to attack on the British flanks, backed up by additional units of militia who had withdrawn into the woods. Tarleton's forced were annihilated at the hands of militia troops and a brilliant commander. With a small group of his men, Tarleton fled the battle, almost being stopped by the leader of the patriot cavalry. He escaped with little more than his life.

The Battle of Cowpens is one of the final turning points of the Revolutionary War. The utter defeat of Tarleton, who lost his entire command, was one event that helped lead to the final defeat of the British at Yorktown in October 1781.

The American forces fought well, as both the organized Continental Army as well as militia forces who volunteered to help give birth to a new country. The British military was one of the best in the world at the time, yet they lost for a number of reasons. The actions of the frontier riflemen had a telling effect against the British, in several stand-up fights but primarily as a skirmishing, irregular force. The use of the long rifle was something the British had not counted on as being a serious threat. They paid for their error with the loss of the North American colonies.

There were difficulties with both the riflemen and the use of the long rifle in organized combat. General Daniel Morgan was one of the strongest supporters of the frontier rifleman companies. He had organized and employed them several times against the British with resounding success. But even he recognized their weaknesses. In 1808, he was quoted by General Graham of North Carolina as saying:

> *"My riflemen would have been of little service if they had not always had a line of Musquet and Bayonette men to support us; it is this that gives them their confidence. They know if the enemy charges them they have a place to retreat to and are not beat clear off."*

The mixing of rifles among the muskets of the military added another difficulty for the fledgling Continental Army. The muskets and the rifles all used a different caliber of ammunition. In 1777, after nearly two years of warfare, the Committee of Safety, who watched over the supply of weapons for the army, complained that they needed to obtain as many as seven different calibers of ammunition just to feed the weapons that were presently in use. The accuracy of the long rifles was needed, and well applied by Morgan as he set his marksmen to take out the British command structure. This was a true sniper's mission, one that would be carried out uncountable times in the future battles of the new United States. But the mix of weapon calibers made logistics a nightmare. At Valley Forge during the winter of 1778, Baron von Steuben of Prussia, who organized and trained the forces under Washington's command, stated that "Muskets, carbines, fowling pieces, and rifles were found in the same company."

It was a supply officer's nightmare. The only piece of ammunition that would fit all of the weapons was the gunpowder. Many of the rifles were loaded with lead balls cast by the frontiersmen themselves. Other odd-caliber weapons forced their gunners to make up ammunition for them, form paper cartridges, fill and seal them, prior to heading into battle. It was not an efficient situation.

The fragility and lack of firepower of the long rifle helped greatly limit its use among the forces of the Continental Army. The long rifle was only seen in the hands of frontiersmen acting as army scouts. The long rifle saw front-line duty again in the hands of militia volunteers during the War of 1812, where it once more proved its worth against the British.

The situation with the long rifles also helped determine part of the Second Amendment in regards to the meaning of "well-regulated." In this case, the meaning of that term can be considered to be the phrase "built to a standard."

The militia, the people of the United States, should possess a weapon that would meet a standard caliber, the same as that used by the military. Meeting that requirement would mean that in time of need, ammunition could be issued for these weapons from regular government stocks. Additionally, the standardized weapons should also be fitted with a bayonet, something the long rifle had never been able to be.

BIRTH OF THE SNIPER

10

TECHNOLOGY OF THE ERA 1810 TO 1860

Developed over a period of years in different countries, the flintlock firing system, though very successful during its age, was far from being a perfect means of firing a gunpowder weapon. The flash pan of the lock ignited the gunpowder in the barrel through a small hole, the touchhole, in the side of the breech. The small touchhole could easily become clogged with debris or plugged with some of the fouling that burning gunpowder produced in such volumes.

When the touchhole of a barrel was blocked for whatever reason, the flash of the priming charge could not ignite the main charge in the barrel. When the priming powder burned, but the gun itself didn't fire, this misfire was known as a flash in the pan. A quick clean-out of the touchhole with a wire pick, a fresh priming charge, and the shooter could try to fire the gun once again.

There was a much worse kind of misfire that could happen that looked like a flash in the pan, but was far more dangerous. This kind of malfunction was known as a hang-fire, where the priming charge had ignited, but the main load of gunpowder in the barrel hadn't gone off—yet.

In a hang-fire, a coal or spark could be smoldering in the touchhole of the weapon. Or the main powder charge could have become just slightly moist. After a few moments, or even long seconds, the gun would suddenly go off, firing uncontrollably. Without proper handling, a hang-fire would send a bullet off in whatever direction the weapon happened to be pointing in. When the priming charge went off, but the gun didn't fire, the skilled musket handler held his weapon pointed in a safe direction for at least a few seconds to ensure that the malfunction wasn't a hang-fire. In the heat of a battle, this safety measure was often forgotten.

Sportsmen also had their difficulties with the flintlock, especially bird hunters. The time delay between when a shooter pulled the trigger and when the gun actually fired, called the lock time, was a noticeable time in a flintlock. The older wheel lock had a very short lock time, which was one of the reasons that style of weapon remained popular with those hunters wealthy enough to afford one. But the flintlock was the far more common weapon, with more than a few ex-military muskets out among the population during the early 1800s.

For hunters, the lock-time pause made it even harder to hit a moving target, such as a bird in flight. The flash from the priming charge igniting could also startle a bird, or other game, into dodging about, throwing the hunter's aim off. It finally took a clergyman who was tired of missing his shots when bird hunting, to come up with a firearm ignition system that eventually completely replaced the flintlock.

THE PERCUSSION LOCK

By the beginning of the 1800s, gunpowder was no longer the only known explosive in the world, but it remained the only successfully employed one. Of the many known explosive compounds and chemicals, almost all were far too sensitive to heat, shock, friction, or impact to be of any practical use. A number of chemicals were simply too expensive to be used for anything except the occasional laboratory demonstration. Some explosive chemicals were so unstable that they would explode if exposed to bright sunlight, or even detonate if just left sitting quietly for a time.

The most commonly known of these dangerous explosive chemicals were the fulminating salts of metals such as gold, silver, or mercury. Some of these very sensitive and powerful explosives were tested as possible replacements for gunpowder. Shattered guns and more than a few serious injuries were the result of most of these experiments.

The Reverend Alexander John Forsyth, the Presbyterian minister of Belhelvie in Aberdeenshire, Scotland, was an avid sportsman and bird hunter who had read the accounts of some of these explosive compound experiments with great interest. The fact that the chemist who had written one of the accounts had decided to investigate much safer materials did not sway Reverend Forsyth from following his own line of experiments.

At first, Forsyth attempted to use mercury fulminate as a replacement for the priming powder in an otherwise normal flintlock. The sparks from the flint did not always set off the fulminate, ending his first investigations. What Forsyth did discover was that striking the fulminate powder with the cock itself would set it off. The flash of the detonating fulminate in an open pan would not communicate with the main charge of the gun through a normal touchhole, but encasing the system with a cover and forcing the flash of the fulminate down a tube did exactly what Forsyth wanted. The system would fire the main charge in the gun barrel every time, and with a much shorter lock time than with the flintlock. With the fulminating ignition system, the flash and smoke of the priming powder was also gone. Instead of a flint striking a steel frizzen, Forsyth's design used percussion, the striking of a rod with a hammer, to crush down on the fulminate powder and fire the gun.

One factor that helped make Forsyth's invention a success was that the system could be easily installed on a standard flintlock weapon. The flash pan and frizzen, with their supporting screws and springs, were removed. In their place was installed a round plug that was threaded into the touchhole of the weapon. Over the plug was a rotating container, one end of which held a quantity of fulminate powder mixture; the other had a flatended rod, the striker. Rotating the container allowed a small amount of the priming composition to fill a depression on the central plug. Rotating the container another half turn put the striker rod over the filled priming cavity.

The cock of the original flintlock was replaced with a flat-nosed hammer. When the hammer hit the striker after the trigger was pulled, the striker in turn crushed the priming composition on the central plug. The flash was guided through a small tube and into the gunpowder charge inside the breech of the weapon.

The rotating container that was central to Forsyth's system gave the whole design the name "scent bottle" lock, after the container's resemblance to a small perfume bottle. By whatever name it was called, the lock was relatively simple, easy to install, and most important of all, it worked. Forsyth received a British patent for his invention in 1807. In spite of the protection of his patent, a number of gunsmiths in England and elsewhere, notably France, quickly copied his design. He spent a good deal of time defending his patent in court. The balance of his time he experimented in developing his design further for use by the British government. Forsyth was even given space at the Tower of London to conduct his researches, as well as leave from the Aberdeenshire presbytery to give him time for the work.

Though his basic idea was relatively sound, it took other people's working with his system to really simplify and advance Forsyth's invention. The strongest claimant to the final design of the percussion lock is Joshua Shaw of England, who later moved to the United States. Shaw's invention was a simple metal cap, about half the size of a pencil eraser. The final cap design was of copper, his first two being steel, then pewter. Inside the cap, shaped much like a miniature top hat, was a small amount of percussion priming composition. The cap was placed over a conical nipple, attached to the top of a drum screwed into the touchhole of the weapon. The cup-nosed hammer would smash down on the cap, crushing the priming composition and firing it. The flash would be guided down the nipple and into the main charge inside the barrel of the gun.

It was an elegant and simple replacement for the flintlock. Shaw received a U.S. patent for his invention in 1822, though he had first applied for the patent in 1817. Within about twenty years, the percussion cap had supplanted the flintlock in U.S. military service. It could fire in the wind or the rain. Once a cap was used, it was just tossed away and another put in its place. It was a sure, fast, and cheap way of firing a weapon and could be fitted to everything from a small pistol to a huge naval cannon.

THE MINIÉ BALL

The percussion lock helped make the musket weatherproof and more efficient as a military weapon, but it did nothing to address the inherent inaccuracy of a smoothbore weapon. Both civilian and military experience had proven that the rifled-bore weapon was far more accurate than the musket, and that trained men could take significant military advantage of that accuracy. Experiments and practical sporting use, especially during the 1820–1850 time period, had also shown that the use of longer cylindrical projectiles with conical tips, the cylindro-conoidial bullet, was much more effective than the traditional round lead ball. The longer conical projectile passed through the air more efficiently with less drag, giving it a generally longer range than a ball of the same caliber. In addition, the bigger projectile was heavier, so it would deliver more impact energy to the target.

The problem with using such projectiles in the military centered again on the rifled-bore guns that were necessary to take advantage of the characteristics of the longer bullets. Taking time to load a weapon properly was fine for the civilian market, where a hunter would only fire a few rounds at most during a hunt. For the military weapon, volume of fire was still the central thought, and that meant a loose-fitting bullet that could be quickly rammed down a fouled bore. The loose-fitting projectile, no matter what its shape, could not take advantage of the rifling of a barrel, so there was no need for the military to even consider the expense of rifling a musket.

A variety of systems for expanding an undersized projectile into the rifling grooves of a barrel were tested and rejected over a period of about thirty years after about 1820. All of the systems examined, including those adopted to a limited extent, had their drawbacks, usually centering on either not having enough accuracy, due to their damaging projectiles, to justify their expense, or the system's interfering badly with the soldier's ability to simply clean the weapon after use.

Working off an idea from Captain Gustave Delvigne, French military captain, Claude-Étienne Minié developed a bullet that appeared to solve the loose bullet question and the rifling engagement problem at the same time. The famous projectile bore the name of its designer and was quickly known as the minié ball, even though Captain Minié insisted that he had developed it from Delvigne's idea.

In connection with another breech design, Delvigne had come up with the idea of an expanding-base bullet. The basic idea of the system was that a cylindrical slug could be made slightly undersized so as to easily slip down the bore of a weapon, even if the weapon was already fouled from shooting. The bottom of the bullet was hollow and filled with a metal ball. The pressure of the burning gunpowder would push the ball up into the soft lead, expanding the sides (skirt) of the projectile enough so that it would grip the rifling in the barrel. The expanded bullet skirt would seal off the barrel and grip the rifling well enough to impart a good stabilizing spin to the projectile.

Minié's modification to Delvigne's design replaced the metal ball in the hollow base of the projectile with an iron cup. The outer cylindrical section of the projectile, the skirt, had several grooves formed in it, grooves filled with a wax and grease mixture to lubricate the bore and keep the gunpowder fouling soft. This system worked even though the iron base plug was occasionally blown clear through the lead bullet when it was fired. In 1849, Minié patented his idea, including in the application such details as base plugs made of wood, copper, zinc, and lead in addition to iron.

The idea of the Minié bullet was a popular success. No modifications to the weapon were needed to employ it; it only had to be made in the proper diameter to match a caliber. Even smoothbore weapons gained some advantages in accuracy when fired with a minié ball, though the projectile was not very stable when fired without rifling.

Within a few years of it being patented, the minié ball was adopted by a number of the world's major military forces. An employee at the American Harpers Ferry armory, James Henry Burton, the assistant master armorer, experimented with the design of the minié ball. The problem of the iron base cup blowing through the lead projectile was a serious one. When the iron cup blew out, it left a lead cylinder, a sleeve, stuck in the bore of the rifle. It had to be pulled out before the weapon could be reloaded. In his experiments, Burton found that a properly designed hollow base allowed the propellant gas alone to open the skirt of the projectile into the rifling. The Burton bullet had an ogival (curved) point with a flat nose as well as three cannelures (deep grease rings) cut around the skirt. It worked extremely well and was soon adopted by the U.S. military. But the Burton projectile was forever known as the "minnie" ball by the shooting troops.

THE FORTY-ROD GUN

The invention of the percussion system and improvements in rifling increased interest in target shooting as a sport in the first half of the 1800s. Originally, target shooting in the United States was primarily a means of teaching marksmanship and polishing that skill. This helped the young hunters of the frontier learn about their weapons and just what they could do with them. It also helped to put meat on the table if they knew how to shoot and not miss their target. By the later part of the first half of the 1800s, a specialized style of rifle had been developed, a design intended, just for competition shooting. These weapons were known by several different names, including forty-rod guns, for the 220-yard distance that was a common one for competition, and slug rifles, for the kind of projectile they fired and to point them out as different from the patched round-ball rifles of the same era.

These massive weapons all shared some common characteristics. They all appeared to be little more than a large octagonal barrel with a buttstock attached. Most models had no front stock at all, just an undecorated flat-sided barrel extending from a very short breech action to the muzzle. The weapons could weigh anything from fifteen to forty pounds, though most commonly they ran

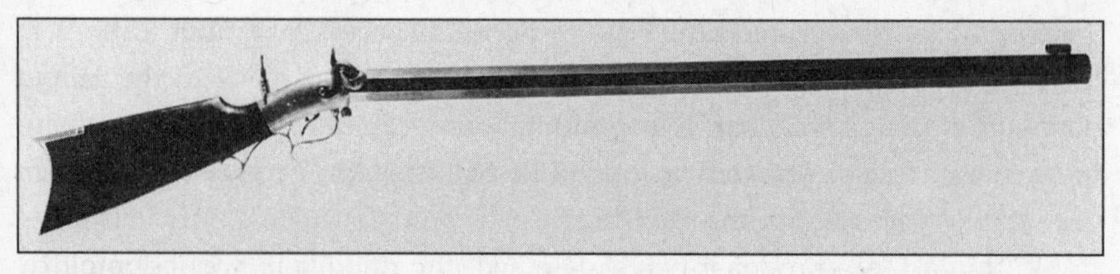

A benchrest percussion target rifle. This specimen has no front stock at all, the bare barrel being intended to lay on a rest for shooting. A finely adjustable set of mechanical sights are on the muzzle and at the wrist of the stock. ***Smithsonian Institution***

from fourteen to twenty pounds, with thirty- to thirty-two-inch barrel lengths. Bore diameters could be anything from .37- up to .56-caliber, with from .40- to around .47-caliber being the most popular. Giant .75-caliber one-hundred-pound slug rifles were known, but not very common.

The makers of the big slug rifles were very limited when compared to the number of people making regular weapons. Only the top gunmakers could produce a slug rifle that would sell, and nearly all of the weapons were produced on a semi-custom basis. The best makers of the most popular rifles proved their faith in their own products by competing with their guns at various matches.

It was these massive rifles and their very particular loading procedures and requirements that helped push forward shooting competitions all over the country, but primarily in the upper northeastern states where the makers lived. The bulk of the gunmakers who produced the slug rifles were based in either New England or New York State.

The main features that separated the slug rifles from all of the other rifles of their era was the bullet they fired and the means used to load it. The best of the slug rifle projectiles was a long, cylindrical bullet with a rounded nose, surrounded by an oiled paper patch. These bullets were made in two pieces, a soft pure lead base and a hard lead alloy nose. The two parts of the slug were swaged together, formed under pressure in a die, to solidly lock the parts into one piece.

The soft lead base of the slug would upset under pressure, expanding to perfectly fit the rifling of the barrel. Between the lead and the barrel steel was a lubricated patch, usually made of paper soaked in sperm whale oil. The hard nose section of the slug wouldn't be easily damaged during the loading process when it was pressed down on a carefully measured powder charge. The bullet was very long when compared to its diameter, a factor that allowed it to be very stable in flight. The long, heavy bullets were also much less sensitive to being pushed off-target by a crosswind. These long, specialized bullets were also known as picket bullets for their shape.

One of the noticeable aspects of all of the slug guns was the number of accessories that were needed to properly employ the gun. Even one of the heavy rifles could be outweighed by the volume of materials necessary to load, clean, and support it. Among other accessories and tools would be a false muzzle for loading, a ramrod, a bullet starter, a cleaning rod, a bullet nose mold, a bullet base mold, a nose swage die, a slug swage die, a patch cutter, a powder measure, cleaning patches, sperm oil, bullet patch material (banknote paper),

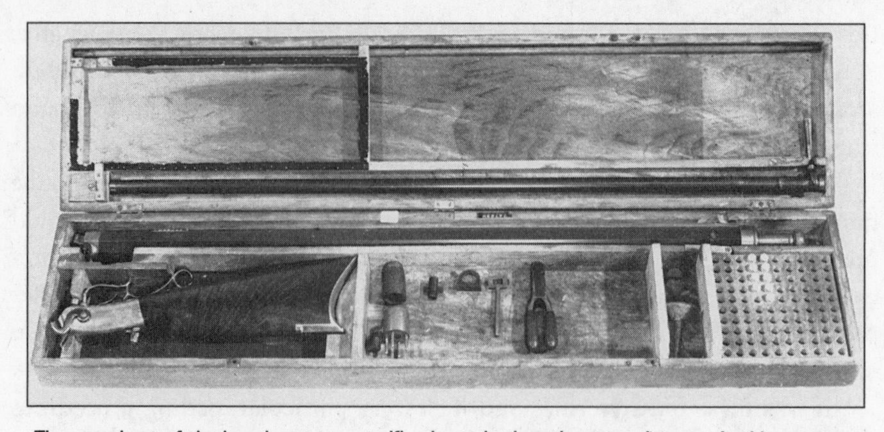

The cased set of the benchrest target rifle shown in the other two pictures. In this case are all the materials and accessories to assemble, load, and care for the weapon, one of the most precise target guns of its era. The long octagonal barrel complete with a false muzzle and starter inserted at the muzzle is in the long compartment at the top of the main case. Along the entire length of the lid to the case is secured the telescopic sight used with this weapon. Other accessories in the case include the precision mechanical sights that could be attached to the weapon in place of the long telescope, and materials for casting bullets and preparing them for loading. *Smithsonian Institution*

lead, lead alloy, powder, caps, and sights. All of these items, along with the rifle itself, were usually packed in a single fitted wooden case.

Loading the slug gun was an even more exacting process than making the bullets. The first step would be to pour a carefully measured charge of high-quality gunpowder down the barrel. A false muzzle, formed when the barrel was made and exactly matching the rifling of this one barrel, was placed over the muzzle. An in-line starter, along with guides for the paper patch strips, would be placed over the false muzzle. Placing a bullet into the starter and then pushing it a few inches into the barrel with a hard smack on top of the starter would get the bullet placed into the barrel exactly in the center of the paper patches and square with the sides of the bore. A shove with a ramrod would put the bullet down onto the powder charge.

Many of the slug guns had a caplock (percussion) action with the hammer underneath the barrel. These underhammer guns kept some of the smoke from firing out of a shooter's face. The weapon itself was usually fired from a benchrest, a long bench with a support for the muzzle of the weapon at one end. The shooter would sit and brace the rifle against his shoulder to fire it. Some light (under eighteen pounds) rifles could be fired effectively by a standing shooter, aiming the gun from what was called the offhand position.

The result of all of this preparation was a supremely accurate weapon that could put its projectiles into a target, very close to the same point of impact, shot-after-shot. Groups of ten bullet holes would often be less than a few inches wide from center to center at 220 yards. Expert shots with carefully tuned, well-broken-in guns could shoot five-round groups that measured three-quarters of an inch across from center to center at 200 yards.

The sights for such accurate guns were also finely produced pieces of machinery. Front "globe" sights usually had a very fine set of wire or bristle crosshairs to be centered on a distant target. The tube surrounding such a sight helped support and shade the delicate crosshairs, as well as providing a superior visual picture. Rear sights were adjustable peep sights with relatively small holes that would circle the view of the front sight exactly.

These weapons were the most precise of their day. Shooting records were set with slug guns that were not broken until the more recent modern age of firearms. Their use and popularity helped develop the science of ballistics and shooting to a very accurate art. They were heavy and massive, but in the proper hands they would also accurately outrange any other shoulder-fired weapon in the United States arsenal until well into the late 1800s and in some cases into the 1900s.

TELESCOPIC SIGHTS

To take advantage of the accuracy offered by the slug gun, they were among the first weapons fitted with telescopic sights. These sights were invented in the United States between 1835 and 1840. The maker of the first telescopic sight was Morgan James of Utica, New York. He produced the sights after having the concept and some details provided to him by John R. Chapman. By the mid-1850s, telescopic sights were also being made by William Malcolm of Syracuse, New York, among others.

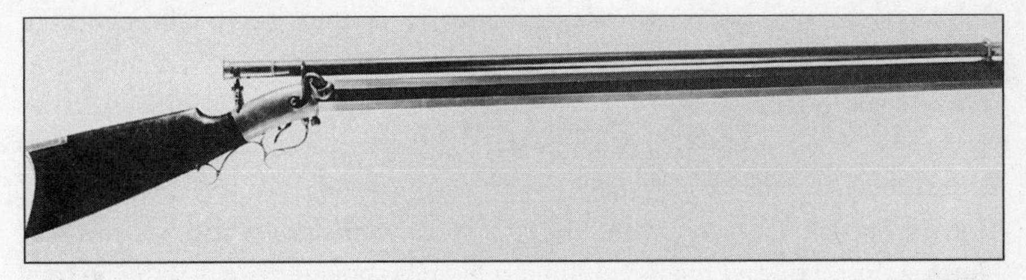

The benchrest percussion target rifle set up with the long telescopic sight in place of the mechanical sights. The front end of the telescope is held in a small hinged mount while a screw piece at the wrist of the stock can be used to adjust for elevation. *Smithsonian Institution*

The first telescopic sights were little more than straight metal tubes with lenses installed. The tubes were often longer than the barrel of the weapon they were fitted to, extending from the muzzle to well behind the action and trigger. Two sizes of telescopic sight were the most common, one being thirty-two inches long, the other being about half that length. Focusing was relatively crude, using a threaded sleeve to move the eyepiece back and forth. The cross-hairs were not adjustable, and the sight was moved for zeroing, adjusting the point of aim to the point of impact, with external hinged and threaded mechanisms. These telescopic sights had a magnifying power ranging from three to twenty times or more.

The main body of the telescopic sight was usually made of brass, occasionally steel, the metal being taken from a sheet and formed into a tube. The diameter of the tube was less than an inch, most often about three-quarters of an inch. The long, thin telescope meant that there wasn't a wide field of view at the target. Only a few feet on either side of a target were visible to the shooter when looking through the scope at a range of one hundred yards. But the magnification of the telescopic sight, even given the relatively crude lenses of the era, helped make a distant target much more distinct to the shooter. The most important factor of a telescopic sight was that it gave a much more precise point of aim than almost any mechanical sight could. Even the finest crosshairs on a globe sight had to be big enough to be seen by the user. Those fine lines would block out a relatively small target at long range. With normal iron sights, a human-sized target could be completely blocked by the front sight at a range of a few hundred yards, especially if the shooter wasn't able to adjust his sights for the range and had to aim well over the target to hit it.

Telescopic sights removed some of these aiming problems. The fine cross-hairs inside the scope tube would leave much of a target visible to the shooter, even at a relatively long range. Telescopic sights were even made by at least one of the major U.S. arms manufacturers of the day. In their 1860 catalog, the Colt Firearms company offered a long telescopic sight for $15.00. The sights are relatively rare, as it appears that only a few were made. But they were made and sold as a commercial item.

BREECHLOADERS

Though Patrick Ferguson brought the first military breech-loading rifle into North America during the Revolutionary War, his design cannot really be considered a success. Only a few hundred Ferguson rifles of all types were even made, and the gun has slipped into becoming little more than an historical

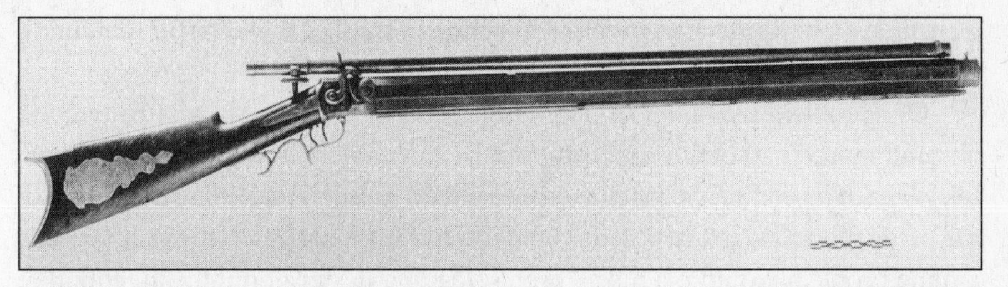

An Edwin Wesson percussion target rifle. This weapon is fitted with set triggers as shown inside of the elaborately shaped trigger guard. Along the top of the barrel is the thin metal tube of a telescopic sight, adjustable for elevation only by the wide wheel underneath the eyepiece at the wrist of the stock. This specimen has been outfitted with a wooden ramrod for field use, the ramrod visible as the light-colored line along the bottom of the barrel. Usually, only target rifles used for field duty are fitted with such ramrods, a military necessity. The cylindrical portion at the front of the barrel is the false muzzle and bullet starter used to carefully load the weapon for maximum accuracy. ***Smithsonian Institution***

footnote. It was an American design for a breech-loading weapon that became the first really successful military breech-loading flintlock rifle.

John Hancock Hall designed a breech-loading action that utilized a tilting breech block to hold the powder charge and projectile. Hinged at the rear, the front of the block would tilt up when a catch underneath the bottom of the weapon was released. With the block tilted up, the open mouth of the chamber was exposed for loading with powder and ball. Pressing the block back down into position locked it in place, sealing off the breech of the barrel and making the weapon ready for firing.

In the final design of the Hall rifle, the flash pan and frizzen were mounted on top of the tilting block, in the center and to the rear of the part. The remaining components of the firing mechanism—the trigger, cock, springs and parts—were behind the back of the breechblock. The entire breech mechanism could be removed for cleaning. Later, some enterprising soldiers realized that the breech mechanism invented by Hall could be carried in a large pocket and used as a crude but effective pistol.

Hall received a patent on his design in May 1811. In 1817, he submitted a military version of his design, a rifled musket with a thirty-three-inch, .54-caliber barrel. After extensive testing, his weapon was adopted by the U.S. military in an improved form as the U.S. Rifle, Model 1819 (Halls). During the working life of the weapon, nearly twenty-thousand specimens were produced, many of them being made at the United States armory in Harpers Ferry, Virginia. The Hall design was effective enough that it made the transition from flintlock to percussion action in the 1830s while still remaining in U.S. service.

One problem plagued the Hall design, and that was the lack of truly good obturation (sealing) of the joint between the breechblock and the rear of the barrel. Without a good seal, propellant gases would escape, weakening the power of the projectile and wasting powder while also being a danger to the shooter. This sealing problem would plague all breech-loading designs until the invention of the metallic cartridge case.

One person who was working at the Harpers Ferry Armory while the Hall rifle was being produced was Christian Sharps. While working at the armory, Sharps became very familiar with the manufacturing processes that produced the Hall and other weapons. The gas leakage problem with the breech-loading Hall was also something Sharps studied. When the manufacture of the Hall designs was finally suspended at the armory in 1844, Christian Sharps had learned more than enough to develop ideas of his own design of a breech-loading mechanism. He received a basic patent for his idea in 1849.

Developed while he lived in Cincinnati, Ohio, after leaving Harpers Ferry, Christian Sharps's idea was for a breech-loading mechanism that didn't tilt, but instead had a rectangular breechblock that slid vertically through a machined channel cut in a solid receiver—what would be called today a sliding breech, such as is presently used on some artillery and cannons. Through a mechanical linkage, moving a lever forward would pull the breechblock down, exposing the open chamber of the barrel. The operating lever was underneath the action of the weapon and was also formed so that it could serve as a trigger guard.

A specially treated linen or paper cartridge containing the powder charge and having a bullet tied to one end was designed to quickly load the Sharps action. The cartridge could be slipped into the breech and the operating lever pulled back, raising the breechblock and locking it into place. The upper front edge of the breechblock was sharp enough to cut off the end of the chambered cartridge, exposing the gunpowder charge. The carefully machined flat face of the breechblock would match with the smooth, flat surface of the breech. Around the center of the breechblock that faced the barrel was a ring of hard metal; in the original weapons it was a platinum ring, surrounding the flash hole. This ring, along with the careful machining tolerances of the breechblock and the channel it slid through on the receiver, would seal off the propellant gases, obturating (sealing off) the breech when the gun was fired.

In the center of the breechblock ring was the open end of the flash hole. On the outside right edge of the breechblock a percussion cap nipple was threaded into the opposite end of the flash hole. A large external hammer on the right side of the receiver could be cocked back, and when the trigger was pulled, the hammer would swing down and crush the percussion cap. The flash from the percussion cap would be guided down the flash hole and into the propellant charge. The burning gunpowder would push the conical lead projectile into the rifling and down the barrel.

Tested by the U.S. Army Ordnance Board in 1850, the Sharps rifle was well received. They issued their report in November 1850. It read in part as follows:

> *This is an arm loading at the breech which is opened and closed by a vertical slide or shear cutting off the end of the cartridge. This arm has withstood all of the trials the Board has considered necessary to make with it. It was fired several hundred times without cleaning, during which the movements of its machinery were not obstructed. The arm is loaded with great ease and rapidity by using a simply prepared cartridge which Mr. Sharps has arranged: and also the ordinary rifle and musket ammunition, with its percussion caps, can be used with facility.*
>
> *The penetration, range, and accuracy of fire from the rifle thus arranged, with the cartridge and conical ball prepared for it, were superior to that of any other breech-loading piece offered to the Board.*

The Sharps Model 1851 design was not adopted by the U.S. military, even after the glowing report it was given by the Ordnance Board. The weapon went

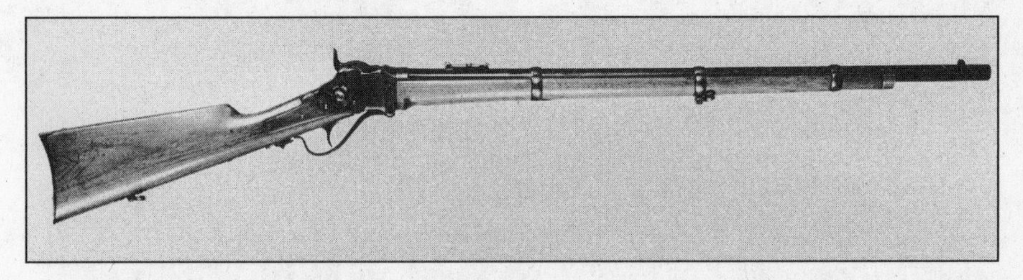

The Sharps New Model 1859 rifle, the preferred weapon of Berdan's Sharpshooters. Though the Sharps Model 1859 carbine was one of the most common carbines in Federal Force hands during the war, the Sharps rifle was made in much smaller numbers and used in spite of a great deal of resistance on the part of some Federal supply and ordnance officers. The Sharps rifle could fire its .52-caliber conical ball projectile at the rate of eight to ten times a minute, more than double the rate of fire for the muzzle-loading weapons used by the rest of the military. Not only could the rifle fire quickly, it could be easily fired and loaded at that rate while the user was in the prone position. Tradition-oriented military supply officers considered such a rate of fire simply a waste of ammunition, a commodity that was already hard to deliver to the front lines in volume. ***Smithsonian Institution***

through several more design changes until the final percussion weapons, the Model 1859 Sharps Rifle and Model 1859 Sharps Carbine, were completed. All of the Model 1859 weapons were .52-caliber with round barrels. The carbine had a twenty-two-inch barrel and was the most popular of the Sharps weapons, accounting for almost 65 percent of the total production of the weapon. About 103,000 Sharps percussion carbines were produced before the percussion model was dropped. The M1859 Sharps rifle had a thirty-inch barrel and was produced in much smaller numbers, only about 9,500 being made. The Sharps was the most successful breech-loading percussion weapon ever produced and saw considerable use with the U.S. military.

11

WAR 1861 TO 1865

The Civil War

Whether it's called the Rebellion, the War Between the States, the War Between the Union and the Confederacy, or simply the Civil War, it was without question the costliest conflict in terms of American lives that the United States was ever involved in. That is the nature of a civil war; both sides of the conflict are from the same country. During those four savage years, over 600,000 Americans lost their lives due to combat, disease, and exposure, the normal hazards of war. But the American Civil War was also so costly in human life because it was such a time of change, of tremendous growth, in the technology of war. The cartridge rifle, the repeating rifle, the balloon, the submarine, the machine gun, mass transit, and armored warships were just some of the technical and tactical innovations that were introduced during the Civil War. And it was the first conflict where the organized employment of what would be called snipers today was used on both sides of the field.

The Union troops of the North faced the Confederate soldiers of the South. The South had seceded from the union of the United States according to what they saw as states' rights. The forces of the Union sought to maintain the United States and the union by forcibly bringing the South back under federal control. It was the heavily industrialized North against the rural, largely agrarian South. The North had industrial production and population. The South had men who had been raised with a weapon nearby much of their lives, a smaller but intensely proud population who fully supported their leaders. It was almost a foregone conclusion that in a war of attrition, the Union would eventually defeat the Confederacy. The Confederate military leadership looked to making the war so costly and unpopular that the Union would eventually sue for peace and they would win by default. It was a recipe for a bloody and drawn out conflict.

Rifled shoulder arms were coming into general issue by the beginning of the Civil War, but marksmanship training in the military, especially the Union military, was still almost nonexistent. It wasn't necessary to train men to aim when the bulk of the military on both sides were still using smoothbore muskets, some of them flintlock arms converted to percussion. Drill was taught instead of expensive and ammunition-wasting marksmanship. Volume through volley fire and speed of reloading were still the primary training results desired by the military. Accuracy was not something that could be quickly, or easily, instructed to raw new recruits no matter how enthusiastic they were during the early part of the war.

The bulk of the military officers, particularly those of the Union, were still firmly behind the idea of linear warfare, large masses of men moving to close range, firing by volleys and closing with a bayonet attack. It was maneuvering the groups of men that would win the day, they believed, and this was the thinking that cost thousands of lives.

Linear maneuver warfare was fine for its day of flintlocks and smoothbore weapons, when the effective range of a weapon was fifty yards. But as the rifled musket became more and more available, the effective range of the average soldier's weapon grew to hundreds of yards. The volume of fire from the soldier hadn't lessened; the adoption of the minié ball had taken care of the problem of loading a fouled barrel. A year into the war, the casualty rates were soaring. Instead of closing to a few dozen yards, firing several volleys, and charging, the ranks of men were approaching groups of men who were able to open fire long before they could close to bayonet range. Charging over hundreds of yards to assault a position where the men were firing from behind cover would just get one side slaughtered, as happened over and over again on both sides.

There was another difficulty for the soldiers besides military leadership failing to adapt to new tactics quickly enough. Corruption was rampant among the politicians and manufacturers who supplied the Union troops. Weapons and materials were being delivered that were completely unacceptable, unsafe, and prone to fail the soldier when he most needed them. To compound this problem, there were very senior officers, well into their sixties, who refused to see the difficulties with the weapons they were arming the troops with. As these antiquated officers wouldn't adapt their tactics to the changing style of warfare, neither would they allow new ordnance to come into their troops' hands. These men allowed their personal tastes to convince them to prevent the adoption or even manufacture of new, innovative weapons that could have actually shortened the war.

Probably the guiltiest of all officers in preventing modernization of the soldier's weapons was Colonel, later General, James Wolfe Ripley, who held the post of army chief of ordnance at the beginning of the war. Ripley had been in charge of the weapons production facility at Springfield Armory, and he knew for a certainty that the M1861 Springfield rifled musket was everything the soldier needed in a weapon—ever.

As chief of ordnance, Ripley actively undermined the actions of his commander in chief, President Lincoln, when it came to the adoption of new weapons, no matter how good they were or if battlefield commanders were directly requesting them. Positive reports on weapons Ripley didn't like were conveniently lost or simply destroyed. Direct orders were ignored and situations manipulated so that the contracts that were finally issued were impossible to meet. Then Ripley could simply state that the manufacturer was in default of the contract and cancel the order. Fighting men died directly because of Ripley and men like him.

There were military people who recognized what technology offered, that there had been growth in fields that had a direct application to warfare. Civilians experienced in these fields were ready to volunteer to join the service and bring their skills with them.

THE UNION SHARPSHOOTERS

In January 1861, when the outbreak of hostilities beginning the active Civil War was still several months away, the Regular Army of the United States consisted of 16,367 officers and men. Hundreds of these soldiers, officer and enlisted together, deserted federal service and went over to the Confederate forces when the war broke out, but in a little over six months, the number of men in federal military service had swollen to over 180,000. In two years, troop strength had reached nearly a million men under arms.

In this rapidly expanding military, a number of people saw their opportunity for personal glory, gain, or just to put into effect ideas that they held dear. Among these men was a thirty-eight-year-old mechanical engineer who was living in New York City at the beginning of the war. Hiram G. Berdan was a very well-known figure in some circles. Target shooting was a popular sport in the northeastern United States in the years leading up to the war. According to

A well-known woodcut of a Winslow Homer drawing of a Federal sharpshooter up in a tree with a telescopically sighted target rifle in the Peninsula Campaign. Under the muzzle of his weapon is hanging a canteen of water, indicating that this sharpshooter intended remaining in the tree for an extended stay if necessary. ***Harpers Weekly***

some reports, Berdan had been ranked the top amateur marksman in the nation since 1846, when he was only twenty-three. Seizing on his popularity and renown as a shooter, Berdan sought to extend his influence into the military and show the professional soldiers how to apply the skills that he knew so well in winning the war for the North.

The idea Berdan proposed was the formation of several military units made up of nothing but the best riflemen available. These units would be armed with the most modern weapons as well, chosen by Berdan himself through his obvious expertise in the field. These troops would act as sharpshooters and skirmishers, operating out ahead of the main body of an army and breaking up enemy formations with the application of aimed rifle fire. This was at a time when the bulk of the military command still firmly believed in volley fire being the only effective way to employ riflemen. The idea of aimed fire being anything but an occasional weapon of opportunity just didn't fit in with the notion of volley fire.

The final result of this attitude being so prevalent at the outbreak of the war would be very heavy casualties. Men would be marched in ranks to close with an enemy, firing only at almost point-blank range with several volleys before charging with fixed bayonets. The only real problem with this scenario that had worked for literally centuries of warfare was that the weapons of war had changed. Instead of both sides having to close to within only a few dozen yards before opening fire, the rifled musket and the rapidly loading minié ball had given the average soldier a weapon with an effective range of hundreds of yards. In addition to the range of the rifled musket, the use of the minié ball had nearly doubled the effective rate of fire of the trained soldier. To close to bayonet range before opening fire was a quick way to have your troops killed. And to charge a position meant that men had to run over hundreds of yards, being shot at all the time, before they could reach the enemy with their cold steel. With this philosophy of combat, there was little need to train the troops in marksmanship. The army spent very little time on the ranges, and only a few men in the average unit could be considered good shots.

None of this was known at the time, or if it was, the few officers who realized it kept their own council. What Berdan wanted to do was simply employ long-range rifle fire, with him in charge of the units. The enthusiasm and reputation of Hiram Berdan was enough to convince the then secretary of war, Simon Cameron, and he gave his official approval of the proposal, signing the order on June 15, 1861. This appointment was later also approved by Secretary of War Edwin Stanton when he was confirmed for the post on January 15, 1862. Initially, two regiments of men would be organized, regiments made up entirely of sharpshooters. While Cameron was secretary of war, corruption in the War Office was rampant. It is very possible that the man made promises to Berdan that he in turn passed on to his men. Whatever the reason, there were a number of very serious problems with broken promises that were soon to arise in developing the sharpshooter units.

A full-strength infantry company for the Federal Army was one hundred officers and men, led by a captain. A regiment was made up of ten companies as well as a headquarters unit and was led by a colonel. Following the custom of the time, when Berdan was given the authority to raise his sharpshooter regiments, he was appointed a colonel of the volunteers, a subservient rank to the active army officers who had attended West Point or otherwise obtained their commissions, but a high rank for someone who had no practical military experience prior to the outbreak of the war.

The next step in creating the sharpshooter units was to obtain the qualified men who could fill out the ranks. Organizing recruiting officers to aid him in manning the new units, Berdan centered his requirements on the men being able to pass a marksmanship test, as well as simply being fit enough to serve. Recruiting officers traveled all over the North, concentrating their initial efforts in the northeastern states. They went to major cities, towns, and small villages to find the skilled men they needed. Because of the patriotic fervor of the time, a large number of men volunteered for the new organization, particularly since Berdan promised them considerable benefits in their serving with his special units. Fliers were put up in large numbers inviting the men to come and join.

Many more men tried to enlist in the new units than could qualify. In some areas, as many as two-thirds of the hopefuls trying out for the sharpshooters failed to pass the practical marksmanship examination. The marksmanship test was a straightforward one; each man could use whatever weapon he wanted, firing it from a rest or from any position he desired, as long as he held the weapon to his shoulder. No telescopic sights were allowed for the test, but that was just about the only restriction. While the public and the recruiters looked on, the shooter had to place ten rounds in sequence into a ten-inch circle at two hundred yards.

The ranks of the sharpshooter companies quickly filled with enthusiastic, skilled men who had almost no military experience among them. Officers were appointed; sergeants and corporals brought in from the ranks. The inducements promised by Berdan included a bounty paid for each man who brought his own rifle along with him. This made available to the unit a large number of heavy target weapons, not normally considered suitable for military service. The bounty for the weapons was $60.00 each, a very good sum for the day. But as time went on, the promised payment was never delivered.

Additionally, the men were told that they would be supplied with the best weapons available at the time. These were to be breech-loading rifles, fitted with telescopic sights and hair (set) triggers. For the men, this meant the Sharps New Model 1859 rifle, but General Ripley, the chief of ordnance, refused to deliver the weapons. The issue rifle of the day cost the military $12.50, including the necessary accoutrements for its use. The Sharps rifle cost $42.50, nearly four times that of an issue weapon. And the new sharpshooter regiments would require two thousand of the Sharps rifles.

This difficulty in supply was completely unknown to the men volunteering and qualifying to join the sharpshooters. As the new companies formed up, they were mustered into service and sent on to a camp of instruction in the countryside near Washington, D.C. The first of these new companies began

TO THE
SHARP SHOOTERS
OF WINDHAM COUNTY

****

YOUR COUNTRY CALLS!! WILL YOU RESPOND?

****

Capt. Weston has been authorized to raise a company of Green Mountain Boys for Col. Berdan's Regiment of Sharp Shooters which has been accepted by the War Department to serve for three years, or during the war. Capt. Weston desires to have Windham County represented in his Company.

The Sharp Shooters of Windham County and vicinity who are willing to serve their country in this time of need and peril, are requested to meet at the *Island House* in Bellows Falls, on Tuesday, the 27th inst., at 1 o'clock P.M., for the purpose of testing their skill in *target shooting*. There are great inducements to join this celebrated Regiment, destined to be the most important and popular in the Service.

No person will be enlisted who cannot put ten consecutive shots in a target, the average distance not to exceed five inches from the centre of the bull's eye to the centre of the ball.

GREEN MOUNTAIN BOYS!

"Rally for the support of the Stars and Stripes!"

YOU ARE INVITED TO BRING YOUR RIFLES.

F.F. STREETE, Supt. of Trial

Bellows Falls, Vt., August 19, 1861.

Phonix Job Office, Bellows Falls

This is an approximation of a broadside announcement used during the early months of the Civil War to induce volunteers to enlist in the sharpshooter regiments. The men who responded to this advertisement, and who qualified, would go on to form the Vermont Riflemen, Company F, of the First United States Sharp Shooters Regiment. They were mustered into the service as a company on September 13, 1861, with personnel strength of 3 officers, 13 noncommissioned officers (sergeants and corporals), and 113 enlisted men. They served in the war from 1861 to 1865.

arriving in late September 1861. There were soon enough personnel on hand to create two regiments. Personally named by Hiram Berdan, the 1st Regiment of United States Sharp Shooters was established, with himself in command, and the 2nd USSS was established soon after. They would go on to create a legendary name for themselves in the annals of the Civil War.

It was during their first weeks at the camp of instruction that some of the men were sent back to their homes. In his enthusiasm, Berdan had authorized the enlistment of 113 men per company. The problem was, this wasn't acceptable to the army, which stated that a company was one hundred men, no more. Thirteen men had to be sent back, only the first of Berdan's promises that were not going to prove true.

No weapons were issued from army stores to the men of the sharpshooter companies until they had completed their training. Only a very small handful of old smoothbore muskets were issued, and those only to arm the men standing guard duty. Instead, the target rifles that so many of the men owned were brought out, and shooting contests arranged between the men, and then the companies themselves. These contests were almost exhibitions for the population of the Washington, D.C., area and quickly became a popular form of entertainment for the yet to be war-weary people. More than a few notable citizens came to see the sharpshooters demonstrate their skills.

So popular had the shooting competitions become that they drew a visit from some very august personages. When General George McClellan, the commander of the Army of the Potomac, came by to witness their skills, only the best shots were brought out to demonstrate the abilities that would make the regiments great. The general was accompanying another famous personality, one who had spent more than a little time firing a weapon himself during his youth. President Abraham Lincoln, joined by several members of his staff, had also come out to see what the sharpshooters could do.

President Lincoln tried his hand at shooting, using the personal target rifle of Corporal Peck of Company F, the Vermont Company. At the end of the exhibition, Berdan demonstrated his skill by firing at a man-sized target six hundred yards away. After satisfying himself that the sights of the weapon were set for the correct range, he made the announcement that his next round would go through the right eye of the simulated confederate soldier. In spite of his showmanship and arrogant claim, Berdan was an excellent shot. He settled in behind his weapon and took a long time making certain of his aim. When the smoke cleared after his shot and the audience could examine the target, they could all see for themselves that the painted right eye now had a black pupil drilled right through it.

But shooting demonstrations and competitions were not what the sharpshooter volunteers had signed up for. Their promised rifles never showed up. Worse still, the bounty that they had been told would be paid for the rifles they brought themselves was considered unauthorized by the military. The sharpshooters felt that they had been betrayed, and they were more than angry at their commander and those above him. Seeking to subdue the anger of his men, Berdan sought the help of President Lincoln in at least obtaining the promised weapons. General Ripley in the Ordnance Department remained steadfast in his refusal to supply the desired weapons. He went so far as to ignore direct orders from his superiors and the President himself.

Finally, the men of the sharpshooters had had enough of the lies and failed deliveries. They threatened mutiny if they did not receive the weapons that they had been promised. Ripley had overstepped himself badly. He recognized the amount of trouble he would be in, but he still tried to weasel out of supplying the desired Sharps rifles. Instead, he offered Colt Model 1855 revolving rifles as a substitute. The weapons were acceptable to Berdan, but not to his men; they considered them little more than mechanical toys.

At least the Colt Model 1855 revolving rifle was not the standard issue Model 1861 Springfield rifled musket. It was a repeating weapon, but only barely. One thousand of the Colt rifles were delivered to the sharpshooters just before they were to go into combat for the first time. The men looked at the Colt as a mediocre weapon at best, a positive danger to the user at worst.

The revolving rifle was effectively a very big, long version of the Colt pistol. A five-round cylinder could be loaded with .56-caliber rounds in the military model of the weapon, the powder and ball stuffed into the front of the cylinder. Percussion caps at the rear of each chamber took care of firing the rounds, at least some of the time. Like the other cap-and-ball revolvers of the day, the Colt rifle could accidentally chain-fire when used. In this spectacular malfunction, the hot propellant gas from the chamber under the hammer ignited the powder in another chamber, one not aligned with the barrel. That chamber in turn fired another and another, until every loaded chamber seemed to go off at once. The result was very close to an explosion taking place right in front of the shooter's face. And, unlike in the pistol, which could also have this kind of malfunction, the shooter held the rifle up with both hands, putting his supporting hand in harm's way, since it would be in front of the chambers in the case of a chain-fire. Because of the power and size of the loads used in the Colt rifle, the chances of a chain-fire were even greater than in a normal pistol.

On top of the problems of safety, the men found that the Colt M1855 rifles just weren't very accurate. The cylinder gap that had to be there to allow the cylinder to rotate caused gases to leak out, changing the velocity of the shot from round to round. The leaking gases could also burn the supporting arm of the shooter. The only way the men would use the Colts to any degree was to load only one chamber to keep from blowing their own left hand off.

Ripley finally got the sharpshooters to accept the Colts on a temporary basis. On April 21, 1862, the order went in for the first five hundred of two thousand Sharps rifles. The weapons would be the New Model 1859 rifles, very few of which would have the set triggers and none of which were expected to be fitted with telescopic sights. It was in May, while the sharpshooters were in the field, that the first consignment of rifles arrived. The men were eager to give up their Colt rifles, and they took up their Sharps with glee. In engagement after engagement, the 1st and 2nd USSS regiments demonstrated their abilities in fighting the Confederates. The firepower offered by the breech-loading Sharps rifles was put to good use in the hands of these expert shots. During some battles, the sharpshooters successfully defeated numerically much greater numbers of Confederate troops armed with traditional muzzle-loaders. Not only was the breech-loading Sharps a faster-firing weapon, the men were able to load and fire it while in the prone position. Often, the only thing the Confederates saw of the sharpshooters was the puffs of smoke rising up from the grass where they lay shooting. As skirmishers, the sharpshooters proved their value over and over again. Berdan was not one of the better commanding officers. In only one incident during the second day of the Battle of Gettysburg did Berdan personally command a scouting of Confederate positions. Otherwise, the man was in Washington and elsewhere. He received honorary promotions to brigadier and major general, but when the ranks did not become full appointments, he finally left the service on January 2, 1864.

Berdan's sharpshooters were not the only such units in the Union Army. There were a number of other companies, battalions, and regiments of sharpshooters raised to help defend the Union of the United States. But the 1st and 2nd USSS were just that, the first units, and they fought during the entire duration of the war. Those shooters set a very high standard for other sharpshooter organizations to try and meet, both in marksmanship and in valor. Of the 2,570 men in the sharpshooter regiments, 1,008 of them were either killed or wounded during the war. The claim of the sharpshooters was that they inflicted more casualties among the Confederates than any other units of equal size in the whole of the Union Army.

THE INDEPENDENTS

The sharpshooter regiments fought primarily as skirmishers. They operated as units at the regimental and company level. They were not snipers, at least not on a massed level. But there were particularly skilled shooters among the sharpshooter ranks who were made special use of, and given special allowance with their weapons.

The massive target rifles many of the men had first arrived with back in Washington weighed between fifteen and thirty pounds on the average. This was a tremendous weight for a soldier to carry on the march, especially over unimproved roads, muddy trails, and cross-country, when added to his regular burden of knapsack, clothing, bedding, and supplies. Besides being heavy, the target rifles were found to be difficult to supply with the proper ammunition. Bullets had to be cast by the user to insure that they were of the correct caliber. And the overall ruggedness of the target weapons just wasn't up to surviving a normal military campaign.

In spite of the difficulties, some of the target weapons were taken along with the baggage train of the unit. The cased target guns were set in a wagon and made available to individuals who had proven that they were very proficient shots. When using their target weapons, these men operated as independents assigned to special service. They were free to move up and down the lines, going to those places where they felt their skills could be put to the best use. Once they had found a place, the independents settled in to fire at selected targets.

One or two independent sharpshooters armed with their target rifles could keep an entire battery of enemy artillery from functioning. As crewmen went to man their weapons, the independent sharpshooters would pick them off with precisely aimed shots. The effective range of some of the best target rifles was equal to the range of the average mobile artillery piece. Only the independent sharpshooter was a considerably harder target.

At the siege of Yorktown, Virginia, over the month of April 1862, the Sharpshooters were first able to demonstrate just what they were capable of when they applied their skills to the battlefield. There were some instances of independent shooters using the target rifles to good effect during the month-long fight. One sharpshooter from New Hampshire, George Chase, effectively captured a major Confederate artillery piece. With his own thirty-two-pound target rifle, Chase would wait until a Confederate gunner went up to the weapon standing several hundred yards away. As the gunner would try to operate the cannon, Chase would fire a shot. The gunner would spin away, falling to the ground, and the cannon would remain silent. For two full days, the Confederate artillery piece was incapable of being fired as Chase completely controlled it with his target rifle.

The independents were not without their weaknesses. In a close-in fight, the heavy precision target rifles were more of a liability than a useful weapon. They were slow to load and couldn't take a bayonet. In close range, the target guns were little more than clubs. It was the supporting infantry units, or the firepower of the other sharpshooters, that could defend the independents against a closing enemy. Together, they made a very efficient fighting force.

These men, never more than a handful in number in any single company or unit, were the forerunners of today's snipers. They stalked their enemy, camouflaged themselves, and tried to take out only the most valuable targets from among the enemy. They were among the most feared soldiers on the battlefield.

Several of these independents became well known in the popular press of the age. Berdan in particular liked one of his men for independent duty, Private Truman Head, a man popularly known as "California Joe." Joe was illustrated in a woodcut published in the August 2, 1862, issue of *Harper's Weekly*. Shown lying behind cover, with his hat off and his long Sharps rifle in his hands, California Joe became something of a symbol of the Sharpshooter regiments. In the article that showed his picture, Joe was credited with taking down a Confederate soldier at fifteen hundred yards, an incredible shot with even a Sharps rifle. It is very possible that the story was exaggerated, but California Joe was well known for regularly taking down his selected targets at six hundred to eight hundred yards, a very respectable distance with iron sights, even by today's standards.

The independent sharpshooters were also targets themselves, particularly of the Confederate sharpshooters, who recognized the danger the other marksmen presented. It is a well-known military rule today that the best way to take out an enemy sniper is by putting your own sniper up against him. It was in the Civil War that this rule was first brought into effect.

During one incident, Private Ide of Company E of the New Hampshire sharpshooters was working against a Confederate marksman. Several shots had been exchanged between the two men and their situation was turning into a personal duel between the two soldiers. Other men in the line, not involved with the conflict between the two men, would turn and watch as they patiently waited until each other exposed himself for a shot.

It was not to be Ide's day and he fell dead, shot through the head by the Confederate marksman even as he was raising his own weapon to fire. But the sharpshooters were all at least competent shots. One of the officers reached over for Ide's scope-mounted target rifle, still loaded and ready to fire. Looking through the telescopic sight, even as Ide's body was cooling, the officer waited. Over on the Confederate lines, a rebel marksman, feeling that enough time had passed to show that he had hit the other man, stood and exposed himself. The officer fired, and even if Ide's hand did not avenge his death, his rifle did.

THE CONFEDERATE SHARPSHOOTERS

There was no Southern equivalent of Hiram Berdan to push for the idea of creating special sharpshooter regiments among the Army of the Confederacy. After seeing what sharpshooters could do, the politicians and military leaders in Richmond were quick to follow the Northern example. By an act of Congress, an official sharpshooter regiment was authorized in April 1862 and General Patrick Cleburne put in command. These sharpshooter units were battalion-sized, each authorized to be made up of three to six companies. Eventually, there would be sixteen of these sharpshooter battalions in the Confederate Army, each assigned to the brigade from whom the initial battalion manpower was drawn.

The sharpshooter battalions, on paper at least, were organized around a commanding officer, 8 commissioned officers, 10 noncommissioned officers (sergeants and corporals), 160 privates, 4 scouts, and 2 buglers. Generally, this quantity of men was divided up into four companies. The battalion would set up in their own camp, usually close by brigade headquarters so that they could quickly receive their marching orders. The sharpshooters were also exempt from the usual guard and other fatigue details that the regular troops had to conduct. This fact alone made serving in the sharpshooter units a very desirable assignment to the average Confederate soldier.

The Confederates who did serve in the sharpshooter units were all accomplished shots as well as resourceful and intelligent men. They had already proved themselves in combat and were a dependable resource for their commanders. The mission of the sharpshooters, like that of their Union counterparts, was to serve as skirmishers, going out in front of the main body of troops and facing the enemy at range, engaging him with precision fire and cutting down on his troop strength before the main battle commenced.

The skills of the Southern marksmen were extremely good. Many of the men had grown up as subsistence hunters and knew how to hold their shot until they were sure of a target. They were every bit a match, if not superior, to the sharpshooters of the Union Army. Where the South couldn't match the North was in being able to supply their sharpshooters with the best weapons available in the numbers which they could be used. The Confederacy had a relatively small firearms manufacturing base, and that was running at full capacity to supply just the general weapons required by the military. The production of specialized weapons, precise long-range rifles and particularly breechloaders, were industrially out of reach. The availability of weapons was so critical that captured arms from the North were considered a major source of supply. It was the Confederate purchasing agent in Europe and England who had to make up the shortfalls in munitions by buying whatever might be available.

A large number of surplus European weapons, both rifles and muskets, were purchased by agents and shipped back to the Confederacy. Among this hardware were around 117,000 British Enfield 1853 Pattern rifles. These .56-caliber weapons became the most popular weapon issued to the Confederate sharpshooter units. It was accurate and long-ranged, capable of hitting a man-sized target regularly at eight hundred to nine hundred yards. With the Enfield, the Confederate sharpshooters became a force feared by those Union Army troops who had to face them.

An Enfield P1853 rifle cost the Confederacy around $150, in gold. There was an additional weapon desired by the purchasing agents, but it was considered frightfully expensive. For $600, the Confederacy could buy the Whitworth target rifle, a militarized weapon that was considered the most accurate production gun in the world at the time. It was cutting edge technology, and that meant it cost. For the best of the best, a cased Whitworth with a Davidson optical sight and a thousand rounds of the special ammunition the rifle worked best with, the price rose to a staggering $1,000. That would be the equivalent of about $18,000 today, even more than the cost of some of the best modern sniper rifles in the world.

But the South wanted the weapons, and the purchasing agents made their buys under the directions of their superiors. Probably less than 150 Whitworth rifles were purchased and delivered to the Confederate forces, but these weapons did a sterling service for the South.

The true snipers of the Confederacy were the scouts of the sharpshooter battalions. These were the men who first received the Whitworth rifles, and used them against the Union Army with telling effect. Only the very best

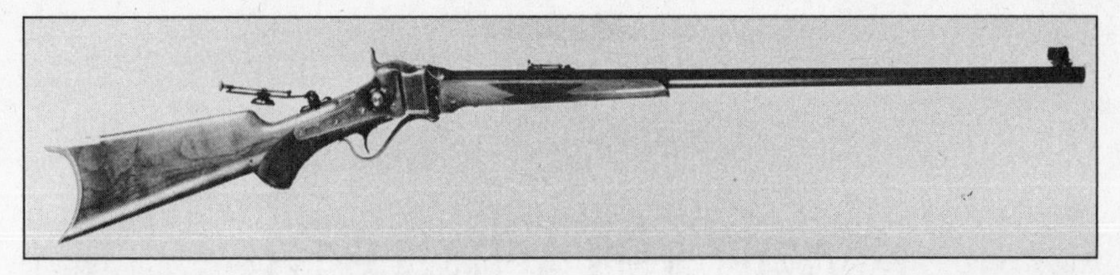

The Sharps Model 1874 Long Range Rifle #3. This was the perfected Sharps side-hammer breech-loading cartridge weapon that was considered one of the finest long-range rifles of its day. The mechanical Vernier scale rear sight at the wrist of the stock, attached to the rear tang of the action, is in the folded position for protection when carrying the weapon. The hooded globe front sight was adjustable for windage, while all adjustments for elevation were made with the rear sight. This was one of the premier target rifles of its era, and shooters armed with it have held long-range records that have lasted until bested only fairly recently. Though the basic action remained the same, the more common hunter versions of this rifle tended to have a shorter barrel. The Vernier tang sight was available for the hunter's weapons as an optional attachment. ***Smithsonian Institution***

shooters were given the best weapons available, and they knew how to employ them. Always in short supply, the rifles were a badge of trust and importance for the men who carried them. General Patrick Cleburne, who became one of the Confederacy's leading generals, received thirty of the precious Whitworth rifles, as well as sixteen of the slightly lesser quality Kerr rifle, for his army. Combined with the best sharpshooters, the deadly rifle made up an independent operator who could move about the battlefield just as his Northern counterpart could.

In addition to their marksmanship, the independent shooters on both sides of the Civil War conflict also acted as scouts, gathering intelligence and reporting back to their respective commanders. They conducted the missions that would be given the modern snipers of today, only they did them for the first time in the U.S. military. In the post–Civil War service, there was no room for the volunteer forces that had served so well during the war. The sharpshooter units were all disbanded and their skilled men returned to civilian life.

12

WAR 1914 TO 1918

World War One

It was in the muddy fields and trenches of Europe where the world, and later the United States soldier, learned firsthand just what kind of devastation could be wrought by enemy snipers across a stagnated battlefield. After a quick series of maneuvers over several months, the war between the Allies on one side and the Central Powers on the other stalled out. The threat of the machine gun and rapid-fire accurate artillery caused both sides to dig into the earth to give their soldiers some degree of protection against the fire coming from the other side. Individual holes were soon connected with one another. Trenches became ever deeper as they were dug down into the earth, and the removed dirt put in bags and piled on the parapets. Now the battlefield became a series of trenches on either side of a ripped and torn area of shell holes, shattered trees, and barbed wire entanglements. The war of massed troops had settled down to one of individual warfare against a new danger.

During the early years of World War I, when the Germans fielded hundreds of new Jäger unit soldiers, armed with telescope-sighted Mauser rifles, the British lost large numbers of men, and particularly officers, to the threat of the German shooter. The German marksmen would target officers whenever they could sight on and identify one, but any soldier who exposed himself for even a few seconds during an unguarded moment ran the risk of falling prey to a precise shot from across no-man's-land. The British newspapers were quick to report on the new threat in the trenches, referring to the German marksmen as snipers. The word quickly caught the public's imagination and has remained in popular use ever since.

The sniper had become a soldier who would carefully hide himself, making his position as undetectable as possible. From this hidden lair, he would use expert skill with a rifle, combining that with an optically sighted weapon for a precise aiming point and target identification. Since a shot could expose a carefully crafted hiding spot, a "hide" as it became known, the sniper would often withhold his shot until certain of a target. With the patience of a hunter after trophy game, the sniper would wait until a target of value showed itself, preferably an officer or other leader, before firing his deadly shot.

The British Army suffered under these losses for months before developing its own sniper program. The leadership of this new form of warfare was drawn from the officer corps, a number of whom had been noted hunters of wild game in their lives prior to the war. It was a number of these men who pushed for the establishment of a British sniper corps, a means to give back to the Germans what they were so ably dishing out along the front.

The initial skills for the British snipers were drawn from the large estates of England, especially those of Scotland. From the ranks of gamekeepers for the nobility, men who controlled and hunted the great forests and hills of Britain, came individuals who knew marksmanship as well as stalking, observation, patience, and particularly camouflage and concealment. The Scottish Highland gamekeepers, deer stalkers and "ghillies," which was the name that has become synonymous with them ever since, were everything that the American frontiersman had been over 130 years earlier, only they had made it a profession in addition to a lifestyle. A ghillie was an individual who attended to sportsmen, the name having been derived from the Gaelic *gille* for "servant." These special men were particularly represented in a unit known as the Lovat Scouts, made up of about two hundred Scottish Highland ghillies.

The Scots introduced the camouflage outfit that would carry their name into the future, and become almost the symbol of the military sniper. Settled on the ground, the ghillie suit appeared to be a pile of rough cloth at most, little more than a rotting compost heap. Properly constructed and used, the suit blended in almost anywhere. The suit was made up of strips and bits of burlap sacking or hessian cloth of different colors, usually browns and greens, attached to the back of a set of trousers and jacket. A hood also covered in tattered cloth strips with a veil of fine netting completed the outfit.

When combined with skillfully chosen local vegetation, the loose outline of a man wearing a ghillie suit would almost completely disappear into the ground. An unsuspecting observer might not recognize a person wearing a ghillie suit for what he was even though the observer was looking directly at him. The veil on the hood would cover the face of the wearer and could easily be seen through from the inside, but from the outside, the veil obscured the face beneath it in shadow. The misshapen form of the ghillie suit did an excellent job of disguising the outline of the body underneath it, as it is the outline of the human form that draws the eye of an observer at a distance. The ghillie suit worked so very well that it remains in use today, virtually unchanged from the World War I version outside of an updating of materials.

But all of these sophistications had to wait until the sniper problem was fully faced by the British military. To arm selected troops, initially "countersnipers," as the marksmen were intended primarily to eliminate the enemy snipers, the British War Office purchased fifty-two sporting rifles of heavy calibers in February 1915. Most of these weapons were large-bore guns designed for taking down dangerous game. Their deep-penetrating bullets would be able to punch through any protection the German snipers might be hiding behind. A small number of other rifles were obtained that had telescopic sights. These were simply issued from general stores to men assigned to sniper duty with no particular special skills or training.

If not for the barrel of the M40A3 rifle and the objective of the telescopic sight extending out from it, this camouflaged mound could be simply a bush, rather than a deadly marine scout-sniper with local vegetation carefully laced through the cloth of his ghillie suit. *USMC*

With no training or any real familiarity with the weapons, the users never completed actions such as zeroing the sights so that the point of aim matched the point of impact at a known range. The act of zeroing is a very basic but critical requirement for the operator to be able to hit with any weapon, but particularly a scope-sighted rifle. Without a good zero, a shooter could have the crosshairs dead on a significant target and not have a prayer of hitting it no matter how great his skills as a rifleman.

Some particular officers who had extensive hunting experience, notably Major Herbert Hesketh-Pritchard, among others such as F. M. Crum and N. A. Armstrong, set out to organize and properly train British snipers. The first British Army sniping school ever was set up at Bethune, France, in 1915. The school did not receive official status until November 1916, more than a year and a half after it had opened training. The First Army Sniping, Observation, and Scouting (SOS) School had a seventeen-day course of training in marksmanship, use of a telescopically sighted rifle, map reading, scouting, patrolling, camouflage, and setting up a sniper's position. The training was successful enough that a second Army School of Sniping was founded soon after the SOS school had begun graduating snipers for duty in the trenches. Now the Germans were facing the same dangers that they had been inflicting on the British for over a year. The British and other Commonwealth snipers had the Germans paying a heavy price of their own in men falling victim to the sudden bullet.

On April 6, 1917, the United States officially entered the war against Germany. The manpower and industrial capacity of the United States was expected to be instrumental in swinging the victory in the war over to the Allies. But the U.S. troops being sent to Europe were very inexperienced for the most part. And the ranks of the military were being rapidly expanded with volunteers who had even less experience with military combat. American forces had to learn from the British and others who had been fighting for nearly three full, grueling years. One of the very new aspects of the trench fighting that had to be learned was how to defend against, and employ, snipers.

The United States Army already had some experience employing as snipers expert marksmen with optically sighted weapons, but they hadn't had to face such a threat since the Civil War. In the Small Arms Firing Regulations of 1904, there was an authorization for the development of a telescopic sight for the new Springfield M1903 rifle. There had been some experimentation and

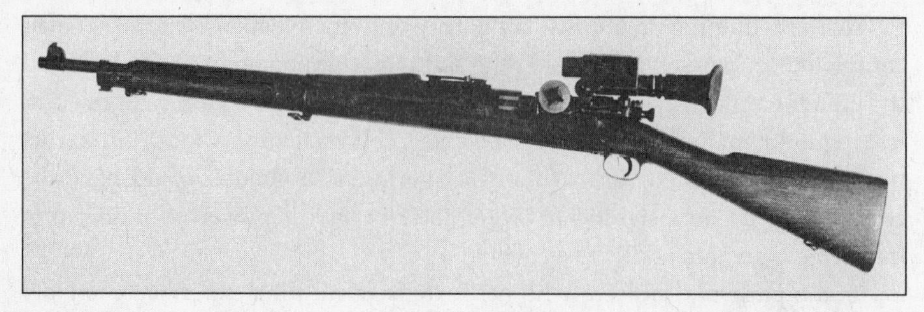

The Springfield M1903 rifle chambered for the .30–06 cartridge and fitted with the Warner and Swasey Model 1913 telescopic sight. This was the primary issue weapon of the Doughboys of the AEF during the United States military involvement in World War I. It is a clumsy, somewhat ungainly device that was mounted to the side of the rifle's breech. The short eye relief of the sight required the user to place his eye within an inch of the back of the sight to aim with it. The large rubber eyecup was intended to protect the shooter from the recoil of the weapon hurting him. ***Smithsonian Institution***

testing of an optical sight for the Krag M1898 rifle in 1900, but no such sharpshooter's weapon had been adopted before the Krag series was replaced with the M1903 and its variants.

The optical sight finally made for the Springfield was the telescopic musket sight, Model 1908. It was a bulky, prismatic device of nominal six-power magnification. Adjustments of the sight for range and windage were limited and poor. Field trials brought out these and other shortcomings, and the sight was later replaced with a slightly improved Model 1913 optical sight of five-power magnification.

Both of these sights were normally referred to by the manufacturer's name, Warner and Swasey of Cleveland, Ohio. These rifles had become general issue by 1909, and the authorization for the telescopic-sight-mounted Springfield M1903 was two per infantry company in both the army and National Guard. It was the experience of using the Model 1909 sight by these units that resulted in the suggested changes for the Model 1913 version.

Prior to the U.S. entering into World War I, the new optically sighted rifles saw some very limited combat use. A number of M1903 rifles fitted with Model 1913 sights were carried by troops under the command of Brigadier General John J. Pershing when he pursued Pancho Villa into Mexico. In addition to the telescopic sights, a number of the rifles also had Model 15 Maxim silencers attached to their muzzles. These special rifles were intended to be used to remove sentries without alerting an enemy camp. The politics of the situation and the world prevented the weapons from being used much before Pershing and his troops were recalled from Mexico. Some few reports indicate that they

A close-up view of the Model 1909 Warner and Swasey Telescopic Musket Sight mounted on an M1903 Springfield rifle. The large dials are for adjusting the elevation of the sight for longer ranges. They worked by tilting the body of the telescope, tilting it downward for increasing ranges. A range scale is mounted on the top of the scope body. ***Smithsonian Institution***

A model M1903 Springfield rifle fitted with a very early Maxim M1910 (Model J) silencer at the muzzle. The silencer could be removed by dismounting the front sight of the rifle. ***Smithsonian Institution***

weren't particularly quiet when fired, possibly a problem with using supersonic ammunition which wouldn't have been known at the time.

One of the reasons Pershing was recalled was that he was put in command of the American Expeditionary Force, the American military contingent going over to fight in Europe. Because of Pershing's insistence that the U.S. troops be fully trained before arriving on the front, few American forces were sent to France prior to 1918.

The first contingent of American troops did arrive in Europe in June 1917. They were followed within a year by nearly 500,000 U.S. military men. Four companies of the new "Doughboys" as they were called saw action as early as July 4, 1917, at the Battle of Le Hamel. But the AEF did not become fully involved with the fighting until the 1st Infantry Division arrived in October 1917. These troops were soon in the trenches and taking part in the fighting. By July 1918, there were a million American troops fighting on the side of the Allies.

Of that huge number of men, there were very few who were trained as snipers in the British schools. Once those men had gained experience in combat and were able to form a cadre of instructors, U.S. organized and staffed sniping schools were established. Sniping was considered a very serious business by the Americans, but the overall number of U.S. snipers remained small. Plans were to have thousands of snipers equipped with a new Model 1918 telescopic-sight-equipped M1903 Springfield rifle, but the war ended before the first handful of new scopes saw duty. Over 5,700 Model 1913 sights were ordered by the U.S. military, but only 1,530 Warner-Swasey musket sights actually saw action. The remaining specimens were delivered so late in the war that they were never issued. So, at most, there could have been barely over 1,500 telescopic-sight-equipped American snipers in the trenches, a tiny number when compared to the overall volume of troops. The small numbers did not mean that a large number of U.S. marksmen did not take part in the fighting, only that relatively few of them were trained snipers.

At Belleau Wood, the rapidly advancing German forces were in hot pursuit of the retreating French when they ran into the U.S. Marines. The unit was the 4th Marine Brigade, attached to the U.S. 2nd Division. The date was June 2, 1918, and the recently arrived marines were not about to join in the French retreat. Instead, they were going to write a new chapter in the history of the Corps.

Utilizing superb marksmanship and discipline, the marines opened fire on the German forces as they came within eight hundred yards. Veteran marine sergeants would call out their estimation of the range to the enemy, and the marine riflemen would set the iron sights on their M1903 Springfields accordingly. The lethally accurate firepower from the marine weapons ended the German advance. Launching a counterattack against the Germans, the marines fought a long and bloody battle, losing more men that day (1,087 casualties) than had been suffered by the Marine Corps in its entire history up to then).

Only twenty years earlier, the feat of marksmanship skill demonstrated by the marines at Belleau Wood would have been almost impossible in the Corps. In 1899, the commandant of the Marine Corps learned that not even one hundred marine officers and men were qualified marksmen with the issue rifle (Krag). He immediately set out to remedy that situation and make being a marine mean that you were also a rifleman, no matter who you were or what your job was.

In 1901, the Marine Corps shooting team finished behind civilian shooters at a number of National Rifle Association–sponsored matches. Learning from the civilian competitive target shooters, the marines invested heavily in studying marksmanship. By 1911, the marines had won an NRA match for the first time. By 1917, every marine sent overseas was a trained marksman. They proved the value of that training against the Germans at Belleau Wood.

Feats of marksmanship were not limited to the Marine Corps. Among the army troops of the AEF was Alvin C. York. On October 8, 1918, within five weeks of the end of the war, in the Argonne Forest, York was part of a small unit of infantry that became lost behind enemy lines. Ambushed by a machine gun while accepting the surrender of a number of German troops, the small unit York was part of was cut in half.

Operating with the eight other survivors of his seventeen-man unit, Corporal York demonstrated American marksmanship to a German machine gun unit. By the end of the day, York and his compatriots had accounted for 25 Germans killed, at least 3 machine guns silenced, possibly more, and 132 enemy troops captured.

Though York never claimed to have operated against the enemy single-handedly, the men who had been with him credited York and his Tennessee backwoodsman shooting skills for their success and survival. The U.S. military agreed and now Sergeant Alvin York became a recipient of the Medal of Honor among other decorations. Two other men among the survivors of York's unit received the Distinguished Service Cross a few years later.

But marksmanship skills are only part of being a sniper. The marines also employed a handful of snipers among their forces. Marine snipers utilized the M1903 Springfield rifle fitted with the Winchester A5 telescopic sight first introduced in 1910. These long, thin steel-tube telescopic sights were relatively fragile for combat use. But they were considered very superior to the Model 1913 sight by the users lucky enough to have an A5. It even saw use with British and Canadian snipers during the early years of the war. U.S. Marine snipers used the A5 scope throughout their involvement in World War I. A number of the M1903 Springfield rifles with the Winchester A5 scope were still seeing active duty with marine snipers on Guadalcanal during World War II, nearly twenty-five years after the end of the Great War, on November 11, 1918.

13

WAR 1939 TO 1945

World War Two

Peace in Europe, or at least not the outbreak of total war, was not quite going to last twenty years from the end of World War I, the Great War. The German forces gathered under the Nazi flag and the leadership of Adolf Hitler had been flexing their military muscles for some years. Appeasement was the word used by diplomats at the League of Nations as the forces of Germany snatched up country after country. Finally, Germany launched its big punch when it invaded Poland on September 1, 1939.

The world saw a new kind of warfare, a coordinated, joint-arms attack utilizing air, infantry, and armor assets to smash through enemy lines and rapidly exploit the resulting breech. It was called "blitzkrieg," lightning war, and it lived up to its name. Poland was taken in twenty-seven days. Early the next spring, Germany turned its forces to the west and the countries that it felt had held it down since the end of World War I. Denmark fell in a day; it took twenty-three days to take Norway, five for the fall of Holland, and eighteen for Belgium to be seized. In France, the country where the trench warfare of World War I had slogged on for more than half a decade, the country was under German control in slightly over five weeks.

On their island fortress, the British held out. The British Royal Navy had made the German naval forces pay heavily to control the waters off Norway. That cost to the German Navy helped prevent the invasion of Britain from ever taking place. Large numbers of German troops assigned to duty in the occupied countries were also unavailable for the combat that Hitler began against the Soviet Union to the east in June 1941.

Standing almost alone against the Nazi juggernaut, Britain held firm in its resolve that its islands would not be taken by the Germans. Across the Atlantic, the United States remained officially out of the war. But everything that could be done to support the British was being done, outside of active combat.

On the other side of the world, conflict had been raging in Asia for years. Imperial Japan had seized large areas of China and Manchuria in the early 1930s. Puppet governments, subservient to the Imperial Japanese Empire, had been installed in the captured areas. By the late 1930s, the Japanese were conducting open warfare in areas of China, threatening to swallow up the whole of the country. In late 1938, the intent of the Japanese Empire is clear. They announced to the world the establishment of the New Order for East Asia. They would continue to expand and become the dominant economic force in the eastern side of the Pacific. The rest of the world could either deal with them, or maintain some small contacts with whatever the Japanese considered as unworthy of their notice.

Britain continued to fight on basically alone but for limited allied troops from the occupied countries, as the war continued through 1940 and 1941. In June 1940, the fascist Italian government of Benito Mussolini declared war on Britain and France. The German and Italian governments had allied with each other, combining to create the Axis forces, with Germany far in the majority position. Combat was taking place in North Africa as the Axis forces worked to consolidate their position there as well as secure needed natural resources. The Battle of Britain raged in the skies, and only with the defeat of the German Luftwaffe did the threat of invasion of Britain really recede.

Desperate to secure badly needed natural resources for their geologically poor country and to maintain their military force, the Japanese launched an attack against the United States Pacific Fleet at Pearl Harbor in Hawaii on December 7, 1941. The stunning attack at first sent the United States reeling, and then it recovered with a thirst for vengeance against Japan. The United States and Britain declared war on Imperial Japan on December 8, one day after the Pearl Harbor attacks. A few days later, on December 11, Germany and Italy declared war on the United States. It was now another world war.

EUROPEAN THEATER OF OPERATIONS (ETO)

The primary U.S. ground combat force against Germany and the Axis was the United States Army. When the United States entered the war, the army was ill-prepared for active combat. The professional army prior to 1941 had been a small force, and it was being rapidly expanded to fulfill the tactical and strategic roles it would have to face in battle. At the outbreak of the war, the United States army had no organized plan for the training or employment of snipers. The only course of training that existed in the U.S. Army prior to the war was a brief introductory training that was offered at Camp Perry, Ohio, much more known for its competition shoots than for the snipers it produced. There was no centralized training of army snipers at all during the war; the details involving the training and employment of such soldiers was left up to the division commanders and their staff.

In August 1942, the U.S. military entered into a new phase of the war, an active phase. The marines landed on Guadalcanal during Operation Watchtower, the first Allied offensive amphibious operation of the war. The experience the marines, and later the army, beginning in October, had in combat against the Japanese convinced army command that there was an immediate need for a sniper rifle to be developed and issued in numbers.

In the years between the world wars, there had been some development of telescopic sights for the military, but these added up to little more than experiments. No finalized device or weapon had been adopted. But sporting development of telescopic sights and mounts to fulfill the civilian demand for such items had gone ahead on a commercial basis and proven very successful. This gave the military a pool of equipment and experience to draw on in coming up with a new rifle. Development had been continued on a low-priority basis for a sniper version of the new M1 Garand rifle, but technical difficulties in providing a suitable scope mount as well as sight prevented that system from

The M1903A4 Sniper version of the Springfield rifle. Having served in the United States military from World War II through the Korean War, this specimen is of the last model issued and is fitted with the M84 2.2-power telescopic sight as developed at Frankford Arsenal late in World War II. ***U.S. Army***

going into production quickly. Instead, the M1903 Springfield rifle, which was entering production in large numbers as the M1903A3, was adapted to a sniper configuration.

Modifications to the base weapons to convert them to sniper rifles were minimal. The iron sights on the M1903A3 were removed completely, though the mounting provisions for them were still in place. The bolt handle was forged to have a curve in it and a clearance cut put in the upper side of the handle. This shaping and cutting of the bolt handle was to clear the eyepiece of the telescopic sight, initially the commercial Weaver 330-C 2.5 power telescopic sight, later designated the M73B1 by the military. The sight was mounted in place directly over the receiver of the M1903A3 with another commercial item, a Redfield "Junior" one-piece scope base and rings.

Production order S-1066 was issued on January 18, 1943, to the Remington Arms Company for them to produce twenty thousand of the newly designated M1903A4 sniper rifles. The production order was accompanied by an urgent request for the rifles, five hundred of which were to be delivered by the end of that month, a thousand more delivered in February 1943, and two thousand per month continually produced by Remington until the contract was filled. On June 20, 1943, an additional 8,365 M1903A4 sniper rifles were ordered produced. The only limiting factor in the manufacture of the sniper rifles was the availability of the telescopic sights. The entire production run of Weaver 330-C scopes from the manufacturer, as well as all available stocks in the country, were obtained by the army for their sniper weapons. The U.S. Army was now assured of a continuous supply of a satisfactory sniper rifle; the employment of the weapons was being left up to the users in the field.

Among the first forces to receive the new sniper rifles were the elite units such as the 82nd Airborne Division already in North Africa. Large unit commanders were setting up their own sniper schools as the new M1903A4 rifles started arriving. In the spring of 1943, active combat in the ETO was limited to North Africa. The units were moving fast to close with the enemy forces, and there was no time at all for setting up any formal school. The terrain the Allied forces were moving through in North Africa was not suitable for sniping operations as the United States Army saw them, which was barely at all at the time. The little use of snipers there was in North Africa for the American forces consisted of individual soldiers with the high marksmanship ability of past hunting skills, operating on their own. A handful of the new M1903A4 rifles ended up in the hands of these men, but their use or effectiveness in combat is little known from the records.

With the fall of Rommel's Africa Corps in May, the Allied forces consolidated their positions with the view to conducting a refit and resupply of their units prior to a further campaign into Europe itself. The M1903A4 sniper rifles that were now being turned out by Remington at a high rate were arriving in numbers and being distributed among the units. During their occupation duty in North Africa, Colonel Sidney Hinds, commander of the 41st Armored Infantry Regiment, showed an interest in applying snipers to his regiment's abilities. While in Tunisia, Colonel Hinds had a formalized sniper training course set up for his unit. Several batches of students completed the five-week course, giving the unit a number of trained snipers prior to their loading up for Operation Husky, the invasion of Sicily.

The primary emphasis on the U.S. Army sniper training during most of World War II centered on the use of the telescopic sights on the M1903A4 and general marksmanship. There was little training in camouflage, concealment, field craft, or scouting. Minimum marksmanship requirements had the army snipers able to hit an enemy soldier in the body at four hundred yards, and in the head at two hundred yards. Many of the shooters were able to exceed these minimum requirements by over 100 percent when the opportunity for such as shot presented itself.

At the Army Infantry School, a sniper course was set up and produced snipers qualified in marksmanship as well as the intricacies of using the M1903A4 rifle and its sight. A special rifle instructor's school was initiated at Camp Perry in Ohio. But there was very limited training offered during the courses in camouflage, concealment, observation, and movement. These were the skills that would allow the sniper to best survive his encounters with the enemy.

After the invasion of Sicily began on July 10, 1943, the terrain of the large island proved itself much more suitable for sniper operations. The snipers of the 41st Armored Infantry Regiment were able to take advantage of the large flat areas of the island and the mountains that surrounded them. The speed of movement for the armored infantry units, most of the men being transported in half-tracks to keep up with the heavy armor, prevented a large number of sniping missions from being performed. But the American Army snipers had been trained to take down high-value targets with their carefully aimed fire, these targets including officers, NCOs, and weapons crews.

Other army units continued their sniping missions as the Allied forces invaded Italy itself and began moving up the Italian Peninsula. One sergeant of the 36th Division took up an M1903A4 rifle and moved out on his first sniping mission along the rocks and hills of Italy. He had spent long hours hunting while living a rural life on a farm in Texas as a young man. The skills he had used in the Texas thickets transferred themselves well to the Italian countryside.

Having taken up a hide among the large rocks on a ridge, the sergeant and his spotter had an excellent command of a dirt road in the valley in front of them. The location of their hide had been chosen to give the men an escape route if the enemy located them and returned fire. They found the enemy as a group of inexperienced German replacement troops started marching along the road in front of their position.

The fact that the German troops were all wearing sharp, clean uniforms with shined boots suggested to the sniper team that these men hadn't seen much in the way of combat. The first target for the sniper sergeant was the officer he recognized leading the troops. Wearing a sidearm was a sign of rank in the German military; it also made a target of the wearer. German forces had faced very little in the way of organized sniper activity from the American forces so far in the war, and these apparently green troops and their leader may not have seen any combat whatsoever.

Sighting on the officer, the sniper shot him in the torso. The man sat back on the road before the sound of the shot reached him. As he fell back, the heels of his boots kicked on the ground and the sniper brought his telescopic sight back onto his target. The officer was dead, and the German troops that had been following him had no idea how to react.

The spotter of the sniper team took down another German soldier before the formation scattered, taking cover in the ditches and craters nearby. The number of troops made the sergeant wary of taking out many more of them; otherwise his position might be detected. When the Germans finally regrouped and continued their march, they ran into another American sniper team farther down the road. It was a leaderless and demoralized troop of German replacements that finally arrived at whatever their destination had been.

This was the general action of army sniping in the European Theater of Operations as the war continued. While preparing for the invasion of mainland Europe in June 1944, the American forces in England included some sniper training in their preparations. Certain units, such as the Army Rangers, included trained snipers at the platoon and even squad level when they were available. M1903A4 sniper weapons were available in volume now, production having continued until the last thirty-one rifles had been completed in June 1944. Further manufacture of the M1903A4 had been canceled in favor of the production of the M1C sniper rifle, based on the standard issue, semiautomatic M1 Garand rifle. "Arsenal America" turned out nearly 6,900 M1C rifles from late 1944 to August 1945, but few if any of the weapons arrived in time for combat use in the ETO. But with over 28,000 M1903A4 rifles having been made, there was no shortage of proper weapons for sniping duties.

The employment of American snipers had become a more common situation in the ETO. Suggestions were put forward on the combat use and training of snipers. Infantry Captain Robert C. Gates made the following suggestions in July 1944 while he and his unit were conducting operations in Italy. His comments included the establishment of "sniper training for the upper-20 percent qualified in marksmanship in each rifle platoon. Organize at least two 3-man sniper teams in each platoon. Allowing for casualties and sickness, you could count on having about 2 two-man teams in each platoon in combat. A sniper's rifle and a pair of binoculars per team is essential."

In the hedgerow country of France around Normandy, it was the Germans who inflicted the worst damage through the use of snipers. A single sniper could hold up a fair-sized military unit until he withdrew, was destroyed, or simply ran out of ammunition. The attitude of the Allied soldiers toward the German snipers was poor to say the least. It was known that General Omar Bradley held a very bad view of the snipers his men were facing and privately he was known to condone the killing of captured snipers, though this was never put into an official document.

As the fighting spread out into the open rural countryside of France, the Allied snipers began working their trade in a much more effective manner than they had while bottled up among the hedgerows. As the German lines collapsed and the retreat to Germany began in earnest, the use of the sniper lessened as the speed of movement picked up. In the open countryside, the long-range shot of the sniper was feared by both sides in the war. In the vicious house-to-house fighting in the cities, the precision of a sniper's shot proved extremely valuable. With a single round, an army sniper could take out a German position inside a building where it would normally take either a major assault or an armored vehicle to blow the enemy free of his site.

THE PACIFIC THEATER OF OPERATIONS

Across the huge expanse of the Pacific, it was the United States Marine Corps that conducted the bulk of the landings during the island-hopping campaign toward the Japanese homeland. Army units operating under the command of General Douglas MacArthur did their share of the fighting, especially along

A U.S. marine scout-sniper scans for insurgents in the streets and buildings along the edge of Fallujah, Iraq, during the first hours of Operation al Fajr (New Dawn) on November 8, 2004. ***USMC***

the islands of the Philippines and on the mainland of Southeast Asia. Even before the war broke out against the Japanese, there were forward-seeing marine officers such as Captain George Van Orden who advocated the creation of sniper units within the Corps, manned with trained individuals armed with specialized weaponry. Along with Calvin A. Lloyd, in 1942 Van Orden wrote a report that pushed for the adoption of an organized sniping program. Along with his other suggestions, Van Orden advocated the adoption of a specialized rifle, one based on a civilian hunting background rather than a converted service rifle. The civilian weapon that showed the most promise in Van Orden's opinion was the Winchester Model 70 Target Grade rifle with a heavyweight barrel. The weapon would be matched to a Unertl 8-power telescopic sight and used with high-quality competitive match grade ammunition.

The marine ordnance officers did not want to bring a new weapon into the supply chain, already overburdened with trying to meet the needs of modern warfare. A service-rifle-based sniper weapon would have parts and support already built into the supply system. Van Orden's words were not heeded by the Marine Corps at the time, but the Winchester Model 70 with the Unertl scope was to see combat duty as a sniper weapon twenty years later in Marine Corps hands.

The use of the Unertl telescopic sight was not an unconsidered one by the Marine Corps. In 1941, the Marine Corps Equipment Board had examined just what was available in the United States in way of a suitable sniper scope. The board tested a variety of sights under field conditions to determine which ones would best satisfy the requirements of the Corps. The telescopic sight recommended for adoption was the target model supplied for the tests by John Unertl. He had made the scope at his own expense and personally submitted it for the testing. His dedication proved itself out. It was an eight-power scope with external adjustments. To adjust the point of aim of the scope, large graduated micrometer turrets on the target-type rear ring mount were turned against their fine half-minute-of-angle clicks. The fine clicks changed the point of aim only half an inch at one hundred yards. The body of the Unertl scope actually slid in the mounting rings under recoil. For each shot, the sniper would have to pull the scope back against its stops to return it to position. The sliding recoil system allowed the fairly sensitive telescope to better withstand the recoil of firing, but it remained a delicate device, especially for use in combat.

When marine combat units landed on Guadalcanal they had very little in the way of sniper equipment in the entire Corps. Weapons that had last seen action in World War I were brought out and issued to qualified marksmen as there were no school-trained snipers available. These weapons were M1903 Springfield rifles fitted with either the original Winchester A5 or the somewhat improved Lyman Number 5A, which was the more common sight at the time. Both scopes were long, slim five-power optical sights that had proven satisfactory in years past. The combat on Guadalcanal and elsewhere during the first marine operations of the war showed the weaknesses of the A5/5A sight, and it was due for replacement.

Whatever reticence the U.S. military may have had in relation to training, outfitting, and fielding formal snipers, the Imperial Japanese forces had none at all. Following the code of Bushido, a Japanese sniper would satisfy his duty if he managed to kill only one of the enemy before being killed himself. What proved more the usual case was that the Japanese sniper took the lives of a number of men before giving up his own life for his emperor.

Camouflage and jungle field craft were very well known by the Japanese troops and particularly the snipers. These shooters would often climb trees to get the best view of the enemy, securing themselves in place with ropes. They had no intention of ever surrendering or of allowing themselves to be taken prisoner. They would have to be killed at their post before they would give it up, and the body might not even fall, forcing the Allied troops to spend even more time making sure that a specific sniper threat was eliminated.

There were many instances of Japanese soldiers acting as solitary snipers against Allied troops. These were the men who would most often fight to the death to hold up the advancing troops for as long as possible. Relatively few of these men were actually trained snipers. As in all of the other military forces of the world, the trained and experienced sniper in the Imperial Japanese military was a valuable commodity. He was not supposed to trade himself for just one or two of the enemy. Though he had the same devotion to duty as the men who tied themselves up in a tree, the trained sniper was to hit the enemy and withdraw to continue his predations at another time and place, until it became his turn to die for the emperor.

The trained Japanese snipers had telescopically sighted rifles available to them. Though the bolt-action weapons appeared to be simplistic to Western ordnance experts, they were particularly suited for the use they were put to. The long barrels and relatively small caliber of the Japanese rifles, notably the Arisaka Type 38 adopted in 1905, made them excellent sniper weapons. Set up with a 2.5-power telescopic sight and a small folding wire mount underneath the barrel, the Arisaka Type 97 was a primary Japanese sniper rifle adopted in 1937.

The thirty-one-inch-long barrel of the Type 38 and Type 97, combined with the 6.5-millimeter (.25-inch) caliber of the weapon resulted in almost no smoke or muzzle flash when they were fired. These characteristics of the Arisaka Type 38 and Type 97 weapons helped make it even harder to spot a concealed sniper in action. A later weapon, the Type 99 Arisaka rifle adopted in 1939 was a larger caliber (7.7-millimeter/.303-inch). It was also made in a sniper version, the Type 99 (1939) and fitted with a four-power telescopic sight. Most true Japanese snipers during the war used the earlier Type 97 weapon; few Type 99 guns were found after snipers had been eliminated.

The U.S. Marine and Army units in the Pacific fought hard against the Japanese sniper threat. What was soon found to be the best way of dealing with a sniper was to saturate an area with firepower. There was little in the way of duels between snipers in the Pacific. Countersniper work involved Thompson submachine guns and Browning automatic rifles. Small artillery in the way of antitank guns would be cranked up and trees completely blown away to eliminate a sniper. It was a brutal but effective way of dealing with the threat.

For employing their own snipers against the Imperial Japanese forces, the Marine Corps established several formal sniping schools, one at Camp Lejeune in North Carolina in December 1942, and the other at Green's Farm, San Diego, California, in January 1943. The Marines also had their new sniper rifles available to them. These were Springfield M1903A1 weapons fitted with the eight-power Unertl scope. Many of the original rifles converted to sniper duty were the very accurate National Match weapons used by the marines for formal competitions during peacetime. The combination was one of the most successful sniper weapons of World War II, only surpassed by the army's M1903A4 weapons because of the huge numbers produced.

The new M1903A1/Unertl rifles were well received by just about every Marine who could get his hands on one. To utilize the new weapons, the official title of "scout-sniper" was given to the men who successfully completed the five-week training course, such as the one run at Green's Farm. On the training site outside of San Diego, the first commandant of the school was Lieutenant Claude N. Harris, previously a competition marksman and winner of the 1935 National Rifle Championship. With a known high-quality marksman in overall command, the fifteen-man classes learned the mission of a modern sniper. The training course included marksmanship, camouflage, observation, map reading, field craft, military sketching, and aerial photograph interpretation. The graduates of the school would be able to fulfill a reconnaissance role, be the forward eyes of the combat commander, as well as a deadly weapons system. Graduates of the school were sent on to combat units, usually three men to a company. The three scout-snipers would make up a team of a sniper and a spotter for combat work and a third man to remain as a reserve.

In some battle areas, the marine scout-snipers had little work to do as active shooters. The very short distances of jungle warfare made their accurate but slow-firing weapons a liability to the trained men. In other arenas, the snipers proved themselves more than worthy of the title they held. As part of the first troops to hit the beach at Tarawa, the marine scout-snipers helped defend the assault engineers by taking out Japanese defense positions. On Saipan, the scout-snipers found a terrain they could operate effectively in, and there were excellent results from their employment.

On Okinawa, there were single Japanese emplacements holding up the advance of entire marine units. In one case, a marine scout-sniper, Private David W. Cass, Jr., took out a Japanese machine gun emplacement through the application of his M1903A1/Unertl rifle. The range was reported later to be as far as twelve hundred yards.

On another occasion, the sniper team appeared to be more of a competition shooting pair than a combat unit.

While on the beach at Saipan, a marine detachment was pinned down by a Japanese machine gun emplacement that commanded the entire area. Every time the marines had tried to assault the position, they had been driven back by the murderous fire of the Japanese, taking casualties with every attempt.

Finally, an inexperienced new officer called back over the radio to have a sniper team sent up to deal with the obstacle.

The sniper team that showed up was a pair of classic Marine Corps gunnery sergeants, complete with the campaign hats that were so well known to a later generation as the "Smokey the Bear" hat. In addition to the incongruous headgear, the two gunnies were not wearing combat uniforms; instead they both had on competition-style shooting jackets. For all of their unusual appearance, the two men set to work with an unhurried expertise.

Unrolling shooting mats on the sand, the two men set up their gear, the observer putting up a spotting scope while the other got into a shooting position with his sling placed around his left arm and his telescopically sighted Springfield held snugly in place. The range to the enemy emplacement was well over a thousand yards. With his spotter calling out corrections, the sniper had his round hitting the bunker within a few minutes. A moment more and a .30-caliber projectile intentionally passed through the firing slit of the emplacement and killed the man firing the machine gun. As another Japanese soldier got into position behind the gun and the weapon opened fire again, another carefully placed sniper round silenced the firing. After the fourth gunner was eliminated, the firing ceased completely from the position.

The marine unit maneuvered into position and destroyed the bunker while under the cover of the sniper team. When the incident was done, the two marine gunny sergeants stood up, put their campaign hats back on their heads, rolled up their shooting mats, and nonchalantly returned the way they had come.

In spite of the excellent service given to the marine scout-snipers during the war, the M1903A1/Unertl weapon was scheduled for replacement toward the end of the war. With the possibility of very heavy action taking place during the invasion of the Japanese home islands, it had been decided that the adoption of the army M1C for use by marine scout-snipers would add firepower and simplify supply matters. The new M1C semiautomatic sniping rifle version of the M1 Garand was examined by the Marine Corps in mid-1945. In early August, the weapon had completed Marine Corps ordnance testing and was being seriously considered for issue. The war ended within weeks of the testing being done, and few of the weapons, outside of field-tested units, saw marine combat use.

The war in the Pacific against the Imperial Japanese Empire finally came to a conclusion after the use of the largest area weapon ever fired against an enemy.

A captain in the Afghanistan National Army Force Battalion, 3rd Brigade peers through the telescopic sight of a U.S. Marine M40A3 rifle. The marines were showing their allies the capabilities of their equipment while serving in Afghanistan. ***USMC***

The atomic bomb could wipe out large areas of a city with a single massive explosion. Two of these weapons showed the world that the end of the war was at hand, and their aftermath suggested that they should never be used again. The marine scout-snipers were the most accurate precision weapon used in the Pacific theater. Their abilities would be called upon again, far sooner than anyone realized.

14

WAR 1950 TO 1953

The Korean War

In the postwar years following the end of World War II, the world situation was tense, but there didn't seem to be much of a chance of the outbreak of total war. The standoff between the Free World and the Communist Soviets and Chinese was stalemated by the existence of nuclear arsenals. In what was now the Atomic Age, large-scale warfare would be conducted with weapons capable of massive destruction. The power of atomic weapons made traditional maneuver warfare conducted by troops on the ground obsolete. At least that was the prevailing military thought of the era.

The art of sniping had been put aside by the U.S. Army, the sniper rifles placed into storage or sold off as surplus. In the United States, only the Marine Corps maintained some preservation of the sniper's skills and mission. There would not be a need for individual riflemen for the foreseeable future, until the middle of 1950.

On June 25, 1950, at 0400 hours local time, the eight divisions of the North Korean People's Army (NKPA) crossed the thirty-eighth parallel into South Korea. The Communist North was moving to capture and occupy the free South. The uneasy peace that had existed between the Communists and the Free World had been shattered. Within three days, the United Nations Security Council had issued an authorization for any member states to aid in repelling the invasion of South Korea. It has been thought that only the absence of the Soviet ambassador to the Security Council, because of an unrelated dispute, allowed the authorization vote to go ahead. The reason the vote passed didn't matter; in Washington, President Truman ordered the U.S. military to intervene, with General Douglas MacArthur in command.

Within a week of the invasion from the North, U.S. Army troops arrived in Korea from Japan. The soldiers had been on occupation duty and were lightly armed, poorly trained, and not in shape for combat. In spite of the odds against them, the army units went up against what the North Koreans would send. On July 12, the 1st Marine Provisional Brigade sailed from California for combat in Korea. Ground units arrived in Pusan on August 2 while the North Korean People's Army was still moving south, pushing army and South Korean units along in front of it. The Pusan Perimeter was established.

Shoved back hard by the North Korean forces, the marines and the few remaining army units fought a long battle to hold the perimeter around Pusan against the oncoming Communist army. Attacks against the Communist positions were conducted by the marine and army units, driving the NKPA back as the fighting went on. By September 15, the tide of the war had turned, as MacArthur launched the amphibious invasion at Inchon, on the far opposite side of the Korean Peninsula from Pusan. The NKPA forces were continually driven back by the onslaught of UN troops, spearheaded by the United States military. The UN forces crossed over the thirty-eighth parallel, moving into North Korea, and the threat from the Communist forces of North Korea appeared to be coming to an end when on October 21 MacArthur publicly announced that the war would be over in a matter of weeks. His optimistic calculations had failed to take into account a significant action on the part of the enemy. Only a week earlier, on October 14, six Red Chinese armies had begun their march into North Korea. The size of the forces aligned against the U.S.-backed South Koreans had suddenly more than doubled.

The Chinese Army in Korea was made up of tough men who had lived through the severities of the Japanese occupation during World War II. They pushed hard against the UN forces, driving them back to the South. It would quickly become a war of wills, the UN forces, supported in the most part by the United States, fighting the Communist armies of North Korea, who were backed by Chinese forces and Soviet material support. The battles would rage across the Korean Peninsula for the next two-and-a-half years.

The Chinese and the North Koreans both utilized snipers to a great extent. Like the Japanese during World War II, many of the shooters identified as snipers by our side were regular soldiers assigned to a harassing duty. But there were a number of very well trained snipers among the Communist forces. The Soviet Union had made very heavy use of snipers in fighting against the

Looking through the telescopic sight of his rifle, this sniper identifies potential enemy targets moving along a rooftop during a search mission in al Fallujah, Iraq. ***U.S. Army***

Germans. Thousands of German soldiers had fallen to the accurate fire of the Soviet marksmen, many of whom were women. That sniper philosophy had been transferred by the Soviets to what they viewed as two of their client states, Communist China and North Korea.

Facing a Royal Australian regiment at a location known as Hill 614 was reported to be one of the best snipers on the Communist side, Zhang Taofang. Propaganda photos of him show him with a standard bolt-action rifle, but the Communist reports of his abilities, even if exaggerated, give him a very high degree of marksmanship skill. Supposedly over a period of thirty-two days at Hill 614, Zhang accounted for 214 soldiers hit. This very high number over a short time is suspicious, but there is no question that some of the Communist snipers were very competent indeed.

There were a small number of Marine scout-snipers in Korea with the first Marine Corps units to hit the ground, but the bulk of these men were the veterans of World War II who were still serving. The venerable M1903A1/Unertl Springfield sniper rifle was again in evidence as the primary weapon the Marine scout-snipers had available to them. The U.S. Army had no snipers in their ranks at the time, and there were no immediate plans in place to build up an army sniper force. It was the marines who demonstrated the efficiency of a trained sniper in the combat in Korea.

Reportedly, when the battalion commander of the 3rd Battalion, 1st Marines had his binoculars shot out of his hands by a Communist sniper, the marine countersniper program really began. A veteran gunnery sergeant who was a competition rifle team instructor was located and ordered to teach the best marine marksmen available how to be even better at their craft. As many M1903A1/Unertl weapons as could be located were brought over to Korea to outfit these new snipers. Within a short time of the new marine scout-snipers having been sent up against the North Koreans, enemy sniping around the battalion commander's position had ceased entirely.

Proper sniper weapons were in continually short supply, and the older Word War II rifles and sights were showing their limitations, but the marines made excellent use of their M1903A1/Unertl Springfields. In spite of some limitations and sensitivities of the equipment, there were a number of long-range thousand-yard shots successfully accomplished by marine snipers using the M1903A1/Unertl combination. Many of the marine snipers considered the venerable bolt-action weapon and its target scope sight to be the best weapon available for their mission.

In the army, the M1903A4 Springfield was still in use, even though its production had been canceled in favor of the M1C rifle more than six years earlier. The Springfield weapons were available in the supply system and put to use by the ad-hoc sniper teams and individual shooters in the army who were able to obtain them. The training and application of snipers in the army was still a question left to the unit commanders to decide.

What the army had available was the new M1C sniper rifles that had been developed in the last years of World War II. The difficulty with producing a sniper version of the semiautomatic M1 Garand was that the weapon was clip-loaded through the top of the receiver. A metal clip that wrapped around the back of eight rounds of .30–06 ammunition would be pressed down into the receiver of the rifle to load the weapon. When the last round was fired, the metal clip would be ejected with a "clang" and the bolt would remain locked to the rear to ease reloading. There was no way to properly mount a telescopic sight directly over the receiver of the M1, as it would have blocked the loading system. Instead, in 1944, a commercial off-set mount produced by Griffin & Howe was obtained that could be attached to the side of the M1's receiver. The mount would hold the telescopic sight, a Lyman Alaskan 2.2-power M73 (later designated the M81, with a crosshair reticle, or the M82, with a tapered post reticle) or the M73B1 telescope as it was available. A conical flash hider that mounted on the bayonet lug and a lace-on leather cheek piece to aid in the proper placement of the head in line with the telescopic sight completed the M1C weapon.

Type classified in June 1944, 6,896 M1C rifles were produced at Springfield Armory from 1944 to August 1945. A second M1 Garand–based sniping rifle was developed at the same time as the M1C and type-classified as substitute standard in September 1944. This other weapon was the M1D rifle. It was virtually identical with the M1C except for the means of mounting the sight. The sight-mounting block on the M1D was at the rear of the barrel, locked up solidly against the front of the receiver. The sight mount itself attached to the side of the block with a single large thumbscrew. It was most often seen with the M84 telescopic sight, a 2.2-power sight similar to the M81 but toughened for military use with much better waterproofing. The weapon also mounted the same flash hider and cheek piece as the M1C. Records indicate that none of the M1D weapons were produced in any numbers during World War II but that the special barrels were produced in numbers at Springfield Armory from January 1952 to February 1953. In addition, Springfield Armory was issued orders to convert 14,325 standard M1 Garand rifles to the M1D configuration beginning in December 1951. An additional 3,087 rifles were to be made into the M1C sniper rifles.

The semiautomatic sniper rifles were well accepted by some users, and discarded in favor of the bolt-action Springfield by others. The semiautomatic action of the M1 did allow for a fast follow-up shot, and gave the sniper a much

A soldier gets down into firing position behind the army's new M107 rifle, the officially adopted version of the Barrett M82A1 semiautomatic rifle. This specimen does not have the large ten-round box magazine fitted into the mag well of the weapon just in front of the shooter's left hand. Army officials expect to complete fielding of the new long-range .50-caliber sniper rifle in 2008. ***U.S. Army***

The .50-caliber sniper rifle conversion of a Soviet 14.5-millimeter PTRD-1941 antitank rifle. This weapon was produced by an ordnance machine shop while serving in Korea in 1951. A .50-caliber Browning heavy machine gun barrel was machined to fit the huge Soviet action, and the bipod is from a Browning M1919A6 machine gun. The Unertl telescopic sight is not mounted in place for this picture; it would normally be on the left side of the weapon so that the operator's face could stay out of the way of the punishing recoil of firing. ***U.S. Army***

greater volume of fire in case of an emergency combat situation. But the act of sniping was considered to be a single, accurate shot placed with care, and the traditional snipers, especially those in the Marine Corps, felt that the bolt-action rifle fit this requirement better, given the greater inherent accuracy of the weapon system. The range of the M1C and D weapons was also limited by the low magnification power of the sights. And a long-range capability was greatly desired in some situations during the Korean War when the fighting was literally from mountain to mountain or across long, flat fields.

Additional weapons were used to increase the range and accuracy of the snipers in Korea. Scope mounts were made up by Marine armorers to adapt the Unertl sights to the M2 HB .50-caliber machine gun. The M2 was capable of single shots and could be set up almost as a small artillery piece with the fine adjustments available from its tripod mount. This weapon was capable of amazing range and accuracy given selected ammunition. Shots from the .50 machine gun were able to reach out to over two thousand yards, taking out an enemy soldier long before he ever even heard the shot.

In the army, sniping was also on the increase, and army personnel were looking to increase range and capabilities. Ordnance Captain W. S. Brophy used his own Winchester M70 target rifle mounted with a Unertl ten-power target scope to take down enemy troops at ranges over a thousand yards. In November 1951, Warrant Officer M. N. Weakley of the 702nd Ordnance

Maintenance Battalion mounted the heavy barrel from an M2 .50-caliber machine gun to a Soviet PTRD 1941 antitank rifle. The huge weapon had a mounting for the Unertl target scope and was used by Captain Brophy to smash into targets well past the thousand-yard range that the .30–06 was generally limited to. There had been experiments with .50-caliber sniping rifles in the post–World War II period. But Brophy was one of the first to put such a weapon to use in combat as a very long-range sniper.

15

WAR 1964 TO 1973

The Vietnam War

In Southeast Asia, conflict had been a way of life for decades, well before the United States entered the picture, from the battle against the Imperial Japanese in World War II, to the expulsion of the French Colonial forces in the mid-1950s. The Southeast Asians had seen war for over twenty years before the first U.S. advisor set foot in South Vietnam. Vietnam had been split into the Communist-controlled North and the more open South by a UN mandate. This situation was not going to be allowed to stand by the Communists in the North, and they set out to overthrow the government of South Vietnam and unify their country under Communist rule.

A military and economic aid treaty between the governments of South Vietnam and the United States had been signed in 1962. United States military advisors and support troops began arriving in the country shortly afterward. The corrupt government of South Vietnam fell to an internal coup, and there wasn't a stable leadership installed until 1965. But the military situation in South Vietnam had increased under continual pressure from the North. In 1965, President Lyndon Johnson of the United States, after receiving Senate and congressional approval, increased the U.S. involvement in Vietnam into active combat.

The watershed event for modern sniping in the United States military was the seven years of active combat in Southeast Asia. The employment of snipers in Vietnam took some time to reach a practical level of acceptance among the military commanders, though the learning curve was much shorter for the Marine Corps as compared to the army. In retrospect, the guerrilla warfare

Armed with a Winchester M70 rifle that had previously served the Marine Corps as a competition weapon, this sniper takes aim through his eight-power Unertl telescopic sight. The M70 was not an official-issue sniper weapon but was pressed into service early during the Vietnam War when few other suitable weapons were available. ***USMC***

that characterized the Vietnam conflict seems almost tailor-made for the employment of snipers on both sides of the battle.

MARINE CORPS SCOUT-SNIPERS

On March 9, 1965, the 9th Marine Expeditionary Brigade arrived in Vietnam with an amphibious landing. The arriving troops were met with flowers held by young silk-wearing South Vietnamese women, rather than by enemy bullets and shells. More arrivals raised the number of American combat troops in Vietnam. Four months after the first landings, there were 18,156 marines in South Vietnam, nearly 10 percent of the entire Marine Corps complement. Army forces also started arriving in Vietnam to help take the fight to the enemy. By 1966, there were 190,000 U.S. soldiers in South Vietnam; by the time of the peak deployments in 1969, their numbers were over half a million.

The fighting force opposing the United States and her allies during the first years of the war was primarily a guerrilla group, the Viet Cong. Unless they were openly carrying arms, VC fighters could simply blend into the local population. Not often given to traditional open combat in large groups, the Viet Cong would attack from ambush and disappear into the jungle. They found the tactic of harassment sniping, simply shooting into an American base without a specific target, a satisfactory way of keeping our forces on edge.

At ranges over six hundred yards, the VC felt themselves relatively safe from U.S. small arms fire. There had been Viet Cong seen moving under arms at ranges of one thousand yards from U.S. positions, without their showing any particular concern about American retaliation. Before a mortar could be fired and corrected onto their position, such visible VC would just move under the cover of the local vegetation. The cost of sending over several mortar rounds every time one or two enemy troops were seen in the open was prohibitive. It was a maddening situation for some marine units.

In 1965, there were no sniper schools in existence in the Marine Corps. The scout-sniper had disappeared completely since the end of the Korean War some twelve years earlier. The staff of the 3rd Marine Division in Vietnam decided to investigate the establishment of a sniper program. Captain Robert A. Russell, a noted marine marksmanship competitor, was put in command of developing and leading the new unit. He was given a free hand to choose his staff of instructors, and he picked five marine sergeants that he knew were all top competitive shots. The men chosen were effectively the best marine marksmen in Vietnam at the time. Four of the men were available in country while the fifth was transferred in from Okinawa at Russell's request.

What helped give Captain Russell an edge in developing a new marine sniper program was that he was a top competition shooter, so he knew exactly what it took to put a bullet into the center of a target at range, and do it every time. What Russell didn't know was how to be a sniper. He had no preconceived notions to get rid of and was willing to learn just what it would take to be a sniper in Vietnam, not World War II, Korea, or anyplace else.

To arm his new unit, a shipment of competition weapons was requested to be sent over to Vietnam. The guns asked for were target-grade heavy-barrel Winchester Model 70 rifles in .30–06 caliber, the best bolt-action weapons available in the Marine Corps at the time. In addition to the rifles, Unertl eight-power telescopic sights were also requested by the new unit. The first twelve weapons had come from the 3rd Marine Division Rifle Team. The telescopes were the USMC-Sniper versions that had been giving sterling service to marine snipers since the early days of World War II. The combination of rifle and scope was almost exactly what had been suggested be adopted as a sniper rifle both during World War II and Korea. It was finally in the hands of marine snipers, or at least competent men who would become snipers.

For use with the target rifles was included target-quality ammunition. M72 match ball ammunition had been produced for the target guns in the Marine Corps, army, and other units since well before World War II. Launching a 173-grain full-jacketed bullet at a nominal muzzle velocity of 2,640 feet per second, each lot of M72 ammunition had to put its projectiles through a seven-inch circle at six hundred yards to be accepted for issue. The very carefully made ammunition had never been intended for combat, but there was no reason it could not be used for such, except expense and availability. The limited number of Model 70 guns in Vietnam made availability of the special ammunition less of a problem.

The new unit was activated in September 1965. With their new sniper weapons, which had actually been in the Corps for decades as extremely well cared-for competition guns, the new sniper school cadre went out to learn just what it would take to be a sniper in Vietnam.

Fifty-five miles from Da Nang were the slopes of the Chu Lai area. Forty-five miles north of Da Nang was an area of mountains and plains around Hue and Phu Bai. For months, the sniper cadre operated in these areas, going out on missions with large and small units, infantry and reconnaissance teams. Broken down into two-man teams, the sniper cadre moved into areas where Viet Cong movements had been noted. The teams would hold in position for hours, all day if necessary, moving in before daybreak and leaving after sunset, to catch a glimpse of the enemy. Their actions were noted by the marine units they accompanied, and the possibilities of their use were considered by commander's in the field.

Two armed VC forward observers were spotted coming into the same area about eight hundred yards from a marine position. The two VC would stop and watch from the same place over several days. When it was confirmed that they were seeing a pattern in the enemy's actions, the marine unit called down to Da Nang and asked for one of the sniper teams to come up to their area of operations. When the snipers arrived, it was not long before the two Viet Cong came back to their habitual position, and met the precision fire of a marine sniper and his rifle.

Once Captain Russell was satisfied with his cadre's skill as well as his own abilities, the next step was to produce a training syllabus as well as a training site. The main need of the site would be a long rifle range. In a small Army of the Republic of Vietnam (ARVN) range a few miles from Da Nang, the sniper cadre found the makings of a satisfactory location. Hard work and the application of bulldozers extended the ex-ARVN range out to one thousand yards. Targets were obtained from the marine mess halls—painted large food cans.

The training program was a hard and fast one, leaning heavily on long-range marksmanship, the use and care of the rifle, and observation. The men who arrived at the school were volunteers from among the best shots in the 3rd Division. They were issued a sniper rifle, and it would remain with them as long as they continued operating as a sniper. Whole days were spent in range time, firing and adjusting until the rifle and its sights became second nature to the marksmen. Classes were taught in safety precautions, how to operate and care for the Winchester Model 70, the Unertl telescopic sight and its care and adjustment, sling use, trigger control, aiming, wind reading, and range determination. Then there was accounting for the effects of weather, maintaining a sniper log, use and care of binoculars, and observation techniques.

As students graduated, more rifles were needed, and they were gathered, prepped, and shipped over from the United States. An additional seventy rifles were located and sent to Vietnam, along with a sufficient number of Unertl telescopes. The use of the marine snipers was increasing within the 3rd Division. Eventually, ninety-two snipers were in use within the division, twenty attached to each regiment. The remaining twelve men went with the division's reconnaissance unit.

Marine Corps headquarters watched the employment of the snipers and the operation of the sniper school with great interest. The results were that the Marine Corps brought back the scout-sniper as a specific military occupation specialty. Officially, the marine scout-sniper was back, and the Viet Cong soon knew about it.

The use of marine snipers grew in Vietnam, and as it did, the problem of obtaining sufficient weapons for them proved difficult. Not only was the Model 70 rifle chambered for a nonstandard round (the .30–06), there had been at least three major changes in the design while the guns were being produced by Winchester. Weapons in use did not have a sufficient number of repair parts available for them, the Unertl scope was showing its age, and the standard military round was the 7.62-millimeter NATO (this was not considering the switch to the M16 and the 5.56-millimeter round). A new weapon needed to be found, and the Marine Marksmanship Training Unit was tasked with obtaining one in December 1965.

Specifications for the new weapon were relatively simple. The new sniper rifle should be made from off-the-shelf components, have a telescopic sight capable of use at a thousand yards, shoot within two minutes of angle with proper ammunition, and be chambered for the 7.62-millimeter round. After some examination of a variety of bolt-action rifles and telescopic sights (a semiautomatic weapon was not considered), the new rifle and sight combination was decided to be a Remington Model 700 with a Redfield Accu-Range three-to-nine-power variable scope.

On April 7, 1966, the new sniper weapon was officially adopted by the Marine Corps as the "Rifle, 7.62mm, Sniper, Remington, M700, w/Heavy Sniper Barrel." A supply contract dated May 17, 1966 first refers to the new rifle as the M40, the designation that it is best known by now. Later, the official designation of the new weapon would be "Rifle, 7.62mm, Sniper, M40." During fiscal year 1966, 250 rifles were planned to come into service, with an additional 450 weapons to come in during the 1967 fiscal year. The weapons were five-shot bolt-action repeaters with wood stocks and special twenty-four-inch-long heavy barrels with a one-in-ten-inch twist rate. The Redfield Accu-Range scope was a commercial item produced in the United States and equipped with a built-in range finder good to six hundred yards. From 1966 to 1971, 995 of the Remington/Redfield rifle combinations were produced for the Marine Corps, the bulk of the weapons (seven hundred) being built and delivered in 1966, and the first rifles arriving in Vietnam the first week of January 1967.

To meet the accuracy requirement for the weapons, they would normally be used with 7.62-millimeter M118 National Match ammunition. Like the M72 ammunition, this cartridge was not originally intended for combat use and came packaged in a white box with red lettering—something that would tend to stand out in a combat zone. Years later, the packaging would be changed to a normal brown cardboard box as the M118 became the "special ball" ammunition specifically for snipers. The M118 load had to meet the same accuracy requirements as the earlier M72 round with a 175-grain projectile launched at a muzzle velocity of 2,550 feet per second.

The marines had their sniper rifle in numbers, and additional sniper schools both in Vietnam and in the United States were turning out numbers of qualified marine scout-snipers. These men very soon made a significant name for themselves among the Viet Cong and in the annals of Marine Corps history.

There were literally hundreds of marine scout-snipers in Vietnam, all of them exceptional men. But even among the ranks of the best, there were some who stood out above all the others. One of these men was the legendary marine scout-sniper Gunnery Sergeant Carlos N. Hathcock II.

In his tiger-stripe camouflage, this Navy SEAL practices with a Remington 700 sniper rifle during his predeployment training. His platoon will soon leave for a six-month combat tour in Southeast Asia. Visible on the ground under his wrist is a white box of ammunition with the word "MATCH" printed on it in large red letters. This is a box of 7.62-millimeter M118 match ammunition, the most accurate 7.62-millimeter ammunition available in the U.S. military inventory at the time. *National Archives*

Hathcock was one of the truly outstanding marksmen of the Marine Corps. Out of a file of two thousand shooters in 1965, Lance Corporal Hathcock won the thousand-yard National High-Power Rifle Championship, the prestigious Wimbledon Cup, at Camp Perry, Ohio. He was a nationally recognized shooting champion but personally a quiet, reserved man.

In 1966, now Sergeant Hathcock was in Vietnam pulling duty as a military policeman in Chu Lai. When Captain Jim Land arrived in Vietnam later that year to set up a scout-sniper school for the 1st Marine Division, Carlos Hathcock was one of the men he wanted as a student and instructor. By October 1, Sergeant Hathcock had arrived at the new school, set up on Hill 55, twenty-nine miles southwest of Da Nang.

If there was a man who could be said to have been born to be a natural sniper, Carlos Hathcock was that man. He built up a reputation as scout-sniper that has been matched by very few individuals in any country. A master sniper by his long-range shooting ability, Hathcock was also an accomplished scout.

He could move slowly through the vegetation, ignoring discomfort and taking hours to cross only a few dozen yards in order to set up a perfect shot. Hathcock had taken to wearing a small white feather in his boonie hat. It was just stuck in the brim of his hat on a whim, nothing more; if he had lost the feather, he wouldn't have bothered replacing it. But the Viet Cong came to know of the sniper referred to as Long Tra'ng—"White Feather."

The enemy in Vietnam placed a high price on the head of the sniper they called White Feather. The bounty bothered Hathcock's commanders more than it did him. One story that came out of this situation has proven so famous that it has been used in movies and on television a number of times, in shows that had snipers of all eras. But it only happened once for real, and it happened to Carlos Hathcock.

When the bounty on his head didn't achieve the results the Viet Cong desired, they sent their own master sniper out to hunt down White Feather. The VC sniper had come to the area around Hill 55 to stalk his foe, and wasn't above taking down more marines while searching for his intended prey.

The enemy sniper was a particularly skilled and dedicated one, at least that was the opinion of Hathcock when he finally decided to leave Hill 55 and go out after the man. Heavy machine gun fire from .50-calibers had failed to dislodge the sniper, who had returned day after day to cut down marines manning their position. Slipping out into the surrounding terrain, Hathcock and his spotter, Lance Corporal John Burke, started to track the enemy sniper. Hathcock knew about where the other sniper was firing his rounds, and that was going to be the place where they started their hunt.

At one point as he was following a faint trail, Hathcock came upon an area where rice had been scattered around. Local birds and other animals were eating the rice, and any movement by Hathcock would have spooked them into the air. It was a simple and elegant early warning system. His opinion of his quarry went up several notches, and he made a point not to underestimate the other sniper.

Setting up at a position overlooking a small open field of fire, Hathcock tossed a branch at the feeding birds. As they flew off into the air, the enemy sniper knew that something had bothered them. It could have been an animal, perhaps a pig or a hunting cat. Or it could be that White Feather had bothered some of his namesakes. The sniper moved out from his position and carefully slipped around the clearing, intending to get above where he thought Long Tra'ng, if it was him, would be.

Hathcock and his partner were waiting patiently at the edge of the small clearing. A sudden sharp noise from the brush off to their side silently startled both of them. They knew that the enemy sniper had circled around to their side, and they had to move, fast and quiet. Scrambling, they got up to move to another position behind thick cover and looked back.

The enemy sniper knew that snagging the small branch and breaking it had warned his prey. Now he was certain that it was at least another sniper out in the jungle with him, possibly Long Tra'ng. He approached the clearing and found the area where Hathcock had been lying just a short time earlier. Now both snipers watched the area in front of them, the clear field that had become the killing zone for their hunt, and they waited for the first one to make a mistake.

Over an hour went by, and neither Hathcock nor his partner had seen anything to indicate the presence of an enemy in the area. Hathcock knew the other was there; he just couldn't find him.

With his own rifle, a 7.62-millimeter Mosin-Nagant Model 1930 sniper rifle with a 3.5-power PV telescopic sight, the same kind of weapon that the Soviets had used against the Germans at Stalingrad, the enemy master sniper carefully scanned the area. The sun was setting as the day was reaching an end and the hunt was not yet over.

From his position, Hathcock saw the glint of something bright. Something was reflecting the sunlight on the far side of the clearing, near where they had been earlier in the day. Burke also saw the glint through his binoculars, and neither man could figure out what it was. Bringing up his Winchester Model 70, Hathcock slowly set up for a shot. A flash of light wasn't natural, and that could be the clue he was looking for, the indication of where the other sniper was hiding in the jungle.

He slowly sighted on the flash of light and decided to risk a shot. The sudden crack of the .30–06 rang out, startling the birds who had once again settled down on the rice. Across the killing field, the body of the enemy sniper bounced back and the man slumped down dead.

When they found the body of the other sniper, Burke picked up the enemy rifle and exclaimed at what he saw. When he handed the weapon over to Hathcock, both men could see that there wasn't any glass in the telescopic sight on the Soviet-made gun. There had to have been glass in the sight earlier; otherwise there wouldn't have been anything to reflect the light back at Hathcock.

With some thought, Hathcock realized what had happened. His shot had gone down the scope of the enemy sniper. One look at the body confirmed that the round had gone through the man's eye. It didn't take long for White

Feather to realize that the only way to have made that shot would have been because the other sniper was aiming directly at him. Carlos had beaten him to firing the shot, a matter of maybe a few seconds at most.

The personal bravery of Carlos Hathcock was proven out by much more than his actions as a sniper. During another tour of duty in Vietnam, the transport he was riding in was hit by an enemy round and burst into flame. In spite of the heat, Hathcock went back into the flames again and again to save his fellow marines. He was severely burned but continued to risk the flames and exploding ammunition to make sure his men had gotten out. He received the Silver Star Medal for his actions that day, and spent years overcoming the scars left by the burns. His partner from that first year of duty as a sniper, Corporal John Burke, also demonstrated what kind of man a marine scout-sniper could be when he later received the Navy Cross, second only to the Medal of Honor, for valor in action in Vietnam.

Hathcock later helped set up the new Marine Corps scout-sniper training program. He brought his skills back from Vietnam to teach them to a new generation of marines.

In his distinctive camouflage-colored beret, this Navy SEAL familiarizes himself with a .308-caliber sniper rifle prior to deploying to Vietnam. Though the SEALs did not have a dedicated sniper program during the Vietnam War, a number of operators did attend the marine training schools. The rifle in the picture, used to a very limited extent in Vietnam, was the SEAL version of the Remington 700-based Marine M40. ***National Archives***

U.S. ARMY SNIPERS

The U.S. Army lagged behind the Marine Corps when it came to fielding snipers for combat in Vietnam. There had been no serious development of a dedicated sniper weapon since the M1D program during the Korean War. The M14 rifle which had been used by the army for a relatively short time did not have any suitable scope mount available for it. The only new development for sniper duty in Vietnam was the addition of the M84 telescope to the old M1903A4 Springfield sniper rifles that had last been made in 1944. A small handful of the M1C and M1D rifles saw some duty in Vietnam, alongside the venerable M1903A4. But these were very few and not considered at all satisfactory.

The 2.2-power M84 telescopic sight was attached to a number of M14 rifles as a stopgap measure to supply a sniper rifle in 1965. But these weapons were produced "in house," manufactured by the unit itself, to give the 11th Air Assault Division some additional firepower when they were called to duty in Vietnam. As far as sniper training went, the last school had closed down in Fort Hood, Texas, a few years after the end of the Korean War. The sniping skills had been lost and would need to be recovered if the army was to field snipers in Vietnam.

In April 1967, a number of sniper systems were tested in Vietnam as an evaluation of sniper operations. A satisfactory method of mounting the M84 scope to the M14 rifle was developed by the Army Weapons Command and entered service in Vietnam with a number of units. Another weapon utilized was a National Match-grade M14 fitted with a modified Redfield three-to-nine-power variable scope with a special ranging cam system (ART). A Winchester Model 70 rifle was included in the evaluations along with the 5.56-millimeter M16A1 rifle with a three-power Realist mounted in place on the carrying handle.

Only ten of the National Match M14s with ART scopes were available for the trial, along with eight of the Model 70s. Snipers were trained in the use of the weapons during June and July 1967, and the evaluation ran from July to October that same year. A total of 259 army snipers from eight different units were trained and sent out into the field with the different weapons. Only 10 men in total used the M14 with the ART scope, while 102 used the M14 with the smaller M84. The largest number of snipers were the 139 men who were armed with the M16 and Realist scope combination. The least were the 8 using the Model 70.

In 124 targets engaged over the length of the evaluation, there were forty-six enemy troops killed in action. Ranges of engagements varied from a few

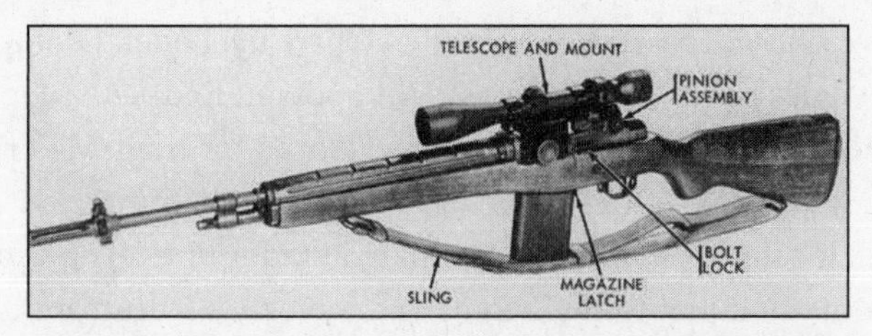

The official U.S. Army illustration of the M21 sniper rifle. This left-side view of the weapon shows the automatic ranging telescope (ART) mounted in place on the receiver. The single mounting screw on this early scope mount is visible as the light-colored disk on the dark square plate on the side of the receiver. ***U.S. Army***

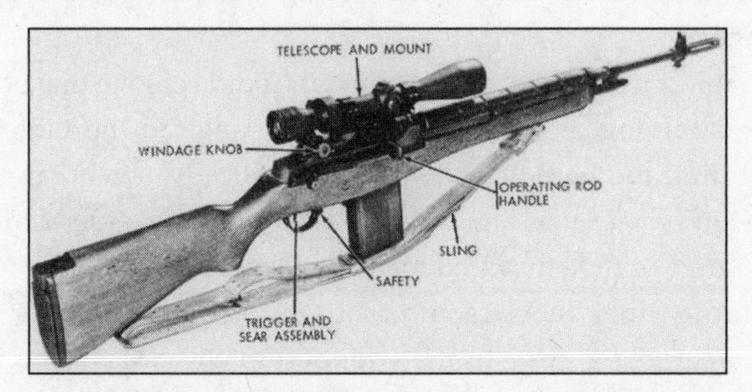

A right-side view of the M21 sniper rifle. In this official U.S. Army illustration, it can be seen that the only difference between the M21 and the National Match weapon intended for competition, is the addition of a telescopic sight. ***U.S. Army***

hundred meters all the way out to thirteen hundred meters for members of the 1st Brigade of the 101st Airborne. The average range of engagement was four hundred meters, still outside the practical range of the issue M16A1 rifle in the hands of the average soldier. Of all of the weapons evaluated, the men who used the M14 with the ART scope considered it 100 percent satisfactory. There were a few men who considered even the Winchester Model 70 rifle unsatisfactory for sniper use, while the other two weapons systems showed serious weaknesses, especially in vulnerability to water and moisture damage.

One of the results of the evaluations in 1967 was the army adoption of the XM21 sniper rifle on September 18, 1969. The new sniper weapon was a match-grade M14, built to National Match competition standards. The telescopic sight was mounted in place, with a large plate attached to the side of the M14 receiver with a single large thumbscrew. The ART (automatic ranging telescope) was based on a principle developed by James Leatherwood. The ART system was derived from a variable-power Redfield scope, the same model as had been adopted by the Marine Corps for their M40 rifle. Turning the power adjustment on the scope allowed the user to bracket a thirty-inch object (wide or tall) between two graduated stadia lines. The thirty-inch distance was roughly equal to the height of an average man from the top of his head to his belt line. Once the distance had been ranged and set with the adjustment knob of the telescope, a graduated cam had also been changed with the movement of the knob. The cam was designed for use with M118 match ammunition, and it tilted the scope as it moved, automatically adjusting the point of aim at the center of the crosshairs to match the point of impact at the indicated range.

It was a complicated ranging system, but it did remove one of the difficulties in sniper training, determining the exact range to the target. Adjusting the point of aim, the "holdover," was automatically taken care of with the system. The new XM21 was big, much heavier than the issue M16A1, and had an effective range of nine hundred meters in expert hands.

Several army units in Vietnam had begun sniper programs with improvised equipment years prior to the adoption of the XM21. The 9th Infantry Division had determined that they had a need for combat snipers soon after they arrived in country in December 1966. By June 1967, the command staff of the 9th had assigned a project officer for a sniper training program. In June 1968 instructors were brought to Vietnam from the Army Marksmanship Training Unit at Fort Benning. The initial group of instructors was made up of a major and eight sergeants, one of whom was an experienced M14 gunsmith.

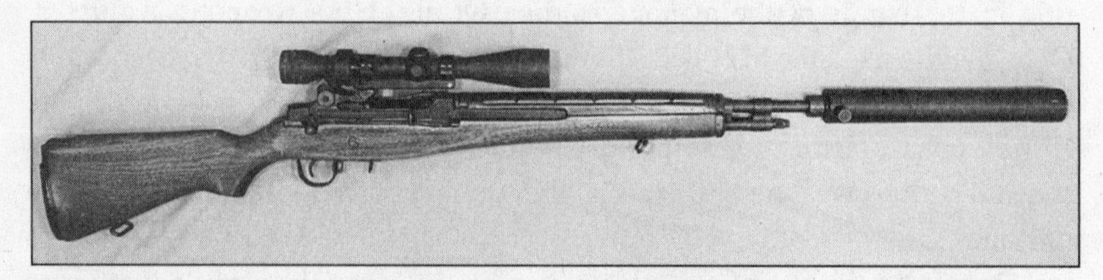

The Vietnam-era M21 sniper rifle fitted with a Sionics suppressor at the muzzle. This specimen does not have the standard twenty-round box magazine locked in place in front of the trigger, though the flaplike magazine catch is visible at the back of the magazine well. This type of rifle saw limited use during Vietnam but did serve to prove that a suppressor could assist in hiding where the sniper was located, even though the sound of the shot was not completely muffled. ***Kevin Dockery***

They initiated a sniper training program that had students qualifying at ranges of 150, 300, 600, and 900 meters.

The students learned field craft, land navigation, adjusting artillery fire, and map reading, among other skills. There were 128 National Match-grade M14 rifles available for the 9th Division training and sniper operations. Fifty-four of the rifles were fitted with the ART scope system, while the remaining seventy-four had M84 scopes. With this equipment and training, army sniping was officially active in South Vietnam.

Army snipers moved as two-man teams, often with a squad-sized security element of infantry troops accompanying them. The Viet Cong were engaged by army snipers at ranges of nine hundred meters. When the use of suppressed weapons came into the sniper program, engagement ranges could also be very short, a few dozen meters or less. Night vision (starlight) equipment came into use that allowed the snipers to work in the dark and still recognize and engage targets hundreds of meters away.

The first kills by the 9th Division snipers were recorded on November 10, 1968. In the four months that followed, fifty-four snipers collected a total of 211 confirmed kills. The trained army sniper was proving to be the most cost-effective weapon of the entire war. Additional units sent volunteers to the 9th Division school to become qualified snipers and return to operate with their parent units. As the XM21 rifle began to be available in numbers, more and

A soldier sights through an SVD Dragunov sniper rifle while providing covering fire for his unit during an attack by insurgents in Sadr City, Iraq. The rifle had been confiscated from insurgent forces during an earlier raid. The soldier had been given sufficient training on the ex–Soviet sniper rifle to be able to put it to good use while assisting his fellows during their operation. *U.S. Army*

more army snipers took to the field. Particularly when they were using a suppressed XM21 in combination with a starlight scope (the ART scope mount could be easily taken off the weapon in the field and the starlight put on in its place), a silent death in the dark came to many Viet Cong who had previously considered the night their friend.

In spite of the military having largely succeeded in their operations in Vietnam, the political situation back in the United States prevented them from ever achieving a real victory. In 1972, no more U.S. forces were sent over for combat in Vietnam. On January 23, 1973, the Paris Peace Accords were signed that officially ended U.S. involvement in South Vietnam. The North Vietnamese Army and the remaining Viet Cong units in the South spent little time in breaking the accords. In 1975, North Vietnam invaded the South with overwhelming military forces. On April 30, 1975 Saigon fell. The country was officially reunited under Communist rule on July 2, 1976.

16

1975 TO 1999

Into a New Century

In the military cutbacks in the post-Vietnam era, only the Marine Corps maintained the hard-won skills of their scout-snipers. Vietnam veterans with large amounts of combat experience behind them returned to a peacetime Marine Corps and set to work to see that their lessons would not be forgotten. In Vietnam Carlos Hathcock had stalked a North Vietnamese Army general, probably his most high-value target ever, for two days and three nights. During his stalk, Hathcock moved only inches at a time, covered bare yards in an hour. Enemy patrols had been moving all about him as he stalked his quarry, one soldier even brushing against his leg as he walked by, but never noticing the sniper in the tall grass. The distance Hathcock covered during his stalk was only two thousand meters of flat ground covered with two-foot-tall grass. He made his shot from eight hundred meters and withdrew while a furious enemy force scoured the area for him.

Such lessons were not overlooked in the layout of the new training curriculum. Students had to conduct eight-hundred-meter-long stalks while under the close observation of a sniper instructor. They learned camouflage, observation, concealment, mission planning, the use of their weapons, and extensive marksmanship. The Military Occupation Specialty (MOS) of 8541, the scout-sniper, did not leave the Marine Corps again. Further sniper schools were opened and the new M40A1 sniper rifle adopted. Instructor training for snipers was conducted over eight weeks at Quantico, Virginia. An additional advanced four-week sniper course was also set up at Quantico. Two additional schools were open, one in Camp Lejeune in North Carolina and another at

Camp Pendleton in California. These were the standard six-week scout-sniper courses that resulted in the student being awarded the coveted 8541 MOS.

In the army, the sniper's position in the infantry became a thing of the past in the mid-1970s. There were selected marksmen for individual assignments, usually those of crowd control within the National Guard and reserve units, but no active sniper employment. In the late 1970s, the XVII Airborne Corps had their own sniper school within the unit. It was the only such course of training in the army at the time and covered four weeks of instruction. The standard subjects of map reading, observation, camouflage, and concealment were all centered on the use of the army's M21 sniper rifle. Some investigations of fielding an improved rifle had been conducted in the early 1970s, and they continued on a very low priority for some years afterward, but with the end of the Vietnam War, all such projects resulted in no practical changes in the available weapons.

The XVIII Airborne sniper school closed in the early 1980s. For some time after that, it was only among the hard-core professional soldiers of the Army Special Forces that any form of sniping school was still ongoing. The curriculum of the SF school was based on the Marine Corps training, but the rate of graduation was so low that the school was closed after only a few classes.

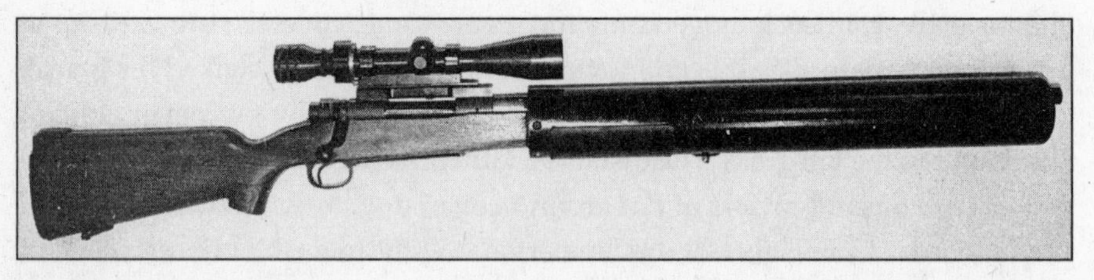

Probably one of the rarest sniper rifles of the Vietnam War, this is a highly modified Winchester Model 70 African, originally a hunting rifle for the most dangerous of wild game. At least five of the weapons were obtained and modified for use as suppressed sniping weapons. The unusual base weapon was originally chambered for the .458 Winchester Magnum, a huge commercially produced hunting round, one of the largest made in the United States. The Model 70 was chambered and fitted to use the .458x1½-inch Barnes, a custom-made "wildcat" round made from a shortened .458 Winchester Magnum. The .458x1½ was intended to be able to deliver a lot of energy on a target without producing the massive recoil of the full-sized magnum load.

Suitable accuracy was given from the small, fat round of ammunition to interest some military designers. Loaded with a 500-grain full-jacketed bullet, the round would launch its projectile at about 1,100 feet per second, less than the speed of sound at sea level. The huge suppressor was designed at Aberdeen Proving Grounds and almost completely silenced the firing of the weapon. But the large rifle and huge suppressor were considered too ungainly for effective use. Though tested in Vietnam, the suppressed .458 Model 70 was not a success, and its existence is almost completely unknown to the public beyond a handful of weapons scholars. *Kevin Dockery*

As the country gradually recovered from the results of the fall of Vietnam, the threat of an old adversary who had never gone away started to rise again. Along with the old danger came a new one on the horizon, one that would overshadow all others in only a few decades.

The Cold War with the Soviet Union heated up as they saw the United States militarily and philosophically weak after the Vietnam War. Communist-inspired conflicts grew all over the world, particularly in close-by Central America. Then the Soviets flexed their muscles even further when they invaded Afghanistan in the early 1980s.

On a new front, Islamic fanaticism showed up as a serious threat when fifty-one Americans were taken hostage in Iran in the late 1970s. Terrorism was growing as an ever-increasing danger to the Middle East and the United States. Under a new presidential administration in the 1980s, the threats against the United States were faced head-on. The U.S. Marines were sent in strength to the Middle East as part of a multinational peacekeeping force for Lebanon. Marine scout-snipers operated in Beirut, but were unable to match the fanaticism of the terrorists. On October 23, 1983, a truck bomb driven by a suicidal individual blew up at the headquarters of the marine detachment in Beirut. Two hundred and twenty Marines were killed and seventy wounded in the attack, along with an additional twenty-one American servicemen killed. That same day, the French forces in Beirut were also attacked by a bomb, with fifty-eight of their number being killed.

Combat operations were conducted by the United States military in Grenada, Panama, Haiti, Somalia, Lebanon, the Persian Gulf, and the former Yugoslavia during the last two decades of the twentieth century. A buildup of the military resulted in the official opening of a U.S. Army sniper school at Fort Benning on July 11, 1987. The sniper would never again be lacking from the U.S. forces in the army, Marine Corps, or even the navy and air force, as their special operations units picked up the sniper skills. New capabilities were brought into the sniper community with the adoption of heavy .50-caliber rifles capable of hitting targets at thousands of meters. During Desert Storm, Barrett .50-caliber semiautomatic rifles in the hands of marine and army snipers were taking out targets at over half a mile.

The U.S. Army switched over from their philosophy of a semiautomatic sniper weapon in the late 1980s. A requirement was put forward in 1987 for a new rifle to be made for the army, a bolt-action rifle based on the Remington 700 action. Several custom makers brought forward their versions of a sniper rifle for the tests, but the final award went to the parent company of the weapon,

In the post–Desert Storm era, Special Operations sniper teams train with their Barrett .50-caliber rifles in the desert. The weapon first saw combat duty with a number of U.S. military organizations as part of the Coalition Forces during Desert Storm. The weapon was not only used to destroy unexploded ordnance at a safe distance, it was also found to be very satisfactory when used on high-value targets such as radar dishes, vehicles, missiles, and even enemy personnel. ***USSOCOM***

Remington Arms. The M24 went into service in 1988 as the new sniper rifle for the army. A number of M24 weapons gradually moved into the other services as they became available. In spite of the precision of fire from the M24 being very satisfactory, experience in desert warfare, especially in Desert Storm, suggested that a semiautomatic sniper rifle also had its place. The M21 was brought back, refitted, and issued as the M25. Through the new rifle, the basic M14 action has remained in army service since the mid-1950s.

The age of the sniper had arrived when the United States entered the new millennium in the year 2000. The whole world would change barely eighteen months later.

17

SOMALIA 1992 TO 1994

In the brutal environment on the east coast of Africa, in the area known as the Horn of Africa, drought and civil war swept through the country of Somalia in the early 1990s. At the forefront of an international relief effort to fight the famine sweeping Somalia, the United States sent in military troops as part of their overall humanitarian effort. The use of the military with their large numbers of personnel, transportation, and organization, had proven effective in past humanitarian efforts by the United States. In the chaos that was Somalia, the initial reception of the American troops by happy Somalis soon dissolved into interclan fighting and attempts by warlords to lay claim to huge parts of the country.

Imminent starvation of the Somali people was put off by the aid brought in and protected by American and other UN troops as part of Operation Restore Hope. With the specter of mass death alleviated, the United Nations turned to restoring the government of war-torn Somalia with the initiation of Operation Continue Hope in March 1993. United States military units would remain under American control while working with the nearly sixteen thousand UN peacekeepers from twenty-one nations.

One of the worst of the Somali warlords in the main city of Mogadishu, the capital of Somalia, or at least what was left of it, was Mohammed Farah Aideed. At his orders, Aideed's militia troops had killed twenty-four Pakistani soldiers and wounded forty-four more during an ambush on June 5, 1993. A warrant was issued for Aideed's arrest, and a $25,000 reward was placed on his head. The situation quickly deteriorated into a civil war between Aideed's forces and the United Nations troops, particularly those of the United States.

The Somali people who were caught in the middle of this conflict quickly sided with the forces of the warlord. The U.S. and UN may have brought food and supplies to end the famine, but eventually, they would leave and the Somali people knew whom they would be left with in the aftermath.

By August, American Special Operations personnel had joined with the marine and army troops on the ground to conduct operations intended to capture Aideed and restore some semblance of a government to the country. These Special Operations forces included U.S. Army Rangers, members of Special Forces Operational Detachment Delta, better known as the D-boys from Delta Force, and U.S. Navy SEALs, along with specialized air assets. The Special Operations forces were organized under the name Task Force Ranger and were on the ground in Somalia by August 28.

The situation in Mogadishu worsened even though raids by the men of Task Force Ranger were making inroads against Aideed's people. Among the men of Delta Force and the Navy SEALs were some of the best snipers in the United States military. There was also the not inconsiderate force of the marine scout-snipers on hand, with more on the way. But the best shots in the world cannot do their job if they don't have a target. Aideed remained in hiding, and his forces remained in conflict with the UN troops. A UN checkpoint was attacked by Aideed's militiamen, backed up by nearly one thousand Somali civilians on September 8.

In early October, the men of Task Force Ranger launched their seventh mission in Somalia, this one to capture two key lieutenants in Aideed's militia. While moving into position over the Olympic Hotel in downtown Mogadishu, two of the birds were brought down by heavy ground fire and rocket-propelled grenades. While a rescue of the crews of the first helicopter was ongoing, several other helicopters took heavy damage from the fierce ground fire. One of these helicopters crashed only a quarter mile from the site of the first downed bird. Convoys of ground troops were immediately organized to go out and rescue the surviving crew members, but it would take them time to arrive on site. By that point, the surrounding Somali and militia mobs would have reached the crashed helicopters and eliminated the crews.

Showing the mark of just what kind of man serves with Delta Force, two of the snipers who were riding in one of the helicopters overhead of the crash site volunteered three times to go down and back up whatever crewmen might still be alive. On their fourth request, the two men were given permission to fast-rope down from their bird, basically slide down a thick rope, to aid the other crew. These men were a sniper team made up of team leader Gary Gordon and his teammate Randall Shughart. They were the first men since the end of the Vietnam War to become recipients of the Medal of Honor, for their actions that day. Their story and their actions are summed up in the official citations:

Master Sergeant Gary I. Gordon

Citation: Master Sergeant Gordon, United States Army, distinguished himself by actions above and beyond the call of duty on 3 October 1993, while serving as Sniper Team Leader, United States Army Special Operations Command with Task Force Ranger in Mogadishu, Somalia. Master Sergeant Gordon's sniper team provided precision fires from the lead helicopter during an assault and at two helicopter crash sites, while subjected to intense automatic weapons and rocket propelled grenade fires. When Master Sergeant Gordon learned that ground forces were not immediately available to secure the second crash site, he and another sniper unhesitatingly volunteered to be inserted to protect the four critically wounded personnel, despite being well aware of the growing number of enemy personnel closing in on the site. After his third request to be inserted, Master Sergeant Gordon received permission to perform this volunteer mission. When debris and enemy ground fires at the site caused them to abort the first attempt, Master Sergeant Gordon was inserted one hundred meters south of the crash site. Equipped with only his sniper rifle and a pistol, Master Sergeant Gordon and his fellow sniper, while under intense small arms fire from the enemy, fought their way through a dense maze of shanties and shacks to reach the critically injured crew members. Master Sergeant Gordon immediately pulled the pilot and the other crew members from the aircraft, establishing a perimeter which placed him and his fellow sniper in the most vulnerable position. Master Sergeant Gordon used his long range rifle and side arm to kill an undetermined number of attackers until he depleted his ammunition. Master Sergeant Gordon then went back to the wreckage, recovering some of the crew's weapons and ammunition. Despite the fact that he was critically low on ammunition, he provided some of it to the dazed pilot and then radioed for help. Master Sergeant Gordon continued to travel the perimeter, protecting the downed crew. After his team member was fatally wounded and his own rifle ammunition exhausted, Master Sergeant Gordon returned to the wreckage, recovering a rifle with the last five rounds of ammunition and gave it to the pilot with the words, "good luck." Then, armed only with his pistol, Master Sergeant Gordon continued to fight until he was fatally wounded. His actions saved the pilot's life. Master Sergeant Gordon's extraordinary heroism and devotion to duty were in keeping with the highest standards of military service and reflect great credit upon him, his unit and the United States Army.

Sergeant First Class Randall D. Shughart

Citation: Sergeant First Class Shughart, United States Army, distinguished himself by actions above and beyond the call of duty on 3 October 1993, while serving as a Sniper Team Member, United States Army Special Operations Command with Task Force Ranger in Mogadishu, Somalia. Sergeant First Class Shughart provided precision sniper fires from the lead helicopter during an assault on a building and at two helicopter crash sites, while subjected to intense automatic weapons and rocket propelled grenade fires. While providing critical suppressive fires at the second crash site, Sergeant First Class Shughart and his team leader learned that ground forces were not immediately available to secure the site. Sergeant First Class Shughart and his team leader unhesitatingly volunteered to be inserted to protect the four critically wounded personnel, despite being well aware of the growing number of enemy personnel closing in on the site. After their third request to be inserted, Sergeant First Class Shughart and his team leader received permission to perform this volunteer mission. When debris and enemy ground fires at the site caused them to abort the first attempt, Sergeant First Class Shughart and his team leader were inserted one hundred meters south of the crash site. Equipped with only his sniper rifle and a pistol, Sergeant First Class Shughart and his team leader, while under intense small arms fire from the enemy, fought their way through a dense maze of shanties and shacks to reach the critically injured crew members. Sergeant First Class Shughart pulled the pilot and the other crew members from the aircraft, establishing a perimeter which placed him and his fellow sniper in the most vulnerable position. Sergeant First Class Shughart used his long range rifle and side arm to kill an undetermined number of attackers while traveling the perimeter, protecting the downed crew. Sergeant First Class Shughart continued his protective fire until he depleted his ammunition and was fatally wounded. His actions saved the pilot's life. Sergeant First Class Shughart's extraordinary heroism and devotion to duty were in keeping with the highest standards of military service and reflect great credit upon him, his unit and the United States Army.

The actions of these men were immortalized in the best-selling book and later movie *Blackhawk Down*. After the incidents of October 3–4 took place, there was a significant increase in the number of U.S. troops in Somalia. But this was a temporary situation. Within months, the bulk of the U.S. forces were withdrawn from the stricken country. By March 25, 1994, only a few hundred marines remained, offshore of Somalia, in case there would have to be an emergency evacuation of the more than one thousand U.S. civilians and military advisors who remained in Somalia. By March 1995, all of the UN and United States personnel had finally left Somalia.

THE **WAR** ON TERROR OPENS

18

SNIPERS INTO THE TWENTY-FIRST CENTURY

The world looked on in horror as civilian aircraft were used as major weapons in an attack against the United States on September 11, 2001. Thousands died as the Twin Towers fell in New York. The ancient granite that made up the walls of the Pentagon shattered when another aircraft was flown directly into them. Only the bravery of a handful of hijacked passengers prevented another such tragedy as they forced their own plane down into a rural field.

Terrorism in the form of Islamic fundamentalism had arrived in the United States. But the government of the United States was not going to simply stand by and allow the freedom that the country had stood for to be changed by the threats of attacks against the innocent. The military was called out and directed to their mission. In Afghanistan, the Taliban government was given every opportunity to deliver the leader of al-Qaeda, Osama bin Laden, to face justice for the September 11 attacks. When the delivery of the terrorist leader was denied, the U.S. military moved in. There was no question that Afghanistan had been acting as a safe haven for al-Qaeda and Islamic fundamentalism for years. They had proven themselves a threat to the world, and that threat was taken very seriously.

In the fighting across the mountains and valleys of Afghanistan, the large-caliber projectiles and long ranges of the American sniper proved valuable over and over. When a heavy weapon system such as air support or artillery couldn't be brought to bear on a distant target, the sniper could come up and eliminate the threat with precision fire. The use of the .50-caliber sniper rifles gave almost unbelievable range to the sniper's shots. The distances that snipers were able to successfully engage targets at were so long that they could have only been reached with artillery during World War II.

A SEAL fire team maintains a 360-degree watch, the four men all facing outward with their backs to one another. To the left is the sniper in his ghillie suit. The outline-breaking action of the ghillie suit is so effective that even in this picture, where you can see the man in plain sight, the eye is not drawn to him, only the other three men in the shot. ***U.S. Navy***

Against the American onslaught, the Taliban and al-Qaeda folded within months. The Afghani people were able to try and rebuild their country and establish a new government. For the first time in history, Afghans voted for their own government, their own leadership. The fight against remnants of the Taliban and al-Qaeda has continued in Afghanistan, but the country is no longer a safe haven where they can move about with impunity.

In Iraq since the end of Desert Storm in 1991, Saddam Hussein had been refusing to cooperate with UN inspectors to prove that he had no weapons of mass destruction. The dictator had used such weapons in the past, gassing his own people with poisons that had been outlawed in warfare since World War I. His dreams of a nuclear weapon had been attacked once before when a reactor deep in Iraq had been destroyed by an Israeli air strike in the 1980s. Biological weapons had been uncovered and destroyed by UN teams during their searches immediately following Desert Storm. But Saddam refused to cooperate fully, finally throwing the UN inspectors out and sealing his country.

The posturing of the dictator was that of a man who is holding cards that he refuses to show. The threat he posed was a great one; even the other Muslim

Soon after Desert Storm, a foreign soldier is given some trigger time behind the Barrett .50-caliber rifle by U.S. Special Operations personnel. The dust cloud kicked up by the muzzle blast of the huge weapon has not settled down completely, but the firer is ready for a quick follow-up shot. ***USSOCOM***

countries around him could not say that the Iraqis had no weapons of mass destruction. During the last months of tolerance, convoys of heavy equipment and sealed containers were spotted by intelligence-gathering assets on the roads across the Iraqi desert. The last opportunity to cooperate was denied by the dictator. At 9:34 P.M. EST on March 19, 2003, Operation Iraqi Freedom was launched.

The objectives of the invasion of Iraq included the elimination of Saddam's regime, freedom for the Iraqi people, the removal of a terrorist base that had been operating for years, and the elimination of support for global terrorism, as well as a possible source for weapons of mass destruction. The United States military at the spearhead of the operation smashed through the Iraqi defenses and defeated the Iraqi Army in record time. The suspected weapons of mass destruction were not located, but there were suspicious materials found that indicated more had been going on than the world had known.

Even though its existence has been vehemently denied by the UN and others, a roadside bomb was detonated that could have been far more destructive than any other. The bomb was a converted artillery shell, a binary agent shell that would have resulted in VX nerve gas being released if it had been fired normally. Instead, the explosion didn't mix the ingredients as well as a spinning shell would have, and all that happened was the creation of a large contaminated

A SEAL sniper team on patrol as part of a public demonstration during the 1990s. The forward man is the sniper and he is carrying a McMillan M88 .50-caliber sniper rifle. The trailing man in the patrol is the spotter, who is armed with an M4/M203 combination carbine/40-millimeter grenade launcher. Both men are wearing ghillie suits for concealment, having just gotten up from a hide where they fired a shot in plain view of the audience, none of whom could locate the sniper team. *Kevin Dockery*

area that had to be cleaned up by U.S. forces. It was a sophisticated weapon and not likely at all to have been the only one that Saddam ever had made.

In the power vacuum left behind by the removal of the dictator, other groups have tried to seize power in Iraq. Identified as "insurgents" by the media, these men have proven to be a constant target for the military snipers fielded in Iraq. The discipline and training of the American sniper allows him to precisely target an insurgent, without putting any surrounding civilians in danger. The military sniper has had to reduce the range of his engagements in the fierce fighting that takes place in the cities, towns, and villages of Iraq. His mission more closely approaches that of a law enforcement sniper today, as the military moves house-to-house and room-to-room.

The sniper oversees his teammates as the fight continues. He is the long arm of the battlefield, any battlefield.

LAW ENFORCEMENT

19

A NEW ERA

The 1960s were a time of upheaval in the United States. Civil unrest, driven by political and economic pressures, grew in violence until it seemed to engulf the country. Crimes that had previously been rather rare took on an appearance of becoming almost commonplace. Part of this was due to the fact that the 1960s were the first time that mass media was able to bring images to the public at large as events were happening. Another aspect was that things were happening more often; once one person had committed an act that brought him even fleeting fame, others copied that act.

The decade began with the promise of a new presidency, vibrant and full of the optimism of relative youth. That promise at least helped bring some balance to daily life that was lived under the shadow of possible nuclear annihilation. An assassin's bullets cut short that life and the promise it held, and began what may have been the downfall of the sixties.

Once so rare in the United States, in the 1960s changing the politics of a situation by assassination began to take place with an almost hideous regularity. The first of these assassinations shocked the world, when a former marine with a confused political background brought down the president of the United States.

"Sniper" was at best a poorly understood word in the early years of the 1960s. Even in the military, most soldiers only knew it as a term for the "shooter in hiding" who cost so many their lives during World War II and Korea. Memories of those conflicts were still fresh among the troops in the military, and there was no active sniper program in the U.S. military to change their opinion of who a sniper was and just what he could do. The name was brought violently into the public's eye when it was used to describe Lee Harvey Oswald after he shot President John F. Kennedy on November 22, 1963.

Not the inconsequential figure as so many have portrayed him, Oswald had tried to assassinate another political figure earlier that same year. He had tried to shoot retired General Edwin Walker, a hard-core anti-communist and segregationist relieved by President Kennedy for his attempts to politically influence the troops under his command. Walker had tried to become the Democratic nominee in the race for the Texas governor's seat but had lost to John Connally. With the Mannlicher-Carcano M91/38 rifle he had bought through the mail under an assumed name, Oswald opened fire on Walker as he sat at his desk in his home in Texas. The shot missed, the bullet impacting on the window frame between Oswald and his target. Walker was hit in the forearm by some bullet fragments and little else. Following his plan, Oswald got away from the site of the shooting without incident. The only evidence of his plan to survive the shooting was a note that had been saved by his wife Marina.

It may have been the failure of that event and the successful escape from it that caused Oswald to improve his marksmanship. In the Marine Corps he had qualified as a sharpshooter at one point, dropping to the marksman level at another. Some reports have him passing the two hundred-yard rapid-fire phase of the rifle qualification course at the sharpshooter's level in 1956, scoring a 48 and later a 49 out of a possible 50 points. But he was never considered much of a marksman by the few marines who knew him and were interviewed during later investigations.

Early on that November afternoon in Texas, Oswald changed the public perception of snipers forever when he fired from the sixth-floor window of the Texas School Book Depository. Three rounds fired in just over eight seconds according to films of the scene resulted in the death of President Kennedy and changed the future history of the United States.

That action was the one that brought the idea of a sniper into the living rooms of the American public as they watched their evening news. In the years to follow the Kennedy assassination, more turmoil threatened the stability of the country. In August 1965, race riots swept through Watts, a neighborhood in the Los Angeles area. In seven days of anarchy, ten thousand people became involved in riots that covered nearly a fifty-square-mile area of the city. By its end, the Watts riots had resulted in the deaths of thirty-four people, with over a thousand injured.

Only a month after the destruction of the Watts riots swept the city, the Los Angeles Police Department found itself involved in another incident, of much smaller scale. When the LAPD responded to a disturbance call, the suspect they were approaching barricaded himself and forced a police standoff. During the incident, the suspect opened fire on the surrounding police. By the time the incident had finally ended, three LAPD officers were wounded, along with a nearby bystander. The barricaded gunman was himself wounded in the shoot-out before surrendering to police.

These were some of the incidents that led several members of the LAPD to seriously discuss the need for the adoption of new police tactics and techniques. They were not the only law enforcement organization in the country to realize such a requirement, but the LAPD were the first to actively try and fill the need. The criminal actions of some were taking on the appearance of being unconventional warfare. A guerrilla war appeared to be rising against American society. A police unit trained to fight back against such actions was going to be needed. And they would need to be much more able to deal with hostage situations, barricaded gunmen, and snipers than the rank-and-file police officer was trained to be.

It was an incident in Austin, Texas, on August 1, 1966 that provided the final push for the formation of such new units. LAPD officer John Nelson came up with the conception of what was going to be needed and made his presentation to Inspector Daryl F. Gates. That suggestion led to the establishment of the first special weapons and tactics units among the Los Angeles Police Department. With those new units came a new assignment for a police officer, that of a SWAT sniper. The law enforcement marksman was made a permanent part of the police community.

20

ANATOMY OF A SHOT AUGUST 1, 1966

It was the actions of a lone gunman, a mass murderer when such nomenclature was almost unknown to the general public, which was the beginning of true law enforcement precision marksmanship. This incident can be pointed to as the single seminal event that established the need for the formation of police special weapons and tactical units all around the country. Late on a hot August morning in 1966, a disturbed young man completed his preparations and followed a detailed plan that would shake a city, state, and country.

The former marine moved a load of equipment and supplies to the observation deck on the top of the Main Building at the University of Texas in Austin. Once in place on top of the clock tower, Charles Joseph Whitman continued a killing spree that he had begun the night before. By the time it was over, Whitman had left sixteen people dead, including an unborn child, and thirty-one wounded, before he was finally brought down by police gunfire. In 2001, one of Whitman's victims died of complications from the wounds he received that day thirty-five years earlier. The death of that final victim was ruled a homicide by the coroner, bringing the toll for that day to seventeen dead.

Whitman had served in the Marine Corps at Guantanamo Naval Base, where he received a Sharpshooter's Badge for marksmanship. Though the sharpshooter qualification made him an above-average rifleman in the Marine Corps, it did not make him the expert shot he later proved to be. By age twelve he was an Eagle Scout, and he enlisted in the Marine Corps a few weeks after his eighteenth birthday. After serving for a year and a half at Guantanamo, Whitman qualified for officer training. Returning to the States, he attended college, majoring in Engineering. It was while at college that he met his future wife Kathleen Leissner. His low grades caused him to fail at officer training, and he was ordered back to active duty. He completed his hitch in the service without incident and left the Marine Corps as a lance corporal. Returning to school, Whitman entered the University of Texas to major in Architectural Engineering, and also worked part-time and volunteered as a scoutmaster.

This was not the profile of what we would now consider a serial killer. But Whitman had taken note himself of his unaccountable rages and fixations on violence. More than four months prior to the shootings, he had become concerned enough about his mental state to seek out medical help. His mother had finally left her abusive husband that spring and moved down to Texas to be near Whitman and his wife. The stress had built up to where he sought out both medical and psychiatric help from the Texas University Health Center.

Never returning for the follow-up treatments recommended by the doctor he had seen, Whitman went back into the population of students at the university. The doctor was concerned for the mental state of the young man he had seen. The rages that Whitman displayed during the session were what had bothered the medical professional the most. He had dismissed as fantasy Whitman's fixation on the desire of going to the top of the clock tower and "start shooting people with a deer rifle."

Later, still suspecting that there might be some medical reason for his fixations and pain—Whitman stated that he had gone through two large bottles of Excedrin in three months—Whitman typed out most of a letter stating his intentions and concerns the evening of July 31, 1966. In what was his last day, Whitman wrote this: "I am prepared to die. After my death, I wish that an autopsy would be performed on me to see if there is any visible physical disorder."

After writing the letter, Whitman went out and stabbed his mother to death. Leaving his mother's apartment, he stuck a note on her door stating that she was ill and wouldn't be in to work the next day. Then he returned to his own home and killed his sleeping wife. After these acts, he appended to the letter a final note by hand. Those last words were "Mon. 3:00 A.M. Both Dead." The letter with the final handwritten postscript was found by his wife's body. She had been stabbed five times with a large hunting knife.

Several sources have stated that Whitman was using, and abusing, prescription drugs, including amphetamines. This does not match up well with his written statement about using simple Excedrin to deal with his headaches. There were also no traces of drugs found in his system by pathologists in their postmortem examinations. The tests were not exhaustive, but adequately complete. Doctors at the University Health Center had given Whitman some drug prescriptions, and some of his irrational behavior can be attributed to a misuse of drugs, especially amphetamines. But he could not have had an excess of any of these drugs in his system and had it not affect the fine motor control he demonstrated during his shooting rampage.

What was found in Whitman by the pathologists was a cancerous glioblastoma tumor in the hypothalamus region of his brain. At the autopsy, performed on August 2, the finding was described as a small brain tumor located in the white matter above the brain stem. This could account for his constant headaches. Some doctors have stated that the existence and location of the brain tumor could have contributed to Whitman's inability to control his emotions and actions. But they also said that this link cannot be established with any clarity.

What is clear is the sequence of actions Whitman took during the last twelve hours of his life. They were methodical, planned out, and very complete. They were also extremely lethal.

Shortly after 11 A.M. on the morning of August 1, Charles Whitman arrived at the University of Texas police checkstand northwest of the clock tower. Dressed as a workman, Whitman told the UT police that he had equipment to deliver to the Experimental Science Building. His story got him a parking permit that allowed him to take his vehicle into the inner campus drive that would take him up to the Main Building itself. It was there that he would park and continue with his plan.

The Main Building on the University of Texas campus held a 307-foot-tall clock tower. In some of the dozens of floors within the tower were stacks of library and research materials. Near the top of the tower was an observation deck surrounding the four sides of the structure, running just below the clock faces on each wall. The building itself was a massive, ornate structure of concrete and rock with thick walls and carved stone decorations.

In 1966, the clock tower was the tallest structure in Austin. The observation deck was 231 feet above ground level. The view from the observation deck's elevation not only covered the entire campus, it gave a clear line of sight into the local Austin community and business area. People shopping along Guadalupe Street, locally known as the "Drag," would soon find themselves trapped within buildings and behind cover as Whitman began the worst shooting incident of its kind in U.S. history.

At 11:25 A.M., Whitman arrived at the ground-floor entrance to the tower. His planning and preparations for the incident were obvious in his unloading

ten Whitman on his perch. But the police were possibly not the first to
ck; the local population did.
idents and students began returning fire at Whitman with their own
weapons. One student, whose wife was researching in the stacks of the
the time, rushed home to retrieve his M1 carbine. The student was an
ceman, a former army airborne trooper, and he had purchased his car-
ough the NRA. Returning back to the scene of the shooting, the student
hrough a police roadblock and approached the area on foot. He com-
fire on Whitman shortly after noon, within twenty-five minutes of the
's opening shots.
oting came in toward the observation deck from multiple weapons.
ports later stated that there may have been ten or more civilians who
the tower under fire. There were no reports of victims of shooting
in from Whitman's fire. What is known is that the heavy incoming fire
Vhitman off the ledge and drove him to seek cover.
re were downspout openings at each corner of the observation deck.
one tubes ran out a distance from the deck and opened into the air. It
ough these downspouts that Whitman continued to shoot; only his
fire were severely limited and his victims were now being wounded
an killed. Incoming police officers now on scene were using rifles bor-
om the locals. In his offense report, one officer mentioned using two
two different occasions during the incident. He went on to report that
btained three boxes of .30–06 and one box of .30–30 ammunition from
ardware store—who he was concerned still had to be reimbursed for
idges.
police and concerned citizens risked their lives to try and move the
d to cover while Whitman continued shooting. Several of these people
ick during their rescue attempts. One ambulance driver was hit in the
bleeding badly from a severed artery. As he held a tourniquet made
belt on his leg to stem the bleeding, the wounded driver was driven to
ital by a second ambulance.
blocks away from the campus, the Brackenridge Hospital had the only
emergency room in Austin. The ER quickly became flooded with vic-
erwhelming the facility. Doctors, nurses, and medical professionals
und the city rushed in to Brackenridge to assist. Triage in the hospital
itients to be lined up along the hallway floor near the ER as the worst
ded to first. The strained staff performed magnificently, some only
ing to the strain after the last of the patients had been cared fo

of a heavy military footlocker onto a rented dolly. The green footlocker was marked "L/Cpl. Charles. J. Whitman," his own from his days in the Marine Corps. Inside the locker were weapons, ammunition, and supplies. At least one of the guns inside the case had been purchased that weekend—a Sears store-brand semiautomatic twelve-gauge shotgun. To make the shotgun a more efficient close-combat weapon, Whitman had sawn off the stock and barrel of the gun. Along with the shotgun was a six-millimeter Remington Model 700 bolt-action rifle with an M8 Leopold four-power telescopic sight, a Remington Model 141 .35-caliber pump-action rifle, a .30-caliber Universal M1 carbine, a Smith & Wesson Model 19 .357 Magnum revolver, a nine-millimeter P08 Luger, and a .25-caliber Galesi semiautomatic pistol. There were hundreds of rounds of ammunition in the footlocker for Whitman's combat arsenal. Along with his tools for mayhem, Whitman had packed large containers of water and gasoline. He had matches, canned food, a transistor radio, rope, wire, knives, a machete and a hatchet, an alarm clock, a pipe wrench, bread, sweet rolls and snacks, a blank notebook, a pen, a flashlight, batteries, earplugs, cleaning gear for his weapons, even toilet paper and a shaving razor. There was also a packed knapsack in the footlocker with survival and comfort materials as well as loaded magazines and additional ammunition.

This was not the equipment list of a criminal involved in a spur-of-the-moment situation or a sudden act of passion. These were the planned supplies of someone going into a siege. Much of the ammunition had been purchased that morning; there was a sales slip included in the trunk that was later found to have been dated the day of the incident.

His footlocker was heavy, and Whitman had to struggle a bit to move it into the building and to the elevator leading to the upper floors. The weight of the footlocker had caused him to rent the dolly to move it about. According to some reports, Whitman didn't know how to operate the freight elevator and asked Vera Palmer, a nearby attendant, to help him. After she showed him how to work the controls, Whitman is said to have thanked her and told her, "You don't know how happy that makes me." Palmer was shortly due to relieve Edna Townsley at her post upstairs, sitting at the receptionist's desk for the observation platform.

Having taken the elevator up to the top of its travel, the twenty-seventh floor, Whitman manhandled his laden dolly up a short flight of stairs to the twenty-eighth floor. There he killed the reception floor attendant, fifty-one-year-old Edna Townsley. It isn't noted if Whitman shot the woman or clubbed her down with a weapon. There is no notice of a gunshot by the only two observation deck survivors, and Whitman may not have been ready to initiate his real killing spree. Either way, Townsley became his first victim of the day. Dragging the dead woman's body behind a nearby couch, Whitman left a smear of blood across the reception room floor. Opening up his footlocker, he began arming himself.

A young couple who had been viewing from the observation deck came down the stairs into the reception room as Whitman was preparing. They noticed the smear of red across the floor, which the young woman first thought was simply paint. No body was in sight, but there was somebody else in the room. This was the point when they met the now-armed Whitman. The young man looked at Whitman and just said, "Hi."

It was the look in Whitman's eyes that told the young man that leaving the area would probably be the best thing they could do. For no known reason, Whitman just let them go. They went out to the stairs and left the room behind them. They had been the only people up on the observation deck at the time.

Taking out his sawed-off shotgun, Whitman was not going to be so accommodating to the next people he met. The shotgun had been turned into a deadly close-in fighting weapon, loaded with number four buckshot; the sawed-off would blast twenty-seven pellets of lead out the barrel, followed in quick succession by four more rounds if Whitman kept pulling the trigger.

A group of tourists coming up the stairs a few minutes after the young couple left met up with a killer much more willing to shoot. Two teenaged boys came up to the top of the stairs and ran into a small barricade Whitman had set up with the desk and other materials from the reception area. As the oldest of the boys craned his head to look past the barricade, Whitman opened fire with his shotgun.

The first blast boomed out, smashing into the boy and driving him back down the stairs. Continuing to fire, Whitman blasted the shotgun through the grates on the stair railing. His gunfire knocked everyone back down off the stairs, killing the first young man as well as his aunt and seriously wounding the other woman and teenaged boy. At the foot of the stairs, the two men of the group were not hit and just stood there in shock for several moments.

Instead of following up on his victims, killing the rest of the group and blocking the elevator, Whitman turned back to his long-range weapons and the observation deck.

It was late in the morning, but if Whitman had held off firing just a quarter of an hour, he would have seen a large crowd of people making up the summer session lunchtime foot traffic around the university. That situation could have

greatly increased Whitman's final body count. As it w
of targets to choose from at random. The first shots
tower at 11:48 A.M. The carnage would continue for ni

The six-millimeter Remington 700 Whitman was
hundred-grain soft-nosed hunting bullets that left th
thousand feet per second. The discharge of a high-vel
of the bullets going by were unfamiliar to most peo
sniper's rig, close to what would soon be used by M
Vietnam. But Whitman wasn't in Vietnam, and he was
was picking off pedestrians downtown in an American
The initial victims of Whitman's marksmanship inclu
bicycle riding across the campus, a young girl he sh
woman who was eight months pregnant. The woman l
shot in the abdomen and her unborn child was killed.

As people started dropping from the gunfire, call
police switchboard. The first such call was made by a F
saw people struck down in the area south of the tower.
by the Austin Police Department at 11:52 A.M. It was f
of additional calls that were reporting the shootings, mo
for help.

There was no apparent pattern to Whitman's victims
old, male and female, white and black. If they were visib
like making the shot, he fired. His accuracy shocked man
dent. The longest distance that Whitman struck a target at
from the tower. He fired between the steel rails of a guar
yards from his shooting position to hit a person on the
any opposition, Whitman strode around the observation de
and firing at will with different weapons, appearing to ol
than one shooter. It was during the first twenty minutes
Whitman killed most of the victims of his rifle fire.

The Austin police and university authorities quickly
reports of shots fired and people down on the campus. But
means with which to deal with a sniper in a fortified positio
and with such clear fields of fire. Back at their station house,
old .35-caliber Remington autoloading rifles. These were th
that had been used against Bonnie and Clyde decades earl
the kind of precision long-range weapon that was needed n
of the police was done with handguns. The issue .38 Special

Citizens who could not help directly lined up for blocks to donate blood at a nearby facility.

One police officer and a local pilot flew around the clock tower in a light plane. The danger to people around the tower prevented the officer from opening fire on Whitman. That restriction did not extend to Whitman, who fired at the plane. The plane pulled back a safe distance and remained circling overhead, acting as an observer for the people down on the ground. Asking for an armed helicopter, Austin police contacted Bergstrom Air Force Base and talked to the military there. Again, the threat of injury to people other than Whitman prevented a helicopter from becoming involved.

The shooting rampage continued until a group of men went out on the observation deck after Whitman. The final group that took Whitman down consisted of three police officers and a volunteer citizen who had been deputized earlier. Other officers were immediately available in the tower when the takedown happened. They were acting as backup, covering the area of the floors, establishing communications, and helping the wounded from the tourist group who'd been the first to fall victim to Whitman's shooting.

One group of the officers, including the deputized volunteer, Allen Crum, a retired air force master sergeant and son of a Chicago police officer, had made their way to the main building by crossing the dangerous open area around the tower. Even with the heavy return fire coming from the armed citizenry firing back at Whitman, if he had spotted the group of men, he could have shot at them while they were exposed.

Another much larger group of police were escorted through the tunnel system underneath the university campus. They came up into the basement of the Main Building without being spotted by anyone. That put two groups of armed and determined men inside the tower building, within reach of Whitman out on the observation deck above.

What was unknown to everyone was that Whitman had brought a transistor radio along with his cache of supplies. On the open deck, Whitman had tuned the radio to the local news station and was listening to the reports on the situation with interest. If there had been any reporting of the approaching officers, Whitman might have reacted differently. But no one, not even the television crew on the campus who had set up their camera and trained it on the tower, had any knowledge of the action going on at the base of the tower.

The camera had been set up earlier and focused on the observation deck with a telephoto lens. Apparently, Whitman had noticed the camera or knew it was there and had refrained from shooting at it. Reports had him waving at the camera at one point, but that's unlikely given the volume of incoming fire aimed at the observation deck. Neither the radio nor the television crew was able to give Whitman any information about what was soon going on only a few dozen feet from where he was shooting.

At the elevator shaft leading up to the observation deck, a group of officers and Crum, the deputized retired serviceman, were deciding their next course of action. One of the officers who had a rifle as well as a sidearm was asked by Crum if he could use the rifle. That was how Crum became armed with one of the department's antiquated Remington autoloading rifles. The old weapon was chambered for the pre–World War II .30 Remington cartridge and had a five-round magazine. It was not a piece of modern firepower, but it was what Crum had available to him at the time. The other officers were mostly carrying their service revolvers, while a few had police-issue pump-action shotguns.

Officer Jerry Day, W. A. Cowan of the State Department of Public Safety, Allen Crum, and Officer Robert Martinez were the ones to finally make the assault on the observation deck. In separate trips, the men had taken the elevator up to the twenty-sixth floor of the tower, deciding to move from that point upward on foot. Covering each other infantry-style, the men moved up the flight of stairs to the twenty-seventh floor. There, they discovered the men from the tourist group physically unharmed but badly shaken. The other four members of the tourist party were in the corridor leading up to the twenty-eighth floor. One of the boys was down on the floor along with the two women. The other young boy was conscious but wounded. He was the first person to tell the men that they were dealing with a single shooter and that he was on the outside of the building.

Moving up the stairway to the reception area, all of the men covered one another as they expected to be under fire at any moment. It was an act of bravery to push at the barricading desk and materials, with the thought that bullets could be coming their way at any moment. No fire came and they entered the reception area. The trail of blood that spread across the floor made the men think that the shooter might have been wounded and was waiting in ambush for them. With Martinez covering him, Crum entered an open room that had the trail leading in through the doorway. Inside the room, the body of Edna Townsley lay in a pool of blood behind the couch where Whitman had dragged her.

Moving back into the main room, the group of men started up to the observation deck itself. Cowan remained back to relay communications between the men going up and the ones to the rear. In the corridor leading up to the observation deck, Officer Martinez and Crum found the doorway blocked. The dolly

Whitman had moved his gear with was braced up against the outside of the door opening onto the deck. Through some windows along the corridor, the men were able to see Whitman's footlocker out near the west wall of the deck, but there was no view of the gunman himself. They would have to force the door to go out and take Whitman down on the deck itself.

By this time, the group had been joined by additional police officers. Officer Houston McCoy had come up, armed with a pump-action shotgun, and Officer Jerry Day had come back from where he had helped move the wounded boy out of the line of fire. It was now about 1:15 in the afternoon and time to end the situation. Lying on the floor to give himself at least some cover, Martinez shook the door as hard as he could, finally dislodging the dolly and freeing the entrance to the deck. The first man out the door, Martinez turned to cover parts of the east and south walls with his revolver. Crum moved away from the door and took up a firing position that let him cover the area opposite Martinez.

With Officer Day covering him, Crum was able to grab up a pump-action rifle he saw lying on the deck next to a spread-out green towel. As soon as he could examine the weapon, Crum saw that it couldn't have been any threat to him or the other officers. Struck by the incoming gunfire, the rifle, a Remington 141, had been damaged so badly that it could not be operated. Someone from the ground had come very close to hitting Whitman and ending the shooting spree before the officers and Crum had arrived on the scene. The only way the rifle could have been hit was if Whitman had been holding it; it was that close a thing, which could have saved a number of victims on the ground.

As the men were breaching the observation deck, Whitman was moving along the walls, stopping and shooting as he saw targets. Suddenly, Crum heard rapid footsteps that sounded as if the gunman were running along the wall where he would soon see Crum and the others. Firing his rifle into the wall, Crum stopped Whitman's advance and drove him back to the northwest corner of the observation deck. There, Whitman squatted down on the deck with his back against the outer north wall. He waited for Crum to come around the corner and expose himself to the fire of the M1 carbine Whitman held in his hands.

The ground fire had helped the group of men assaulting Whitman, as it had pinned him down on the west wall of the observation deck. The lack of communications meant that neither the officers nor Crum could tell the people on the ground when to quit shooting. They just kept their heads low and got ready to take the gunman down. Any letup of the incoming fire would have let Whitman know something was going on even before Crum had fired.

Officer McCoy had moved up to cover Martinez with his shotgun as the other officer moved along the wall. Martinez was trying to get to a position where he could see the area of the deck hidden from view by the tower. Once he reached the corner, he looked around and saw Whitman sitting at the far corner, facing south where he would be able to kill anyone coming around that corner. Seeing that Crum was in danger, Martinez opened fire with his revolver, hitting Whitman in the head.

Wounded, Whitman tried to turn the carbine in his hands toward the officer who had just shot him. Crouching, Martinez emptied his revolver toward Whitman. Officer McCoy, who was much taller than Martinez, stood up behind him and pumped two rounds of twelve-gauge buckshot into Whitman where he sat. Seeing that Whitman was still moving, Martinez grabbed the shotgun from McCoy and ran up to the gunman. Whitman could still have been a threat; there was no way of knowing how he was armed or what he could still do. But when Martinez fired the shotgun point-blank into Whitman's head, the ordeal was over.

Waving handkerchiefs and the green towel, the men out on the deck signaled to the citizens below to stop shooting. In the final tally of events, Whitman had killed sixteen people during his spree, including his wife and

A U.S. Army sniper team scans for enemy activity from an Iraqi police station in Mosul, Iraq, November 17, 2004. ***U.S. Army***

mother, and wounded thirty-one. Law enforcement professionals all around the country examined Whitman's shooting incident and the responses of the local law enforcement. The involvement of what they called citizen shooters was not going to be allowed to happen again, no matter what value they may have been in controlling the situation. The success of individual action by police officers, who had never trained together, backed up by the local citizenry, could not be depended on to deal with such a situation. The civil unrest of the latter half of the 1960s was on the rise; the Watts riots in Los Angeles had already taken place. To deal with barricaded gunmen, hostages, and unusual incidents that the police were not normally trained to face was going to take a new organization. The Los Angeles Police Department established the first official special weapons and tactics unit in the United States. They were soon followed by other major departments.

21

TECHNOLOGY OF THE ERA

Law Enforcement Sniping

Unlike the military sniper, the law enforcement sniper has a different requirement in the selection and application of his weapon and ammunition. The military sniper regularly sets up for much longer-range shots than his police counterpart. In the military, snipers want to stalk as close to the target as practical while still remaining undetected—both before and after the shot. A compromise in this difficult situation is the taking of long shots, ranging from three hundred to six hundred meters, to minimize the exposure of the sniper. Targets of opportunity in a battlefield can pop up from close range to the extreme range of the rifle and the ability of its shooter. The military sniper normally goes for the simplest shot, that of aiming at the center of mass in his target.

For a police marksman, the range of engagement, how far they can be expected to shoot, can be between one hundred and two hundred meters on most callouts. Often, this range is shortened even more due to the demands of urban areas, to fifty meters or less. Even though the range is short, the police marksman must operate at an extreme level of accuracy, far more than that demanded of a military sniper. The police marksman must hit exactly the point of aim on his target; anything less could easily cause the injury or death of a hostage, an innocent bystander, or one of his teammates operating in harm's way. The average target area for a police marksman is much smaller than that for his military equivalent. To end an incident that could otherwise result in the death of innocents means an instantly disruptive shot. A killing round to the central nervous system is a precise head shot aimed for the cranial cavity, specifically the base of the brain and the upper part of the spinal cord. That's a target roughly the size of an apple. And, like William Tell, the police marksman must hit that apple with his shot. The collateral damage of a missed shot is completely unacceptable both to society and the police marksman himself.

It is the cranial cavity shot that the police marksman has to train to hit the first time, every time. If the situation allows for a different shot placement, that simplifies the marksman's job to an extent. But it is only the destruction of the central nervous system (CNS) that insures an instant incapacitation of the target. This catastrophic destruction of the brain can stop the subject from being able to pull the trigger of a weapon. A center-of-mass shot that strikes the heart will kill a subject, but he may take ten to fifteen seconds to fall unconscious from the bleeding. So the police marksman must take the more difficult CNS shot to insure the end of an incident.

Above all, the police marksman must make his shot the very first time he pulls the trigger, what is called the "cold shot," the first round out of the barrel after an indeterminate time in storage or transport. That cold shot has to be absolutely dependable, every time the weapon is used. This requires a dedication to training, consistency of equipment, and an almost religious attention to maintenance and cleaning.

Caliber choice is also something that has to be considered for the police marksman. For the military sniper, the choice is overwhelmingly that of a standard issue caliber. The 5.56×46-millimeter NATO round is a very high-velocity slug that is particularly destructive to tissue for its size. Police marksmen are not limited in the choice of their projectiles as military snipers are. There is no Hague Convention outlawing hollowpoint bullets in the civilian community. There are only the problems of overpenetration of a target and the need to be able to reach that target.

For the 5.56-millimeter projectile, even the partially steel-cored military M855 round does not suffer from overpenetration. The projectile can normally be expected to stop within a human-sized target or lose enough velocity to be minimally dangerous on the far side. The 5.56×45-millimeter ammunition has less recoil than most of the other rounds a marksman has to choose from, resulting in possibly more accuracy and a faster follow-up time to get back on target after a shot. What the 5.56-millimeter projectiles suffer from is a lack of penetration through hard targets. The projectiles can disintegrate or deflect when going through window glass, especially car safety glass. They also cannot penetrate heavy apartment building windows or if they do, they have no penetrating power on the far side of the glass.

The most common police marksman ammunition choice by far is the 7.62×51-millimeter NATO round. The 7.62-millimeter NATO cartridge was once the standard rifle round of the U.S. military. Now it is primarily used in light machine guns, or the weapons carried by military snipers. Available in a wide variety of loads and bullet types, including both hunting and military projectiles, the 7.62-millimeter NATO round has proven itself accurate over years of competition use.

For the weapon type, the overwhelming choice for most law enforcement agencies and their marksmen is the bolt-action rifle. The minimal number of moving parts in the bolt-action, virtually none after the trigger is pulled, gives it a superior edge in dependability and accuracy over a semiautomatic design. At a secondary level of importance is the fact that, given the same caliber and level of accuracy, a bolt-action weapon is far less expensive than a semiautomatic one. One fact that is more significant to the police weapon than the military one is that there must be a deliberate, intentional act on the part of the shooter (working the bolt) in order to put another round in the chamber for firing. Liability is of great concern to law enforcement agencies, and a bolt-action weapon is a much safer one on this score than a semiautomatic design. It takes a conscious effort and action on the part of the officer to chamber another round in a bolt-action weapon, and the action can be left open, or the bolt completely removed, to demonstrate visibly at a distance that the weapon is safe.

When the first police agencies and marksmen had to decide on a model and caliber of rifle, they turned to the military and the experience they were developing in Vietnam. Though the U.S. Army was fielding trained snipers by the late 1960s, it was the Marine Corps that held the greatest pool of experience in combat sniping. The Marine Corps adoption in 1966 of a modified Remington M700 BDL short-action rifle as their M40 sniper rifle deeply influenced law enforcement agencies' selection of a marksman's rifle. Remington released what was essentially a civilian version of the M40 in 1967 when they put their 700 BDL Varmint Special on the market.

The Model 700 BDL Varmint rifle was a heavier but more precise shooting version of the hunting rifle. It had a twenty-four-inch-long, slightly tapered, heavy barrel, no iron sights, was drilled and tapped for a telescopic sight mount, and could be purchased chambered for the .308 Winchester round, the commercial equivalent of the 7.62×51-millimeter NATO round. This weapon quickly became the rifle chosen by the vast majority of police marksmen and their departments. The M700 bolt-action rifle is still the favorite by far of all police marksmen, and is the action used by specialist manufacturers for their own sniper rifles.

In the late 1960s, there were relatively few other weapons in use as marksman's weapons in the law enforcement community. Those that were used remained generally bolt-action rifles. A handful of police officers around the country utilized what were basically privately owned hunting weapons that the individuals were very experienced with. Two other commercial weapons that were purchased in some numbers by police departments were the Winchester Model 70 Target rifle and the Remington 40-XB Rangemaster repeating centerfire target rifle.

Both of the Winchester and the Remington target weapons had been originally designed for longer-range match shooting competitions. As match-quality weapons, they were made to exacting tolerances, much tighter than those of the general hunting class rifles. They also had crisp, relatively light triggers, as well as hand-fitted actions and stocks. There was no question that the weapons were as accurate as a commercial weapon could be at the time. But that accuracy came at a cost of field serviceability. With tighter tolerances, a match weapon cannot operate in a dirty, dusty, or sometimes even a very wet condition. The looser spaces between parts in military and hunting weapons, measured in just a few thousandths of an inch at times, allow a certain amount of dirt and fouling to build up while still allowing the weapon to operate dependably. This extreme tightness of the action prevented the Model 70 or the 40-XB from becoming as widely accepted as the Remington Model 700.

The Winchester Model 70 Target weapon had a five-round internal magazine and had a twenty-four-inch heavy barrel. In police and some military use, the weapon was consistently found chambered for the .30–06 cartridge. Some long-range versions of the weapon were chambered for the more powerful .300 Holland and Holland magnum, but these were rarely seen in law enforcement's hands and were primarily long-range competition weapons. The Model 70 Target rifle was first put on the market in 1935. It had been in use with both the U.S. Army and Marine Corps as a competition weapon. The Remington 40-XB was from its beginning a competition weapon, the majority of which were chambered for the rimfire .22 long rifle round. In 1964, the Model 40-XB Rangemaster centerfire repeater model, made up from a perfected Model 700 receiver with a five-round internal magazine, was placed on the market. The heavy-barrel Model 40-XB was made in a wide assortment of chamberings; the most effective for police use was the 7.62×51-millimeter/.308 Winchester. Scope-mounting blocks on the receiver and barrel allowed for the easy mounting of a precision target telescopic sight. These long, externally adjustable telescopic sights had been used by both the Marine Corps and the army during and since World War II. They were supremely accurate but sensitive to dirt and rough handling.

OPTICS

The telescopic sight is an absolute must for the law enforcement marksman. Not only does he need the precise point of aim offered by a quality telescopic sight, but he absolutely requires positive target identification before a shot can be fired. The clarity and magnification of a telescopic sight offers the marksman both the needed precision and visual identification.

The selection of satisfactory optical sights in the later 1960s was somewhat limited. Some very good designs were available on the open market, development of such sights having been pushed forward by the demand of hunters and target shooters. Extreme accuracy and high magnification were available in the form of target sights. This extreme is exemplified in the Lyman Super Targetspot, a telescopic sight that was available in magnifications ranging from ten-power all the way up to a thirty-power scope. But the Super Targetspot was also very long and thin, being slightly over two feet in length. And the high magnifications gave a very small field of view, only four feet wide at one

Looking through the scope of a sniper rifle helps to enlarge targets and to ensure that the sniper has a clear shot. ***USMC***

hundred yards for the thirty-power model. Combined with its very fine external micrometer adjustment controls, the Lyman Super Targetspot was suitable for the range and not for use in the field, even by police marksmen who would not normally expose their weapons to the rigors of active combat.

The law enforcement marksman required his weapon and its sight to be rugged enough to stand up to the rough movement and handling of a callout, not the invasion of an enemy-held beach. The rig that would be commercially successful for a hunter could easily meet the requirements of the law enforcement community. The added weight of the Varmint-style heavy-barreled rifles was acceptable to law enforcement for the additional accuracy of shot placement they offered.

The demands of military sniping actions in Vietnam made their results known on the sight market in the United States. The Marine Corps made the Redfield Master Variable model scope their general sight of choice. The Marine Corps mounted the unmodified Redfield scope on their M40 rifles while the army used it as the basis for their ART (automatic ranging telescope) sight used on the M21 rifle. In both cases the Redfield three-to-nine-power variable scope provided generally excellent service, though it was hard to repair in the field.

For the law enforcement marksman, the Redfield Master Variable telescopic sight was the most common model mounted on department weapons. Two other variable telescopic sights, the Leupold Vari-X II and Bushnell Scopechief II, both three-to-nine-power variable magnification designs, were also used in a number of departments. Fixed power telescopic sights, usually those in the four-power range, were also found in use, the four-power magnification being found to be the best magnification for all-around use at ranges of one hundred yards or less.

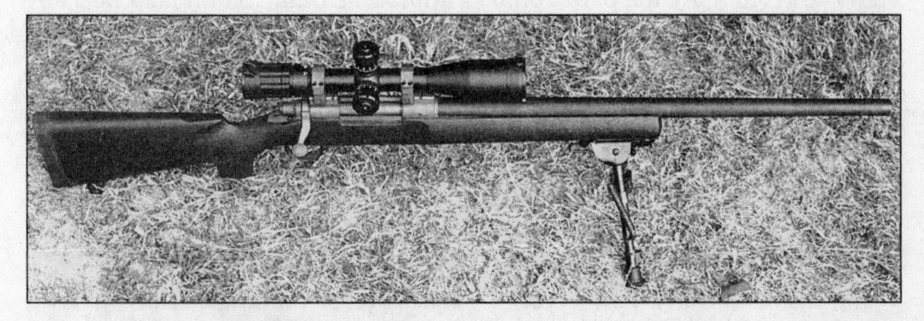

An example of the .308-caliber sniper rifle used by Navy Special Warfare (SEAL) snipers. The five-shot repeating bolt-action weapon is based on a Remington 700 action with carefully selected and assembled parts. The black synthetic stock is almost unbreakable and impervious to moisture, a great improvement over the wood stocks of the Vietnam era. *Kevin Dockery*

22

ANATOMY OF A SHOT DECEMBER 2, 1992

There are occasions when the time the police sniper has to set up, acquire the target, assess the situation, and decide to shoot can be measured in seconds. On a cold December evening, twenty-year-old Marshall Smith, Jr., forced the development of one of those situations in Albuquerque, New Mexico. Earlier that evening, Smith had assaulted his live-in girlfriend, the mother of his child. Taking his infant daughter up, Smith left the beaten woman and rushed from his home, driving off into the night. Notified of the situation, the Albuquerque Police Department responded quickly, trying to cut off Smith's flight in his speeding car.

On the road approaching the I-40 Bridge over the Rio Grande, Smith struck a stopped police cruiser with his car, but that did not end his flight. Snatching up his baby daughter, the man abandoned his car and ran from the police. Once out on the bridge, Smith went to the side guardrails before any officer could reach him. As the patrol officers approached him, Smith held his baby daughter out over the side of the bridge, the infant dangling helplessly over a drop ending in jagged concrete and the sharp ends of steel rebar. As the officers backed away from the possibility of watching the infant fall to her death, Smith drew the child back from over the drop and hugged her against his chest. Just when the police thought that they might be disarming the situation, Smith leaned far out over the guardrail, now threatening to throw himself and the child down onto the rocks below.

The situation had now become a stalemate, as Smith continued to hold officers at bay by loudly threatening to drop the baby or jump over the railing of the bridge with her. The very real possibility of witnessing a murder/suicide kept the patrol officers from committing any action that might force the desperate and unbalanced Smith to carry out his threats. For thirty minutes, they tried to calm the man down, to talk him into giving up the child, but Smith refused to allow any of them to approach. When officers moved forward, he would once again hold the baby out over the drop. Several officers made a desperate attempt to place themselves under the bridge with a blanket. If Smith did throw the child over, they could at least try to catch the infant in the blanket. The darkness and uneven footing of the concrete and exposed ends of the rebar made such an attempt hazardous for the officers under the bridge, and they had to avoid being seen by the unhinged man on the bridge above them. If Smith saw what the officers were considering, all he would have to do is toss the child out away from the bridge. All the helpless police under the bridge would be able to do is watch the child die.

The on-duty Albuquerque SWAT unit was immediately called out to help deal with the situation. It was unquestionably a hostage incident, with an immediate possibility of harm coming to the baby. The bridge was closed, and traffic along the busy I-40 highway diverted even as other SWAT officers sped to the scene. Roadblocks were put up by the police to control traffic, which quickly built up as the busy bridge and its approaches were cut off.

Caught up in the traffic was Steve Rodriguez, a veteran Albuquerque police officer and SWAT team sniper. The freeway leading to the bridge was at a standstill, clogged with a half-mile traffic backup, when Rodriguez decided he could cover the rest of the distance to the bridge faster on foot. With his sixty-five pounds of personal equipment, including his body armor, web gear, ammunition, and sidearm, the sniper grabbed up the additional twenty-five pounds of precision rifle in its own protective bag and started out running.

As Rodriguez arrived at the bridge after running a half mile through stalled traffic, he did not have communications with the other officers on the scene. The level of physical fitness demanded by his job and training allowed the sniper to control his breathing and heartbeat to some extent as he quickly surveyed the situation. Kneeling down behind a parked police cruiser, he quickly set up his weapon. The image that Rodriguez could see through his telescopic sight was a chilling one.

It looked as if the negotiations with the hostage taker were either at a standstill or had broken off completely, as Rodriguez watched at least one officer walking away from the confrontation on the bridge. The view through his sight did nothing to change that opinion. Smith was holding the baby out at arm's length over the side of the bridge. Rodriguez knew that the drop from where the man was standing on the bridge was at least thirty feet or more down to concrete construction. Certain death for the child, and the man holding the infant would probably follow her over the side.

The situation was one of the worst that a SWAT officer could face: an innocent hostage in what had all the appearances of a homicide/suicide situation. What further complicated the situation was that the danger to the hostage was what the police referred to as a "dead-man switch." If Rodriguez fired to stop Smith, Smith would just drop the child. If he fired to wound the man, the child would still go over the edge, probably with the hostage taker right behind her.

There was no weapon visible to the sniper, but there didn't have to be. Smith wasn't holding a gun, knife, or bomb; he was going to let gravity do his killing for him. The threat was an obvious one, and there was no question that an innocent life was in imminent danger of serious bodily harm or death. In such a situation, the sniper was legally and morally authorized to use lethal force to end the situation.

As the SWAT sniper watched, Smith drew the baby back from over the edge of the railing and again hugged her against his chest. This was a possible window of opportunity for Rodriguez to make his shot and end the situation without exposing the infant to more danger. He quickly considered the various possibilities of a shot in his mind. A center-of-mass shot would put the bullet through the chest of the perpetrator, killing him quickly and making for an easier shot in the poor light at the bridge. But with a chest shot, there would be the danger of the bullet going all the way through the target, a through-and-through shot, and possibly striking the infant. The power of the sniper's rifle could also put the very small child in danger from bone fragments kicked up by the bullet's impact. It would have to be a central nervous system (CNS) shot, the ultimate stopping shot. A correct CNS shot would end the situation immediately by cutting off all control to the hostage taker's body. The sniper's target would be the head of the hostage taker.

All of this went through Rodriguez's mind as Smith started to move along the bridge, going away from the officers and farther into the shadows. Within moments, the opportunity to end the situation safely for the infant would be gone. Smith could move into darkness, taking away the sureness of a shot, or he could just extend his arm out and put the infant once more out over the railing.

With his target selected, Rodriguez tracked his moving target carefully; the last thought that went through the sniper's mind as he squeezed back on the trigger was "Just don't mash [the trigger]." He didn't.

Through his sight, the sniper could see a spray come off the back of Smith's head. The danger to the child was over as Smith crumpled to the ground. Three other members of the SWAT team who had not been able to effect the situation on the bridge ran up to where Smith lay. They picked up the child and ran her back to the rescue unit nearby, where medical personnel could examine her. She was going to be all right.

From the point when the sniper arrived on scene to when Rodriguez fired the shot was ninety seconds. The sniper had a minute and a half to set up, assess the situation, and decide on his course of action. There was never any question in his mind that he was about to witness a homicide/suicide. The visible intent of Smith's actions was that he was going to kill the child and then himself to make a point that would be understood only by him.

The child survived the ordeal without any physical trauma. After an investigation, a grand jury ruled the event a justifiable homicide. Rodriguez had acted in defense of an innocent life and committed no errors in judgment or actions. Smith's family did not feel the same way, but it took them five years to sue the officer, police department, and the City of Albuquerque. A jury decided in Rodriguez's favor, coming to the same conclusion as the grand jury, that the shooting had been a justified one.

The Austrian-made Steyr Mannlicher SSG PII rifle. Made particularly for police work, this weapon is in use in a number of countries around the world as well as seeing good commercial sales. Part of the accuracy for the rifle is due to the large bearing surface where the barrel meets the receiver. Instead of being threaded in as most barrels are, the receiver of the SSG is heated and then the barrel inserted into the expanded receiver socket. When the receiver cools and shrinks around the barrel, the resulting joint is as solid as if the two parts were made from a single piece of metal. No sights are fitted to the barrel or receiver of the PII, it being intended solely for use with optical sights. The sight on this specimen is a Heinsoldt ZF 10×42 model set in quick-release mounts. The mounts used allow the sights to be easily removed and another sighting system, such as a starlight or thermal-imaging device, to be installed while the zero of the original scope is not changed. ***Kevin Dockery***

23

THE LAW ENFORCEMENT SNIPER MISSION

Tactical and Legal Considerations

One of the difficulties for a police sniper does not directly involve the activities of his job, but the public's perception of it. The vast majority of callouts for a police sniper involve long hours of watching a situation unfold in front of his rifle barrel. After an extended period of observing and reporting what he sees, the situation is resolved, usually without major tactical action on the part of the police. The sniper will then pack up his gear and leave the scene for a debriefing.

In a relatively small percentage of incidents, the perpetrator will convince either the sniper or the overall police commander of the situation that there is an imminent chance of the loss of a human life. It does not matter if that danger is to a fellow law enforcement officer, a hostage, or an innocent bystander; once the danger has been confirmed, the use of lethal force is authorized. It is that term "lethal force" and the finality it brings with it that causes some of the reaction of the public, and forces some authorities to pull back from the situation.

The mission of the law enforcement sniper is not always a physically demanding one, not when compared to other activities in a tactical unit. What the sniper must do that is different is maintain a very high level of mental alertness for an extended period of time. He cannot allow himself to become fatigued, or worse still, excited, while reacting to an incident. Either of those physiological states would affect the sniper's carefully practiced accuracy and his ability to squeeze off a single precision shot when called on to do so.

Part of the difficulty for the sniper is that the situation can change so drastically for him within seconds. The use of deadly force has to follow strict legal guidelines, and the sniper must constantly apply those guidelines through every moment that he is actively sighting his weapon during an incident.

A sniper may witness a sudden immediate lethal threat, very likely not one that endangers him personally but instead an action that threatens another officer, hostage, or bystander, with death or serious bodily injury. This is a situation where the sniper and only the sniper might see a person aiming a weapon or possibly even firing. In the normal course of events, in a situation like this, the sniper is legally justified in using lethal force to end the threat.

The imminent threat is a slightly more difficult situation, where the police sniper may shoot, or even must shoot, when he recognizes the threat and it is one of serious bodily harm or death. This is the kind of incident where the sniper can see a dangerous situation developing, or is told that such a situation is taking place outside of his view of the scene. If given the green light to fire by higher authority, or if a temporary opening in the situation is seen by the sniper—a break in events where a lifesaving shot may be fired—then he may apply lethal force. But the imminent threat situation, the "may shoot" scenario, can suddenly change into a "must not shoot" event.

In addition to the pressures of the threat level of a situation—the "must shoot," "may shoot," "must not shoot" scenarios—there are additional unrealistic demands that have been pushed at the sniper by an unknowing public or self-serving political leadership. These are the result of situations where people have questioned the results of a shooting well after the fact, and even after the event has been thoroughly investigated by higher authority and legally declared a justifiable application of lethal force.

Unrealistic suggestions, or even demands, in some cases include the requirement that the sniper shoot to wound or shoot to disarm, rather than directly apply lethal force to the person instituting the threat. For a public who gets the bulk of their information from unknowing news sources, the popular media, or films and entertainment, it seems that simply wounding a subject should stop them from committing any further action, rather than going to the extreme of using lethal force. And these same well-meaning people think that it would be better still if the subject wasn't directly fired on at all, if the sniper simply shot the weapon out of the individual's control. Knock the gun from the bad guy's hand, just as the public has seen the cowboys do for decades now in the movies and on TV.

For a public brought up on television dramas and Hollywood creations, such suggestions seem completely logical, and much preferred to the horrible option of death to an individual. In the very real world of the police sniper, these options are a practical impossibility—besides of the fact that they just don't work most of the time. As another very serious reality for the police sniper, nowhere in the United States legal system is there an allowance for the use of lethal force in an intentional nonlethal manner.

During the infamous Whitman tower incident in August 1966, one of Whitman's weapons was struck by a bullet and damaged so badly that it was rendered useless. The rifle could still fire, but another round could not have been fed into the chamber through the damaged action. Whitman simply switched to another weapon and continued shooting. It is so very unlikely as to be nearly impossible that the weapon-disabling shot was aimed and fired intentionally at such a small target, but it did happen.

Due to a very few, rare but well-publicized incidents, the act of shooting a weapon from a person's control has been given a name—tactical disarmament. There is no equivalent term for the even more difficult idea of shooting to wound. The only true answer to shooting to wound or tactical disarmament is the development of less-than-lethal weaponry. This technology is forthcoming, but its application at this stage of development is limited. The police do not have a *Star Trek* phaser capable of being set on "stun," though they would very much like to have one.

The technical question of tactical disarmament has been examined by a number of officers, departments, and outside authorities. The manner in which a weapon can safely and effectively be destroyed or rendered safe by a gunshot changes with the situation and type of target struck. Tests have been conducted on dozens of types of weapons where they were safely held in a secure manner and fired on by police marksmen. In many cases the result of a high-velocity projectile striking the hard metal of a weapon was exactly what was expected—not desired, but expected. The bullet smashed into the metal and rebounded off, screaming away in any direction as an uncontrolled ricochet. Other shots had the target weapon disintegrating and large chunks of metal flying off at very dangerous velocities, penetrating wooden witness panels around the site of the test weapon. In many weapons, but revolvers especially, the ammunition in a struck magazine or cylinder exploded from the impact of a rifle bullet, adding the target weapon's own projectiles to the lethally moving mix.

As a last aspect of the use of tactical disarmament, it was found that many times the impact of a shot on a target weapon would not result in the weapon becoming nonfunctional. As was noted in the case of Whitman's rifle, the struck target weapon could still fire at least the round that was already loaded in the chamber.

The result of firing at a weapon in a subject's hand can be a lot messier, and legally liable, than an unknowing public realizes. In one shooting incident, a sniper fired on the weapon in the hands of a man holding the police away at gunpoint. The criminal was armed with a semiautomatic handgun and was holding the weapon up to his own head. The police commander on the scene told his sniper to shoot the weapon, to disarm the individual, if he had a clear shot. The sniper did as his commander requested and shot the gun out of the man's hand at a range of twenty-four yards.

The bullet from the sniper's weapon struck the magazine well of the pistol in the hand of the criminal. The tremendous impact of the rifle bullet detonated at least one of the rounds held in the magazine. The result of the rifle bullet's impact and the ammunition exploding was the criminal receiving a broken left arm and two amputated fingers. The legal implications of such a shot are a litigation attorney's lifeblood. In another incident, a situation where the application of lethal force was legal but not used directly, involved a gunman holding a weapon and threatening to use it on himself as well as others around him. The sniper fired a shot that wounded the man, causing him to lose his weapon in the process. But before the surrounding police could subdue the gunman, he picked the weapon back up with his other hand and shot himself.

24

ANATOMY OF A SHOT
JULY 3, 1982

Shoot to Wound

The use of police marksmen during a hostage situation was still relatively new in 1982 when the hijacking of a Trailways bus near Jasper, Arkansas, and the seizure of sixteen people as hostages forced a new consideration in the employment of police snipers. The bus had been traveling from Little Rock, Arkansas, to Wichita, Kansas, when it was taken over by a young armed couple. The couple, Keith Haigler, twenty-six, and his wife Kate Haigler, twenty-four, had been followers of little known religious leader Emory Lamb. His worshipers considered the fifty-three-year-old Lamb, a resident of Jasper, to be the messiah of their religious cult, the Foundation of Ubiquity (FOU), also called the Father of Us.

The two hijackers had made their message clear from the very beginning of the incident: They wanted media attention and were more than willing to put innocent people in jeopardy to get it. They had seized the bus at gunpoint, holding the hostages in place but not directly threatening any of them. Stopped out in the open, the bus sat at an angle on a small bridge over a culvert, completely blocking the road. There were no concealed avenues of approach that could be made on the vehicle as the Haiglers had a clear field of view in almost all directions.

Initially, the two hijackers had a single demand and that was to meet with the news media. Besides the implied threat to the hostages on board the bus, the Haiglers said that they would "blow up the bus and destroy Jasper," if there was an attempt by the police to take the bus. The threat of a bomb had to be taken seriously, even though there was no evidence of explosives being available to the couple. The state police as well as local law enforcement had the area around the bus secured, and they finally allowed the news media to approach the bus in an attempt to defuse the situation.

Dressed in a decorated Levi vest over a yellow T-shirt, Keith Haigler sat quietly at the front of the bus and startled the reporter and cameraman with what he had to say. The couple was convinced that they were the two witnesses spoken of in the eleventh chapter of the Bible's Book of Revelation. That was a direct quote from Haigler during his interview. His wife Kate was at the rear of the bus along with a number of the hostages. Neither hijacker could be assaulted without excessive risk to the hostages. And there was no practical way to disarm them, as each of the Haiglers had taped their revolvers to their right hands.

Keith continued to say how they had completed the one thousand two hundred and threescore days (1,260) stated in the Book of Revelations since their awakening on January 21, 1979. Their fate was a straightforward one—they were supposed to die that day, July 3, 1982, in order to lie in state for three days before the spirit of life would enter them and they would stand up to give witness to the final time of judgment. Their bodies were to be laid in state, without any funerary preparation, on Emory Lamb's property, until July 7, when they would once again rise. These same statements were delivered to the local sheriff, Ray Watkins, in a letter signed by both of the couple.

The demands were relatively simple but bizarre. They wanted the news media to broadcast the story of their group to the world. And they wanted the police to kill them.

Without question, this was a "suicide by cop" situation well before that term had come into use, and the Arkansas State Police and the local sheriff were not going to accommodate the Haiglers if there was any way to avoid the use of lethal force. The hijacking of the bus and seizure of the hostages had been intended as an attention-grabbing stunt, to get the media to focus on their story. It might have been enough that the newsmen had filmed them.

The situation wore on under the hot July sun. Options to the death of the couple had been suggested by the local sheriff as well as the news reporter on scene. It was said by witnesses that Keith appeared to consider the options while Kate remained set on their holy mission. It may have been a mention by the reporter that the deadline to make the evening news broadcast was approaching that did more to end the standoff than anything the negotiators said. Kate Haigler wanted her husband to "get on with it," according to witnesses. Having received the promise of the reporter that their story would be aired to the world, and that their remains would be treated according to their wishes, the Haiglers released the hostages.

The news camera filmed the people climbing down from the hot Trailways bus, their Fourth of July holiday weekend having become a more memorable one than they would have preferred. The news crew pulled back and kept filming the now almost empty, huge, white-and-red-trimmed bus sitting on the concrete. State Police marksmen armed with scope-sighted hunting rifles kept their aim on the vehicle.

The same news crews that were filming the drama at the bus recorded the orders of the local sheriff to the state police sharpshooters. If the Haiglers forced an end to the situation by firing on the police, the sharpshooters were only to shoot to wound. They were to try and hit each of the two hijackers in the right shoulder. It was thought to be a limited application of lethal force, something not really allowed under law, but the sheriff did not want simply to give in to the couple's demand that they be killed.

The couple had arranged their situation carefully. With their guns taped to their hands, they couldn't put them down. They felt they would remain a threat to the police until they were killed. Climbing down from the bus, they faced the small line of police cruisers about seventy-five yards away. Having walked slowly a few steps away from the front of the bus, the couple knelt down on the concrete. Keith put his left arm around his wife and drew her close, kissing her for the last time. Then the two settled back, sitting on their legs, and looked at the police in front of them. They did not raise the weapons at their sides, simply looked.

Powerful shots boomed out from the rifles of the marksmen. Hit with the first round, Keith was knocked backward, the cloth at his right shoulder fluttering where the bullet impacted. As Kate was also struck in the shoulder, she immediately raised her weapon and fired a shot in the general direction of the police even as she was knocked back. As the woman spun around and fell, she continued to pull the trigger, firing three more rounds off wildly in all directions.

Lying on his back, wounded and in shock, Keith Haigler extended his pistol, aiming it toward where Kate lay apparently so that he could shoot her in the head. There is no way of knowing what the two may have tried to communicate with each other in their last moments, or if Keith's wound prevented him from being able to pull the trigger. It is very likely that the couple had a mutual suicide pact they intended to carry out if the police simply wounded them, as had happened.

Kate sat up, swung her pistol around, and fired into her husband. Then she turned the gun on herself, bending her arm and firing her last round into her chest. The two still bodies lay on the concrete as the sheriff continued to helplessly shout "cease fire."

Outside of the first rounds fired by the marksmen, none of the police forces surrounding the scene discharged a weapon toward the couple. Medical personnel on-site immediately ran up to the bodies, but it was far too late for Keith Haigler. He was announced dead at the scene. His wife Kate wasn't declared dead until she later arrived at a local hospital.

The situation had been carefully planned and orchestrated by the deranged couple. There had never been an intention of them negotiating a release of the hostages and bringing the incident to a close without their death at the hands of the police. The only thing that didn't go according to their plan was their resurrection.

Following the wishes of the Haiglers, the authorities arranged to have their bodies turned over to their respective parents. The three-day wait was conducted as their letter to the sheriff had requested. Four days after the incident, Keith Haigler was buried and Kate was cremated.

25

ANATOMY OF A SHOT AUGUST 16, 1993

Shoot to Disarm

It was an amazing shot, one witnessed by the public all over the globe. The shot took skill on the part of the sniper, almost perfect conditions, and a very large amount of luck on the part of everyone involved. Very well filmed by news crews live at the scene, the footage of the shot was shown all over the country and on news broadcasts overseas. And it was a shot that most police snipers and SWAT officers wish had never been fired.

A despondent, suicidal thirty-seven-year-old male in a Columbus, Ohio, residential area held police at bay for hours as he sat literally in the middle of the street with a revolver at his head. It was a hot, August day in the upper Midwest, and the tattooed man was wearing a black T-shirt with the epithet "Fuck Off" in large white block letters on the front as he sat drinking from a can. Every time police tried to approach him, the man would lift up his weapon and place the muzzle up underneath his chin.

Gathering information on the man, identified at Douglas J. Conley, the police learned that he had broken up with his girlfriend the day before. Calling the woman up that morning, he told her that he was going to commit suicide. At about twelve-thirty on Monday afternoon, with a snub-nosed .38-caliber revolver in his hand, Conley took a chair from his parents' home, set it down in the middle of the busy intersection of Snouffer Road and Bent Tree Boulevard on the northwest side of Columbus, Ohio, and sat down. Several witnesses, including one off-duty police officer, called for assistance. While Conley sat in the street prior to the police arriving on-scene, nearly twenty motorists passed the man by. Some witnesses reported that Conley had fired his revolver into the air once before police arrived to take charge of the area.

By one o'clock that afternoon, the police had evacuated several houses, an apartment, and a local swimming pool around the area where Conley had placed his chair. Several times, the man placed the barrel of his pistol inside his mouth or pointed it at his head.

The threat was an obvious one. If the police came too close, the man would pull the trigger and kill himself. This was in the middle of what normally would have been a busy residential street intersection. The man had brought out a white plastic lawn chair and had sat down in plain sight of everyone in the neighborhood, the police covering behind their cruisers less than twenty yards away, and news cameras set up across a wide grassy field overlooking the scene.

On the far side of where the man sat, as the cameras were looking, were brick single- and two-story family houses. As the incident went on, camouflage-wearing SWAT snipers were moving through the backyards of homes, their long guns an incongruous sight against the backdrop of children's swing sets. Near the base of a line of evergreen trees, three police snipers had their .308 rifles set up on attached bipods, the men trading off with each other as they observed the drama unfolding on a white plastic chair less than a football field away.

It was not just a wait-it-out situation for the police on the scene. The man sitting in the chair was considerably more than just a threat to himself and an inconvenience to the local motorists. There were active roads close enough to be within dangerous range of the man's handgun. The tension of the police increased each time he waved his weapon in the direction of the passing cars nearby.

The negotiators on-scene were within twenty yards of Conley as he sat and threatened mostly himself. During the two-hour-long standoff, Conley even stood up several times, and approached the negotiators at least once when they supplied him with a can of soft drink.

The suspect would just sit in his chair, taking an occasional drink from the soft drink can he would pick up from the pavement. If the police made a threatening move, the weapon went up to his head. Other times, the man sat forward, slumping down in his chair, the short-barreled revolver in his hand dangling down between his legs.

As time wore on, negotiations did not appear to be accomplishing anything. Conley was becoming more and more agitated as the situation continued, even waving his pistol in a threatening manner at the police negotiators. When Conley's ex-girfriend refused to assist the police in talking to Conley, SWAT Lieutenant Peter Tobin made a decision. The sniper had a clean shot with a solid backstop for his round, a concrete curb at the side of the road and the green grass-covered dirt of a neighborhood yard beyond. Conley did not appear to be willing to give up his position or his weapon. That choice was going to be taken away from him.

The SWAT commander told the sniper that if he had a clean shot at the weapon, to take it. The green light was given for one of the snipers to fire, but not on the suspect. Instead, the sniper was going to target the weapon in the man's hand. Lieutenant Tobin's announcement went out over the radio to all of the officers at the scene and was overheard by the news reporters nearby: "I've authorized officers to fire to shoot the weapon away from him if possible."

Never before in the history of the Columbus SWAT department had a sniper fired his weapon during a callout. This was going to be the first shot ever done since the team was first organized in 1974, and it would be one of the most unusual ones a sniper could even consider. SWAT sniper Michael Plumb was going to take the shot. The forty-three-year-old police sharpshooter had practiced on hitting targets as small as golf balls and even smaller, successfully hitting them at a hundred yards on the range. The conditions were near perfect for the shot; the snipers were in a slightly elevated position up on a grassy knoll overlooking the intersection. Sixty yards away, Plumb settled in behind his weapon, a .308-caliber Steyr SSG PII sniper rifle. The Austrian weapon was considered by many to be one of the most accurate production sniper weapons in the world at the time. As Conley settled back down in his chair, he let his pistol dangle down between his legs. Plumb squeezed the trigger.

The 168-grain full-jacketed .308-caliber bullet, moving at over twenty-five hundred feet per second, smashed into the revolver in Conley's hand. The powerful bullet blew the steel pistol into pieces as the startled Conley jerked his hand back. Fragments from the weapon and the sniper's bullet sprayed out in all directions, one chunk of metal tearing into a leg of the plastic chair. Outside of being stunned for a moment, Conley simply sat there in his chair. He leaned forward, resting his hands on his knees, as a number of SWAT officers rushed up to him. They tossed the man roughly over on his back, dumping him to the ground, where he was quickly secured with plastic riot handcuffs.

Outside of some minor cuts to his face and chin, Conley was unhurt by Plumb's shot. At first, the cuts were thought to have come from fragments of either the weapon or the sniper's bullet, but it was later decided that the injuries were a result of Conley being roughly pushed down onto the pavement and secured. He was considerably less injured than he would have been if the sniper had been forced to raise his point of aim a few feet to end the danger to others around the scene.

The shot was unique and quickly seen by millions of people as it was replayed over the national news broadcasts. A standard police investigation followed the shooting, but it was considered a formality. Supervisors of both Tobin and Plumb praised the men's decisions and the results of their actions. Conley was arrested and charged with aggravated menacing and inducing panic. He was also put through a psychological evaluation by a judge's order.

26

ANATOMY OF A SHOT APRIL 4, 1991

In what has since become known as the "Good Guys Incident" a mass hostage seizure by a group of disillusioned Asian gang members was a classic example of how everything can be done right by the police sniper, but everything can still go very wrong. In the incident, a Good Guys retail electronics store at the Florin Mall in the south area of Sacramento, California, became the site of the largest criminal hostage-taking event in U.S. history up to the date of this writing. Around 1 P.M. on a quiet April afternoon, four young Vietnamese refugees ranging in ages from sixteen to twenty-one years old, left their 1982 Toyota Corolla and entered the busy store at the mall. The young men were armed with three pistols and a shotgun, a fact that was demonstrated to the shocked customers and employees of the store as the weapons were fired into the ceiling.

The sudden violence of the invaders' entrance froze most of the customers and employees in place. A few individuals in other parts of the store who weren't shocked into immobility made their escape through back doors and fire exits in the immediate confusion following the shots. Those people who remained in the main part of the store were quickly gathered together under the control of the weapons in the hands of the young men.

Calls started coming in to the Sacramento Sheriff's Department as the people who had escaped the situation in the Good Guys store reached telephones. Thinking they were responding to a robbery gone bad, the Sheriff's Special Enforcement Detail (SED), their tactical response team, were immediately told to respond. The SED was part of the department's Narcotics/Gangs Division, and they had been preparing to go out on a drug raid operation when the call came in for them to respond to the Good Guys store. Being already prepped for their planned operation, the SED were able to respond to the call very quickly.

According to some accounts of the incident related afterward, the four gang members had seized the Good Guys store after a botched robbery attempt. This was not the actual situation. Instead the reason for the action on the part of the four young Asian men was much more dangerous for the hostages than a robbery gone wrong. The youths were all members of a violent Asian street gang known as "The Oriental Boys" that had been preying on the Asian refugee population in the Sacramento area. The gang was known for conducting a wave of home invasions where they smashed their way into a private home, holding the residents until they would disclose the location of whatever valuables they might have.

The four young Asians had decided to use their skills learned during the home invasions to gain some notoriety for themselves. After later investigation, their intent appeared to be just to take hostages for the media attention it would give them. This was a particularly bad situation for the police as the gang members didn't have any real demands or plans for escape; they wanted the show, the public spectacle of them holding the police helpless and at bay, while the media showed their actions to the area, and the country at large. There was nothing concrete that the gang members wanted to negotiate for, not even their lives. They simply wanted to draw out the situation as long as they could.

Not knowing what the real situation was, the sheriff's department responded as they knew how. As the SED headed over to the scene at the store, they took along the department's Critical Incident Negotiation Team (CINT), trained and experienced individuals who would attempt to disarm the situation and bring it to an end with the least amount of danger to the hostages, or even the perpetrators conducting the seizure. The remaining off-duty men of the SED were paged to report in, too, as the tactical unit left for the incident site. Other local and state law enforcement units were also converging on the scene. Alerted by the police response, media organizations quickly sent in their own representatives and film crews to cover the unfolding drama at the mall. Their arrival on scene was exactly what the hostage takers wanted, and added a great deal of pressure on the already tense situation that the police had to contend with.

There was no question that the hostage takers inside the store knew at least part of what was going on outside the building. It was an electronics store; they had a whole wall full of televisions that showed them everything that the news cameras were transmitting.

Within thirty minutes of the call first coming in, the SED had a command post up and running in a nearby bank. As the command post was being set up, the initial action of the SED commander was to send out his two snipers to oversee the situation. The negotiators of the CINT began communications with the hostage takers inside the Good Guys store. The employees and customers who had escaped from the store were talked to to see what information they could give the police on just whom they were facing. Off-duty store managers were called in to help give the police an overall view of the interior of the store, while building engineers brought in floor plans of the building and the mall area around it.

The commander on-site was particularly worried about what the store might have available to the hostage takers. Specifically, did the large electronics store sell police scanners and did they have any in their inventory. If such scanners were available to the perpetrators inside the building, that could seriously compromise police communications in the area. The store managers were able to confirm that Good Guys did not carry such items in their inventory. The only way the hostage takers could have a police frequency scanner was if they had brought one along with them. That possibility was one of the reasons careful control was kept on just what information went out over the police communications frequencies.

In spite of the lack of availability of police scanners to the hostage takers, their taking refuge in an electronics store soon proved a liability. Snipers viewing the situation from their vantage points could see into the building through the large glass doors at the front of the store, doors that were on either side of an opaque center panel. The snipers reported back to the command post that through the glass, they could see a number of the television sets inside the building, and that they were all tuned to the local news broadcasts. The gang members had a perfect view of the police preparations being made on the outside of the building, courtesy of the news media.

The white front of the building looked out onto a wide mall parking lot with a scattering of civilian vehicles parked for the duration of the incident. It was a bright day, but through the glass doors, the interior of the building looked dark and cavernous. By nightfall, the inside and outside lights of the Good Guys store brilliantly illuminated the tense situation visible to the news cameras through the heavy glass doors.

Meetings between the media representatives and the police minimized some of the problems with the news broadcasts being made live from the site. Instead of using helicopters or other means to extend their view of the situation, the media representatives agreed to only show the front of the Good Guys building, and to minimize their broadcasts of the police preparations. This meant that the public could see at least some of what was going on, but rescue preparations could also take place without the gang members inside the Good Guys store knowing about them as quickly as the police could put them into place.

Inside the store, the four gang members had quickly consolidated their position after securing the hostages. Using wire and coils torn from installations, the gang tied up the hostages and put them on display for the media. Through the thick plate glass at the front of the store, hostages could be seen lined up, some kneeling down while others stood, all facing out through the glass toward the relative safety that was out of their reach.

Moving along the hostages, one of the gang members continually tormented the helpless people. He taunted the hostages, poking at them with his handgun and apparently relishing their discomfort. Some of the hostages had their hands tied behind their backs, while others had them secured in front of them, and all of the hostages were securely bound with tough, strong wire. The police could only stand by and watch the gang members as they walked in front of the glass, their backs turned to the officers outside.

Negotiations conducted by the CINT personnel continued with at least one of the gang members inside the store. While the negotiators continued to try and defuse the situation, the tactical officers and their commander investigated other more active possible remedies. Their investigation and information gathering had shown the SED officers that only one entrance to the store did not have alarms on it. That entrance was at the freight receiving area in the rear of the store. A seven-man entry team moved into position to try and take advantage of any opportunity that might show itself to them.

On the north side of the Good Guys store was a fabrics store that would cover the entry team's approach. Inside the evacuated fabrics store, the officers were able to move into position in time to hear the gang members shout "Stay away from the door!"

Immediately the officers froze, thinking that they had been discovered. But there were no more shouts directed their way. Through the shared ceiling area between the two stores, one of the entry team officers was able to look down into the receiving area of the Good Guys store. The officer saw one of the gang members directing some hostages to stack large boxes against the back entrance door. Once the barricade was built up enough to satisfy the gang member, the hostages were directed away from the area. The back of the store was open to the entry team if they wanted to try and conduct a dynamic entry.

On arrival at the scene, the trained negotiators of the CINT began to establish a dialog with the hostage takers as soon as communications could be made. An individual who identified himself as "Thai" was the spokesman for the gang members inside the store. For over two hours, the negotiators continued to talk with the Asian criminals inside the building, the demands of the four men becoming more and more outrageous. The hostage takers never gave the negotiators a clear list of demands, not even those for their own escape. Various materials demanded by the hostage takers included $4 million in cash, thousand-year-old ginseng roots for making herbal tea, more weapons, and a massive military troop transport helicopter capable of moving more than fifty men through the air. Among the most irrational demands was one for transport to Thailand, where the hostage takers, all Vietnamese refugees, felt they could launch a fight against the Viet Cong and help free their homeland.

It was these odd demands, appearing almost random, that told the negotiators that the hostage takers were becoming increasingly irrational. At one point, shots were heard coming from the interior of the store. Hostages were still being held at the front of the store, which prevented the police from launching an immediate dynamic entry of the building.

One demand from the gunmen that remained consistent was their desire for bulletproof vests. This was something that they insisted on so heavily that the negotiators were able to use it to gain the release of a few hostages. One vest was exchanged for several hostages who were immediately interrogated by the police on just what the situation was in the store. The entry team in the fabric store had slipped a small camera device to a point where they could see at least part of the interior of the Good Guys store. But the layout of the store prevented the small camera from getting a full view of the situation. The released hostages were able to fill in a number of important details.

The shots the police heard being fired earlier were not hostages being killed or wounded. The gunmen had gone around inside of the store, shooting out the security cameras. For the moment, none of the hostages were injured.

The negotiations had resulted in some positive action on the part of the gunmen. This gave the CINT officers more hope that the entire situation could be brought to a peaceful conclusion without anyone being hurt. Thai even agreed to surrender at one point, but only on the condition that the gang was allowed to keep their weapons and vests, even while in jail. This was an extremely unrealistic demand, but the negotiators had been dealing with stranger ones on this callout. Normally, in such a hostage situation, the gunmen were able to be talked down from their position and the situation resolved without using lethal force or a tactical assault. But this was not a common incident; the hostage takers had no set demands, and there were no grievances that the negotiators could use as a tool to win over the gunmen. These gang members wanted to put on a show for their own reasons, something the negotiators had very limited means to deal with. When Thai left the phone to speak with his partners, the negotiators felt that they might be finally approaching the end of the incident. Then the phone line went dead from inside of the store.

Following their normal procedures, the SED had been preparing an assault-and-rescue plan for the hostages in case the negotiations failed. The seven-man entry team in the fabric store felt that they had a reasonable means of getting into the store undetected, and the assault plan centered on this. Before any go-ahead was given, the negotiation team managed to reestablish contact with the hostage takers, only to learn that the situation inside of the store had apparently changed considerably.

At first, when the CINT officers tried to call back, all they had heard was a busy signal from the store phone. On their second attempt, the phone was answered, this time by a new voice, a gang member who identified himself as "Number One." He told the negotiators that he was now in control and that they would have to deal with him. There would apparently be no nonviolent resolution to the incident.

The situation at the standoff now started to go downhill. More shots were heard coming from inside of the building, but neither the police snipers nor the entry team in the fabric store could identify just what, or who, was being shot at. As it turned out later, the gang members were once again shooting at the security cameras, a system that the police had no access to.

The seven men of the entry team would have very much liked to have access to the interior security cameras of the store. They had slipped into a storeroom at the rear of the Good Guys building. They were actually inside of the structure, but still separated from the main showroom by a box-blocked door and two metal roll-up windows. In position to be able to fairly quickly assault the main showroom of the store, the entry team settled in to wait for the go-ahead command. In case one of the gang members decided to check on the inside of the storeroom, two of the officers carried suppressed MP5SD3 submachine guns. If discovered, they would cut down the gunman with the suppressed weapons and immediately conduct an emergency assault on the main area of the store. It was not a comfortable situation for the entry team, and the incident was long from being over.

The situation had been going on for nearly eight hours, and the gunmen inside the store felt that they were not getting the reaction they had wanted from the police or the media. The news programs were just showing the front of the store and a very limited amount of the police preparations, little more than shots of uniformed officers pointing guns at the store. The gang members decided to stir the situation up by sending out a hostage, someone who could specifically give their demands to the news media. At the same time, they would make an example of the hostage to force the police to take the young men seriously. They had released several hostages and only received one of the promised bulletproof vests. They would force a change in the attitude of the police.

A male store employee was picked out. He was told that the gunmen's demands were simple; they wanted three more bulletproof vests, a helicopter, and more weapons. That was it, nothing unreasonable in their view. All they were trying to do was draw the world's attention to the plight back in their native Vietnam. To keep the other hostages safe, the released man was to make sure that he told the news reporters exactly what the gunmen had told him to. And to show their seriousness, they shot the store employee in the leg—after asking him to agree to being wounded.

The gunmen shot the employee, risking killing him if they had struck the femoral artery in his leg. He would have bled to death long before any help could have reached him. The wound was not a lethal one, and the man was allowed to crawl past one of the glass doors, leaving a bright red smear across the concrete as he painfully made his way to safety.

The police were not about to give the gang members more firearms, and the helicopter wasn't going to show up any time soon either. The bulletproof vests might be a bargaining chip that could be used by the negotiators, but an attempt would be made to resolve the situation another way first.

Allowing the released hostage to talk to the press, the police hoped that all of the gang members would move to the area in front of the TV screens. They knew that there were only four gunmen in the store. If they could get them all in one place, the entry team could move in and take them down. In addition, the snipers might be able to end the incident with several well-placed shots.

But the Vietnamese gang members inside the store did not take the bait. The situation continued on into the night. Hostages remained visible through the glass doors. One of the gunmen kept moving among them, torturing the helpless people. It was later said by some of the hostages that the gunmen would toss a coin into the air to decide whether a person would live or die. The situation was deteriorating, and the hostage takers were losing patience.

One older hostage started suffering a medical attack. As far as the gunmen were concerned, he had just volunteered to be the next person shot. The ill man was shot in the leg and pushed out the door to make his way to the police. The gunmen had wanted to be taken seriously, and now the police were going to have to deal with them in the most serious way possible. Innocent victims had been shot; more were in danger of serious injury or death. The use of lethal force was authorized.

An authorization was given for a dynamic entry to rescue the hostages and bring the situation to a close before more people were shot by the gunmen. It would be a particularly rare kind of dynamic entry, a sniper-assisted hostage rescue. Inside of the storeroom, the entry team was informed of the command decision, and they were prepared to make entry into the main area of the building. The snipers were also given the green light. When a selected sniper had a clear line of sight on a gunman, his shot would be the signal to initiate the assault.

To draw out one of the gunmen for the sniper, another bulletproof vest was delivered close to the front door of the store. One of the female hostages was sent out to retrieve the vest, the short young woman walking with her hands tied behind her back, a wire tether connecting her to a hood-wearing gunman standing in the doorway. The heavy glass door was only open a short way, with the gunman holding the wire tether just inside the building. It was not a good shot—the sniper would have to put his bullet through the glass of the door—but it was going to be the best one he would get. He pulled the trigger.

As the young woman was crouching down next to where the bulletproof vest lay on the sidewalk, the heavy glass door behind her shattered into thousands of pieces. Inside the building, the gunman staggered back, unhit by the sniper's round. On either side of the center panel of the doorway, more than half a dozen hostages stood stunned at the sudden action after so much time had passed. Holding his face, the hooded gunman staggered away from the widows and into the building. Standing hostages fell to the ground; one man who had been just on the other side of the glass door when it exploded jumped up and dove through the now open doorway.

Just as the one hostage escaped, the police and news cameras witnessed a horrible action. Running from right to left, the hooded gunman moved quickly across the opening, shooting down at the helpless hostages with the pistol in his hands. Before the police could react to the action of the gunman, he was out of sight, leaving only the wounded and dying behind him.

Inside the store, some of the other gunmen opened fire on the hostages. The command went over the police net for the entry team to make their move. Moving in the coordinated manner that they had practiced over and over, the entry team opened the door and shoved away the boxes that had been stacked there. It was not a time for sudden shooting at obscured targets by the police. There were four active gunmen in the building, but there were also several dozen hostages. Stray bullets from the police could kill a hostage as easily as an intended shot from a gunman.

The entry team split into several units, some of them remaining as rear guards to prevent any of the gunmen from getting to the back of the store. The rest of the entry team moved to the right and left, coming around both sides of the main showroom in a controlled but fast-moving manner.

The first gunman encountered by members of the entry team was the one armed with a shotgun. He was firing into the hostages as he spotted one of the police officers. The gunman raised his weapon and fired, but the officer dove to the floor and the blast of shot passed harmlessly over him. The other officers opened fire, cutting down the shotgun-wielding gunman with accurate bursts of suppressed fire from their submachine guns.

The other gunmen didn't know what had happened to the man with the shotgun. They hadn't heard the shots from the suppressed weapons, or if they had, they hadn't recognized what they were. In their confusion, one of the gunmen ran right into the line of fire from the police and was cut down. Two more gunmen were cut down by the controlled fire of the entry team as more officers started to enter the building through the shattered front door.

In the immediate aftermath of the incident, only three of the four known gunmen could be accounted for. In the screaming confusion of the panicked hostages and firing gunmen, it was possible that one man had either escaped or was hidden, possibly waiting in ambush somewhere in the store. Finally, the fourth gunman was found, unarmed and huddled on the floor wearing the one bulletproof vest the hostage takers had received. When he was checked by the very wary entry team officers, the young Asian was found to be seriously wounded. It was Thai, and he was the only survivor of the four young Asian gang members who had started the incident long hours earlier.

In their violent spree of gunfire, the gang members had wounded eleven hostages and killed three more. Three of the hostages had been injured by the broken glass from the shattered door. One of the hostages had been pregnant and shortly afterward lost the child due to the ordeal. Forty-one hostages had been involved in the incident. The one surviving gunman was convicted and is serving concurrent life sentences in prison.

The situation had been a tactically difficult one. Once the gunmen had started shooting hostages, and threatening to start killing more, the decision to initiate the assault was the only option left to the authorities. The sniper shot that struck the door had established that shooting through glass was the most difficult manner in which a sniper could engage a target. Since the incident, new ammunition has been developed to assist the police sniper in making such a shot.

In the Good Guys incident, the police on-scene had done what they could to control the situation and bring it to a satisfactory end. The decision to use lethal force had been taken only after there was an unmistakable threat of the possible loss of life of the hostages. The situation is looked back on by others as a classic example of everything being done right by the sniper who initiated the assault, but of things going completely wrong in spite of that.

TECHNOLOGY

Suppressors

Throughout the United States there are manufacturers of shooting equipment used by snipers, civilian competitors, and sports shooters. This gear can run from firearms and rifles to ammunition, uniforms, web gear, electronics, maintenance gear, telescopic sights, lasers, and night-vision devices. To fully investigate and show the range and breadth of the materials used by both the law enforcement and military sniper would take a good-sized volume of its own.

One piece of equipment that is sometimes used by the sniper is more esoteric and unusual than most. Its applications are almost as misunderstood as the mission of the sniper himself. Shadowed in secrecy, the silencer, or suppressor as it is also known, brings with it an image as a tool of the assassin or the secret agent. These ideas have been instilled in the public by partially known and misrepresented factual stories and complete fabrications. The modern media has reinforced the image of the silencer through popular books, movies, and television which show the silenced weapon primarily in the hands of the murderer and criminal.

The history of the suppressor and its modern use is considerably different from that public image. Instead of being veiled in a shadowy world, the devices were once openly sold to the public in this country, the place where they were first invented. They still are sold fairly openly in some countries overseas.

There are some companies, both large and small, that specialize in making these unusual devices for the law enforcement community, the military, and public sales in states where they are allowed.

28

EQUIPMENT MANUFACTURER

Dr. Philip Dater, Gemtech Suppressors

Currently I build silencers of different models for various firearms for law enforcement, military, and civilian use. "Suppressor" is technically the correct term, but "silencer" has entered common usage, and the terms are completely interchangeable for all practical purposes. Technically these devices are sound suppressors, and this is the term I prefer. Silencer would imply that it actually truly silences something, which it doesn't do; rather it reduces the sound level by a certain amount.

Some of the technology of suppressors dates back to the nineteenth century, when there were silent weapons systems. These were in Austria, I believe, and the weapons were military-grade air guns used from the late eighteenth century into the nineteenth century. The air guns, rifles really, were about .50-caliber, and in place of the buttstock had a large replaceable air chamber that was pumped up with hand pumps. The pressure and volume of the air in the chamber was enough for the user to get a number of shots off before he had to replace the chamber. Due to their relative silence and lack of smoke, these air rifles were considered such a dastardly weapon that the mere possession of one was considered grounds for summary execution.

The modern silencer pretty much started with Hiram Percy Maxim of Hartford, Connecticut, back in the early part of the twentieth century. He is responsible for introducing the term "silencer." His patent of 1909, I believe, was the first patented muzzle silencer for firearms. He was also the father of amateur radio. Incidentally, it was a different Hiram Maxim who invented the Maxim machine gun.

The Gemtech Stormfront suppressor being tested on an M107 Barrett .50-caliber rifle in Tikrit, Iraq. The big suppressor aids in cutting down on the huge blast and noise that comes from firing the big fifty rifle. It also greatly reduces the dust plume kicked up by the weapon when it is fired, helping maintain the security of a sniper's concealment. *Private Collection*

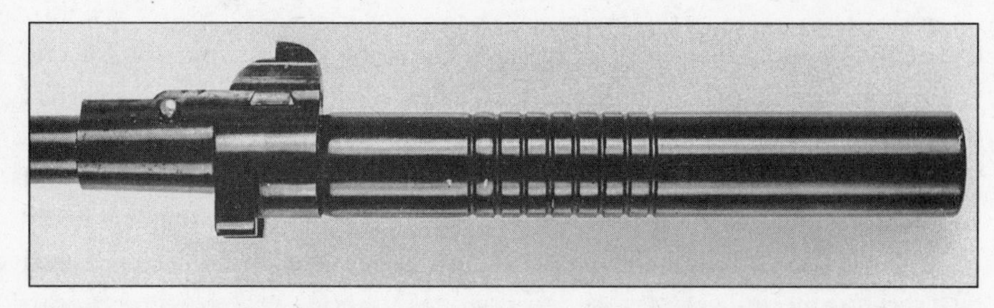

A close-up of the Maxim M1910 (Model J) suppressor. It is recognizable from the later Model 15 U.S. Government silencer (made specifically for the M1903 Springfield) by the smooth, tapered rear part of the mounting sleeve. The Model 15 silencer has a six-sided "nut" at the rear of the mounting. This silencer slips over the end of the barrel and is held in place by the cross-pin that normally secures the sight. The circular rings on the body of the silencer are to hold the internal baffles in place. *Smithsonian Institution*

Maxim aimed at the civilian market. It was considered acceptable in polite society to shoot .22s with silencers. Some of Maxim's early catalogs had pictures of young ladies in ankle-length dresses shooting rifles with his silencers on them. Though some Maxim silencers saw military duty in World War I, probably the first serious usage of the silencer was in World War II. I'm not certain of the exact sequence of their appearance, but the weapons that come to mind are the Sten Mark II S (for silenced) submachine gun, the Welrod, and the DeLisle. The Welrod was a manually operated, bolt-action, .32-caliber assassination pistol with an integral silencer. The DeLisle carbine, designed and built by William DeLisle, was an integrally suppressed .45ACP carbine built on a modified Enfield rifle action. The DeLisle carbines were notoriously inaccurate, but one of the tales that is told about the weapon regards a demonstration put on by DeLisle.

DeLisle was shooting at a target during the demonstration of his weapon, and there was a chicken walking across the field. He missed the target he was actually aiming at, but managed to hit the head of the chicken, killing it instantly. Everyone watching the demonstration was impressed with the accuracy of the exotic silenced rifle. The truth was he wasn't even aiming anywhere near what he hit.

Those weapons were all produced by the British. Here in the United States, the Office of Strategic Services, the OSS, started using the High Standard Model HD Military, a .22-caliber semiautomatic pistol, which was fitted with an integral silencer. I believe those silencers were designed by Bell Labs. There was also the M3 Grease Gun submachine gun produced in a silenced version, and we also produced a small number of Welrods; ours were chambered for 9 millimeters.

After World War II, silencers stopped being very prominent anywhere in the country. It wasn't until the Vietnam War and a man named Mitch WerBell and his company, Sionics, that a number of silencer designs were produced that were relatively advanced for the era. WerBell teamed up with Gordon Ingram, who designed the Ingram submachine gun, and they developed a silencer for that compact weapon. He also made Sionics suppressors for the M16 rifle and the M14 rifle, an integral suppressor for both the Ruger .22 pistol and 10–22 rifle. WerBell popularized the use of the suppressed weapon in Vietnam. He was pretty much a showman and a character, but he made sound suppressors respectable and, in a way, mainstream. His company became part

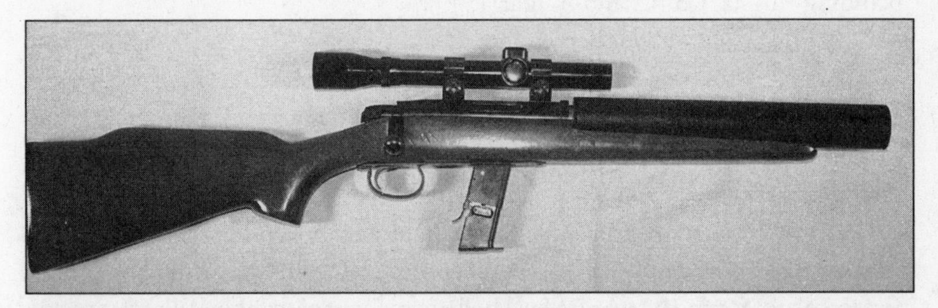

The Vietnam-era Remington 788 M1 rifle produced by the Military Armaments Company and modified to use P-38 pistol magazines and chambered for the standard nine-millimeter round. Made by the Military Armament Corporation, these weapons were produced as a compact suppressed sniper rifle for the U.S. military and other customers. Fired with special subsonic, heavy-bullet loads, the weapon is very quiet and was intended to be used by snipers within one hundred yards of a target. *Kevin Dockery*

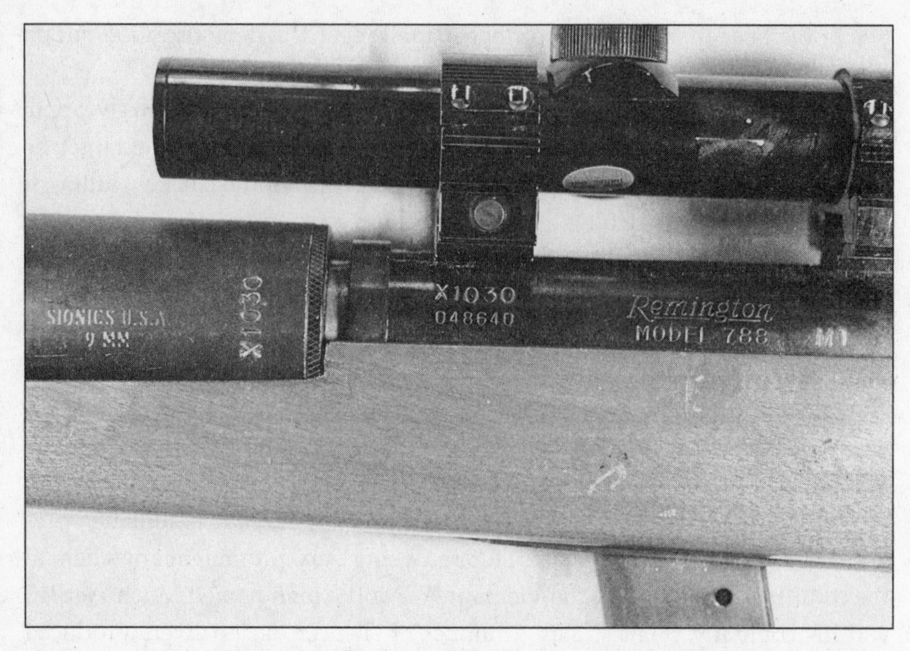

A close-up of the markings on the Remington 788 M1 suppressed nine-millimeter sniping rifle. Both the rifle and the matching suppressor bear the serial number X1030. The Sionics suppressor utilized a wipe-type design, which meant that the exiting bullet passed through a flexible rubber baffle as it left the muzzle. This cut down on some of the accuracy inherent in the weapon and made for a necessary replacement part (the end cap of the suppressor with the plastic wipe). New suppressor designs do away completely with the need for wipes. This is also a particularly rare weapon as only four are known to still exist in museum collections. *Kevin Dockery*

of Military Armament Corporation in Georgia. The Cold War spurred development of a number of interesting designs from the Communist block nations, including Russia, China, and Bulgaria.

After the Vietnam War had ended, the first person to put a serious mark on the commercial market for suppressors was Jonathan Arthur Ceiner. I would easily say that he was the father of popular silencers in the civilian market. He ran ads in *Popular Mechanics, Popular Science,* and similar periodicals starting in the mid-seventies.

For myself, I started working with silencers in 1976–1977. Reed Knight became involved during the same general era with the Hush Puppy, a suppressed pistol for the Navy SEALs. I believe his model went on a Beretta pistol, while the navy had used an older Hush Puppy design that they had developed for the Smith & Wesson Model 39 pistol.

The civilian market for silencers through the late 1970s and into the early 1980s was primarily Ceiner's. My own suppressor work progressed during that time. I started off with the Ruger pistol, making an integral suppressor that surrounded the barrel.

My start with silencers began with an integrally suppressed Ruger pistol that had been made by Military Armaments Corporation (MAC), the company that WerBell and others had started to market their weapons and suppressors. That pistol worked wonderfully, but after about four or five hundred rounds, it wasn't very quiet anymore and my neighbors were starting to complain about my shooting in my backyard in Albuquerque, New Mexico.

Calling Military Armament Corporation, I was told that the suppressor could not be cleaned, could not be rebuilt, and could not be made to work better. I was told flatly that the weapon was made to have an operational life of 150 rounds and was intended to be thrown away after that. The people at MAC didn't know how that weapon could be rebuilt in the field. Well, with a lot of work, I managed to take the thing apart and found that there were a lot of copper screen washers that were packed along the barrel, like the old OSS High Standard HD pistol and M3 submachine gun suppressors. This screen was considered the packing material inside the suppressor. It slowed and cooled the escaping propellant gases that were bled from behind the bullet through holes along the sides of the barrel. Slowing and cooling these gases was what helped eliminate the loud "bang" of firing.

Through my experiments, I found a substitute packing material, a common copper scouring pad, and figured out a way to repack the suppressor. It wasn't hard for me to realize that I could improve on the basic design. I teamed up with a Class II weapons manufacturer in Albuquerque (S&S Arms Co.). We made some suppressed .22 pistols and .22 rifles that were an improvement in both size and performance on the MAC design. In 1978, I got my own license and started making some suppressors for the M16 and the Smith & Wesson Model 76 submachine gun, and progressed from there.

My first suppressor business was called Automatic Weapons Company, a company I founded in 1976 in Albuquerque. In about 1983, one of my customers was Lynn McWilliams in Friendswood, Texas. He suggested that I let him manufacture and market the suppressors while I did design work. That was the basis of a good working relationship. His company eventually became AWC Systems Technology, which is still manufacturing suppressors, now in Phoenix. We worked together for about six years before amicably going our separate ways. Lynn and I remain friends. We simply had different goals in mind.

I continued my original Automatic Weapons Company, having moved to Boise, Idaho. About 1993, I met up with a young guy who was a gunsmith and silencer builder from Washington named Jim Ryan. Jim, his partner Mark Weiss, and I formed Gemini Technologies, Inc. (Gemtech) in 1993. Greg Latka became part of it in 1994, and we worked together for a number of years.

In late 1998, we decided we had different goals, and Greg and I continued with Gemtech while Ryan and his partner went on to form Tac-Ord. In about 2000, Kel Whelan joined Gemtech for marketing, and that was probably the best move we made.

For a long time, our suppressor market was primarily a civilian one. We did some law enforcement but never did any military that I knew of. Some of my civilian customers were what you might call just a little different; they didn't talk too much about what they did, but they traveled extensively in Central America in the early and mid 1980s, in places like Panama. They would come back and tell me that the silenced Ruger pistol that I made was absolutely wonderful. When I asked what they were doing with it, all I was told was a little target shooting here and there. When I would try to call these individuals a few weeks later, I was almost always told that they were out of the country. I suspect that they were probably working for the United States government. But they were licensed Class Three dealers and their ownership of my products was completely legal and properly papered.

Kel's coming on board initiated a large push for the government market, both law enforcement and military. The military market for suppressors is different from the civilian. Knight's Armament, which has excellent .223 suppressors and a number of other outstanding designs, has had contracts with the Crane Naval Weapons Support Center, which acquires all of the suppressors for the Navy SEALs, among others. Knight suppressors have national stock numbers and are items that are kept on the shelves for military acquisitions. The KAC units are what are available through normal supply channels.

A number of military units have preferred features that we have on our products. Either the individual personnel have purchased them, or their unit has, in both small and large lots. By this means, we have marketed suppressors to special military units who are working clandestine areas and doing things that we would probably prefer not to know about.

Suppressors are desired by these units for a number of reasons. Historically, in Vietnam, the suppressed weapon was used to take out sentries, guard animals, and sometimes to eliminate political targets. David Truby in some of his books talks about episodes of such actions that WerBell was involved with. Probably Colonel Robert Brown was also "out in the woods" so to speak on such missions.

Operators would sneak up on a village meeting where there was a Viet Cong leader standing up on a platform in front of the crowd, haranguing the village. A single shot to the head with a silenced 10–22 rifle and the guy would just drop. The people in the audience might think nothing of the action for a few minutes while the target was just lying there, bleeding out the side of his head.

Situations like that call for the use of a suppressed weapon for the protection of the unit. They can fire their shot and withdraw before anyone even realizes they were in the area. Guard dogs, ducks, and geese are also targets for suppressed weapons, to keep a unit from being discovered. It is because of targets like this that the SEAL suppressed Smith & Wesson pistol was the first weapon named the Hush Puppy. Any situation requiring neutralization of an enemy combatant without attracting attention to yourself, and permitting escape afterward, is an application for a suppressed weapon.

There are also a lot of other justifications for a sniper to use a suppressed weapon. By suppressing the muzzle blast of the weapon, the sniper will conceal his location. While the ballistic crack of a supersonic projectile passing through the air is moderately loud, it is almost impossible subjectively to tell where the shot came from.

I didn't realize just how true this was until a number of years ago when I was in New Mexico and a friend of mine shot a silenced Springfield M1903 .30–06 rifle from about fifty yards away over my head while I was lying in a little ditch downrange of him. I was trying to learn just what I could hear and if I could tell where the shot had come from. I heard none of the muzzle blast. What I heard sounded like somebody had just popped a small ladyfinger firecracker on the downrange side of the ditch.

What I was hearing was the ballistic crack, the sonic boom, of the supersonic projectile as it passed downrange, the sound reflecting off the far side of the ditch. We have found while coyote hunting that if you miss the animal with a silenced shot, he can't tell where it came from. He will usually run toward you since he perceives the shot as having come from downrange, behind him. He's not hearing the muzzle blast.

The sound suppressor is also a very good flash hider. As well, it helps reduce the dust signature raised by a weapon that is fired close to the ground, such as a sniper's rifle when the operator is in the prone position. With a long-range suppressed weapon, firing in a military situation, you conceal the location of your sniper so that he can either get more targets, or have an easier avenue of escape afterward. In the case of the law enforcement market, the use of a suppressor is even more important. Although there will still be a ballistic crack when firing, the sound level of the shot is so significantly reduced that the average citizen will not perceive the sound as being close by, or even a shot.

When driving and there is an accident off to the side of the road, even though no lane is blocked, traffic will slow down almost to a crawl because it seems that everyone has some morbid curiosity. In the case of shooting, the average citizen wants to see what is going on. They won't stay in their homes; they look out the window, they come out into the street, and they place themselves in harm's way simply because of curiosity. This is human nature. If one of these curious folk is struck by a projectile or a ricochet from either the good guys or the bad guys, it doesn't matter who fired the shot—the city will get sued, and they are going to lose. When a sniper uses a sound-suppressed weapon, the citizen doesn't realize that the action is close by.

Sound-suppressed submachine guns, which are not sniper items of course, also have their use for law enforcement. For example, when raiding a drug house, officers don't have to use hearing protection or risk serious hearing damage from shooting in a confined area. This allows them to hear communications, their command and control, and potential threats much more easily and with little distortion. Prior to entry, they can shoot the guard dog, shoot out the streetlight, and put a hole in the guy's car so that he can't get away during the raid. The use of suppressors helps give the officers a much higher success rate in catching the bad guys.

One very unusual aspect of using a suppressed weapon is the reduction in flash for a flammable environment. If the officers are raiding a known methamphetamine laboratory, the atmosphere is often highly inflammable because of the chemicals being used. The average drug chemist is rarely observing chemical safety procedures, although he probably would not be actively smoking while there's propane or ether in the air. For a team going into this situation, a suppressor allows for the firing of weapons with significantly reduced chances of causing an explosion, especially if a little bit of water is inserted into the suppressor. Firing a can "wet," as it is known, not only further lowers the muzzle blast significantly, it reduces the chance of igniting combustibles.

There's another issue that's coming up in both the military and law enforcement arena, and that's one of hearing protection. This is resulting in increasing workman's compensation claims and litigation over permanent hearing loss. Wearing earmuffs frequently makes it difficult to hear commands or hear the radio, and an earpiece on the inside of the earmuff is far from a satisfactory solution. Failure to use hearing protection will result in significant hearing loss in a short period of time. There are a number of communities that are starting to see hearing-damage-related workman's compensation claims from law enforcement officers.

Probably one of the worst situations that really needs universal sound suppression is the military. We as taxpayers, through the Veterans Administration, spend an average of about $3,000 per year per veteran, on hearing damage claims. Some of these young kids coming back from combat tours have suffered serious hearing damage. One of the kids in my neighborhood did two tours in Iraq with the Marine Corps. He is all of twenty-three and is going to be wearing hearing aids within a few years because of damage suffered in a combat zone. He was sitting directly underneath a .50-caliber machine gun in the vehicle he was assigned to. Actions in a combat environment are always dangerous, but some of the dangers can be reduced. Hearing loss due to muzzle blasts can be one of these reductions. It is far easier to reduce the sound levels at the source than later try and correct the hearing loss.

Of course, we would like to see more people buying more suppressors; after all, that is my business. It is my personal belief that virtually every rifle used by the military should have some degree of sound moderation. Perhaps not the level of suppression that we have available currently, and suppressor design has come a long way, but at least something to knock the sound levels down by about twenty-five decibels so that you are at or below the threshold where the military presently says you have to have hearing protection.

The military standard (MIL-STD-1474D) says that hearing protection is required when the sound pressure level is over 140 decibels. I can tell you that an M4 carbine has a nonsuppressed level of 164½ to 165 decibels. That is a significant level for hearing damage. Smaller, lighter, but maybe not as efficient as some of the current designs, suppressors would be an ideal tool to mediate this noise problem. At Gemtech, we have several designs that can meet this requirement; there just hasn't been the demand to produce them yet.

We believe that most law enforcement organizations should go to sound-suppressed weapons. This would be difficult to accomplish for the general officer's handgun because of size. But then again, most of the time an officer is not drawing his handgun and shooting. It's just a weapon that is available to him. However, when the team is going on a callout, such as taking down a drug house, all members should be using sound-suppressed weapons—it keeps the neighbors happy, lets the officers hear what's going on, and lets them hear threats that may be coming up.

If somebody shoots a round inside a building, these guys are temporarily deafened. Their ears are ringing, and they can't hear the threat of someone racking the slide on a shotgun, they can't hear their radio, they're in trouble. So I think that law enforcement should be using suppressors for almost all tactical missions.

Suppressors are not magic or unduly complicated devices. They operate on relatively simple principles, though the practical application of those principles effectively seems more like an art, especially reducing size and weight while increasing efficiency.

There are several basic principles as to how a suppressor works. The barrel of the firearm is a tube. When the cartridge is fired, you're forcing a hunk of lead and copper through essentially a piece of pipe that's a bit smaller than the projectile. It takes a lot of pressure from the propellant gases to drive that projectile through the barrel and make the bullet reach an effective muzzle velocity. The moment that the bullet exits the muzzle of the barrel, there is a very sudden release of the pressurized propellant gases that were driving it. This rapid escape of gases is the muzzle blast that you get from the shot. It's the "bang."

High school physics and the General Gas Law tell you that you can reduce pressure by increasing volume or decreasing temperature. The suppressor is a chamber off the end of the muzzle that will allow the volume of the propellant gases to expand, which will then reduce their pressure at the exit of this chamber. If you put in a number of baffles that will absorb heat rapidly, you reduce the temperature, which also reduces the pressure, and sound is pressure. Additionally, by delaying the gas exit, and spreading out the time curve of its escape, you change the release of this energy.

Imagine taking two children's party balloons filled with air. One of them, you stick a pin in and let the air out instantly. It makes a loud pop. The other balloon, you undo the valve at the bottom and release the air inside. You're getting rid of the same amount of energy and the same amount of pressure, but over a little longer period of time, perhaps several seconds. The noise is significantly less, somewhat distinctive, but it isn't a "pop" and it isn't particularly loud.

In a suppressor, generating turbulence to where the gas exit is delayed spreads out the time curve of the gas's escape. You're controlling the same amount of energy, just getting rid of it over a longer period of time. You change the loud "bang" of a gunshot to a dull "thud" or "pop."

I firmly believe most firearms in the general civilian market should be suppressed. In a lot of European countries suppression is encouraged. In France and Finland, almost all .22 weapons sold come threaded for a suppressor. Not only is it neighborly, but you're saving people's hearing. There's nothing unsporting about using a suppressor. In fact a number of states even permit hunting with one. I'm not completely convinced that that's good for big game, because as a hunter, you need to have other hunters know where you are. But it sure does help with cutting back on hearing damage. I shoot a lot of small varmints, and even a .22 Magnum or .17HMR is very loud, at 155 decibels. If I'm shooting, I use a suppressor or a sound moderator (less reduction than a standard suppressor) to drop the sound level below 140 decibels. That way, I can go out and shoot six or seven hundred rounds in a day and not have my ears ringing.

Sound levels can be a difficult comparison. A normal human conversation can be at around sixty to seventy decibels. OSHA says that if you have continuous noise levels over eighty-five decibels in an industrial environment, some sort of hearing protection is necessary. With firearms, we are dealing with a very high peak sound level, but it is an instantaneous peak and doesn't last very long. As a result, there's not as much energy under the curve in a couple of gunshots as there is in listening to a boiler or a locomotive running. Hearing damage is associated not only with the absolute sound pressure level, but also the cumulative dose. The sound levels to an engineer running a locomotive all day are much higher than even a number of gunshots, and the ubiquitous MP3 music players' ear-buds glued in teenagers' ears are equally as damaging.

If you look at the instantaneous peaks or the sound curve, a lot of CO2 pellet guns have sound pressure levels in the 120–125 decibel range. A Daisy

The Navy Special Warfare (SEAL) McMillan M88 .50-caliber sniper rifle. The big weapon is a single-shot, removable-bolt, shell-holder-type weapon. The triple-baffle curved muzzle break helps cut back on the terrific recoil developed by the powerful .50-caliber machine gun round the M88 is chambered for. The metal section on the wrist of the stock, just behind the action, is a joint where the stock can be removed to cut down on the overall length of the very large weapon for easier transportation in cramped submarines and other vehicles. *Kevin Dockery*

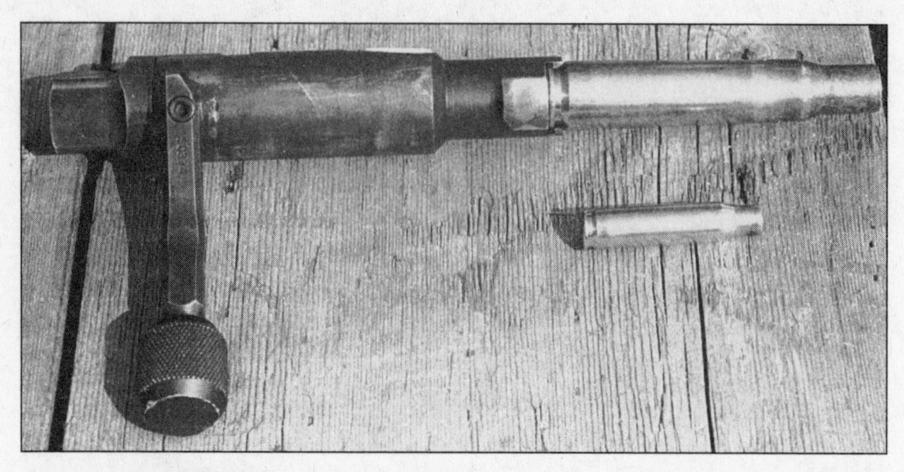

The bolt of the M88 McMillan sniper rifle removed for loading. Inserted into the face of the bolt is a fired .50-caliber casing. The smaller casing below that is the brass from a standard 7.62×51-millimeter round. The head of the big .50-caliber round fits into a holder machined into the face of the M88 bolt. This type of action, purely a single-shot, is referred to as a "shell-holder" system. ***Kevin Dockery***

BB gun is around 110–115 decibels. Some of the pump-up air rifles are getting up around 134 decibels. Yet, subjectively, they're fairly quiet, because they are short-duration sounds. In a standard sniper's .308 rifle, the sound level will depend on the barrel length, and a shot will vary from between 165 to 168 decibels. A .22 pistol will be at around 152 to 159 decibels depending on barrel length, and a 9-millimeter pistol will be 160 to 163 decibels varying with ammunition. These are all loud. A .22 rifle, depending on the barrel length, may be just under 140 decibels and not need any real suppression. The big guns, such as a .50-caliber rifle, have a massive sound peak, in the 175 decibel range, as I recall. A good, well-made suppressor can drop these peak sounds twenty decibels and more.

THE SHOOTERS
LAW ENFORCEMENT

Note: The opinions given in these interviews are those of the individual and are not in any way the official position of any department or organization.

LAW ENFORCEMENT SHOOTER

Deputy Vickers

Working for the Ada County Sheriff's Office in Boise, Idaho, I wear two hats when I'm working. I am a patrol deputy first, and then one of my specialties is that I'm on the SWAT team as a sniper/marksman. We used to call it Emergency Services but have gone over to calling it SWAT now, officially the Ada County Metro SWAT because we have another small PD in our county that is part of the SWAT team now. There isn't a full-time SWAT team in my department; everybody who works on the team is a deputy or officer full-time, with the SWAT duties being a separate, secondary job.

The Ada County Sheriff's Department is an excellent organization, and I feel pretty privileged just to be a part of that group. The SWAT team is an even more professional unit than the line deputies; to use an old line, being part of it is to be one of the best of the best. I don't know who said that, but the line fits pretty well. You feel like you've made it into pretty good shape when you become one of them.

The position I wanted on the team was that of sniper. For a long time, they had two snipers in the unit. There was a need felt for an additional two more snipers, so that we could actually have two teams of two snipers each. When the position became available, I was very excited and honored that I was able to apply and test for it.

My start in law enforcement was back in 1975, and I spent a couple of years doing that. Basically, I was involved with instruction, I never had anything to do with guns until I was about twenty-one, but I've always had an interest in shooting. Over the years I've been in and out of law enforcement. About ten years ago I got back into it after having been in the private sector. I've been a shooter in three-gun competitions—that's shooting a pistol, rifle, and shotgun—and am ranked as an expert in international three-gun competition. It's really just something that I've gravitated to over time.

Changing over from shooting paper targets to possibly firing at a person was something I thought about quite a bit, never having had anyone in the crosshairs until one of my first callouts. It wasn't very glamorous at all; I was in a ditch and there was a gal who had taken over a motel room. They had told her to check out and she was obviously on drugs, but she said that she had bombs in the room. So we were called out to deal with the situation, and I was in a position at the rear of the motel by myself. My other team member hadn't been able to make the call.

The first time that I had her in the crosshairs, it was a very sobering experience to know that all I had to do was touch that trigger to take someone's life. That passed very quickly because I went back to thinking that I had to protect my team and the public. That's why I was there and that's why I was chosen to do that job.

My primary job as a team sniper is to protect my teammates and feed back information as an observer. Most people would use the word "intelligence" to describe the information that a sniper gathers from his observations. I don't feed intelligence back to command; I feed them information. They in their turn determine what makes up intelligence and what doesn't. As snipers, we try to deploy before the rest of the team arrives on-site. That allows us to set up our positions and start to observe the situation. We don't have to wait for a go sign from a superior officer. If I roll into a situation and there's somebody holding someone hostage, if I can protect that hostage by taking a shot, then I'm allowed to do that. Another point is if my team is on a perimeter, or they are making an entry and there's a situation that comes up where I have to act to protect them, then I do that also. By far, the majority of the job is information gathering.

Being able to take a shot is not a special situation; it's no different than what I do on the street every day as a deputy. If I roll into a situation, say responding to a 911 call, and come into a situation where someone's life is being threatened and someone else has the motive, opportunity, and means to bring harm to that person, I can choose to stop them as necessary. That situation, using lethal force, is no different whether I'm on the street as a deputy or behind the rifle as a SWAT sniper.

I don't think the situation would be any less intense if I were to walk into a domestic battery situation where he's holding a knife to her throat, maybe he's already cut it and she's screaming, he screaming. He's already injured her and now he's threatening to cut her throat—that's a pretty serious and intense situation. I don't think it's any different that I'm up close and going to apply lethal force with a pistol as compared to over a long distance with a rifle. The situation is probably more intense, because I'm physically close to the action.

The legal ramifications of what I may do as a sniper are something that I try not to dwell on. We all want to be tested; we train and train and train, and are 99.9 percent sure that we're going to do the job when we are called on to perform our mission. As I saw in a movie, we're not a life-taking organization; we're a lifesaving organization. But if I have to take someone's life in order to save an innocent person, that's what I'm delegated to do.

Sure, the thought is in the back of my mind that if I pull the trigger, I'm most likely to have to go through a departmental review, possibly a coroner's inquest. Then once the legal mumbo-jumbo is done, there are the civil things that can happen if, say, a perpetrator's family decides to sue me or anything like that. We've had a couple of incidents like that take place in Boise recently with the city on the receiving end of the lawsuit.

As an example, a seventeen-year-old was out of control after he had been taking drugs. He had been collecting military weapons, complete with bayonets. One of the weapons he had was an old rifle with a bayonet. The father couldn't control the boy and called the police. As the police showed up and the officer started getting out of his car, the boy charged him with a bayonet-mounted rifle. The officer apparently had his clothing pierced with the blade before he shot the kid in self-defense. The boy died as a result. The father kept saying that he told the police that the rifle wasn't loaded. It was interesting at the coroner's inquest that they were making a big deal about the man announcing that the weapon wasn't loaded. You don't have to load a bayonet to make it lethal, and the pointy thing was on the weapon. That didn't really come out very well in the inquest.

I try not to dwell on that aspect of the job, but sure, it's in the back of my mind. But I cannot allow those kinds of considerations to cloud my perceptions of a situation. If I am not mentally prepared to use lethal force properly to save another's life, then that's the time I need to find another position because I cannot do my job as a sniper.

I'm still fairly new to the sniper position, so I try and study different shootings to learn techniques or what not to do. None of them jump out at me as being dramatic. For the majority of encounters with law enforcement, the ranges that the LEO snipers will work at are much shorter on the average than those fired by our military counterparts. Law enforcement sniper work has much more of a duty if we fire a shot that it be an incapacitating one. Obviously, if we're shooting somebody who's armed with a knife or a gun and he is threatening someone, we can't just wound him. We have to stop him from pulling that trigger or using that knife by completely incapacitating him.

I don't personally know that much about being a military sniper, but I do know that it is much more of a long-distance targeting kind of thing than we do in law enforcement. And if a military sniper wounds a guy, it's still a good shot because you're taking him out of play. Those are just some of the aspects that are different between the two applications of what really can be the same job.

In police sniping, we have questions that would never come up for military snipers. There was one very publicly shown law enforcement sniper shot where a guy in Ohio was sitting in a chair, threatening to kill himself. He had the chair in the middle of a residential street, and the police had set up a perimeter to contain the situation. At one point, the man was holding his pistol loosely, letting it hang down between his legs. The police sniper shot the weapon right out of the guy's hand, ending the incident. There's been so much interest in this incident that there's now even a name for this kind of shooting, they call it "tactical disarmament."

Thanks to the news networks, the public has seen this shooting over and over again. The general public just says that they saw that gun go flying, so now they question every incident by asking, "Why didn't you just shoot the gun out of his hand?"

There are a number of reasons this action is almost never taken by police snipers. One of those reasons is that a person's torso is a relatively small target at even a hundred yards. The head is an even smaller target. But a pistol is positively tiny. If you try and shoot a pistol out of someone's hands, what's to say that it doesn't go off from the impact? Or if you miss the weapon, then what happens? And if you hit that weapon and your bullet deflects off the metal, where is it going to end up? It's just going to fly off in an uncontrolled direction.

That shot in Ohio was a very unique situation. It worked out for everyone involved, but everyone was also lucky. You can't depend on luck when you are applying lethal force to a situation.

When I finally went in for law enforcement sniper training, I personally found it very interesting. I had been a shooter; I knew how to shoot and I was good at it. Before I became a dedicated marksman, I was a firearms instructor for the agency. When I went to the sniper school, I only picked up a small portion of the knowledge that is available. But for me, I learned a huge amount of information.

The first thing that I got out of the school was that if we put any more snipers on the team, they should not go out on callouts before they'd gone to the school. I had gone out on a few callouts as a sniper, a dedicated marksman, thinking I knew what I was doing and feeling fairly comfortable at my ability to perform my job. That feeling lasted only until I attended sniper school.

At the school, I realized that I didn't know what I was doing at all. The ability to precisely place a single shot is only a small fraction of the job that a sniper is expected to perform. Because of my lack of knowledge as to just what an actual dedicated marksman did, had I been called on to make a shot, I probably would have been able to make it, but it could have been called into question over certain training aspects, which of course opens up the possibility of civil lawsuits. That's something that is always a concern, and such actions call in the training records of the shooters involved. That's one of the reasons we keep such meticulous records.

In the dope book we keep for our rifles, we note down what day we went to the range and that my cold-bore shot was here. Follow-up shots from the cold-bore one are also noted down. So is the fact that no corrections had to be made to the scope between the shots. The temperature, ammunition, weather, and range are all noted down as part of the dope book records. These records would be important in case they are called in to question during either a criminal or civil legal case.

The cold-bore shot, or just a cold shot, is very important to know about because that is the shot that you're going to be making on a callout. Making a cold shot is where you pull your rifle out of the pack, so it's at the ambient temperature of whatever it is like inside your car trunk. So that first round out of the barrel is the cold-bore one. If you are called on to make a necessary shot, the chances are about 100 percent that it will be a cold-bore one.

If you're out training, sometimes when you are shooting, your point of impact will change a little bit as you continue firing. Some weapons do this a little less, some change a little more, as they warm up. But the cold-bore shot will be the most important one you make during that entire training session. That's the shot that will have to count if you fire a round during a callout. You always have to know where your rifle hits on its cold-bore shot.

Our agency put out the request for applications as marksmen/observer. There are supposed to be two-man teams, and you aren't supposed to remain behind the rifle during a callout; the accepted standard is twenty minutes.

With their spotters peering on through scopes and making notes, these snipers check the zeros on the AM/PVS-10 sniper night sights mounted on their M24 sniper rifles. ***Private Collection***

After that time, you trade off with your partner and switch over to using either binoculars or a spotting scope, and he goes behind his rifle.

On the motel callout I went on early in my career, I learned the value of working in a two-man team and not staying on the rifle scope. For me particularly when I'm trying to get behind a long-range scope, I close my left eye to concentrate on the view through the lens. After a while, my vision gets tired and I start having a little bit of a wandering eye. During that first callout, I spent three-and-a-half hours behind the scope without a trade-off partner. For the next two days after that incident, I couldn't move my neck because I had been in that craned-head prone position for so long.

There are numerous advantages to having a partner with you on a callout. As a sniper, you need to be behind the scope, ready to go at an instant's notice to make that lifesaving shot. Your observer is providing you with information. He's telling you what your windage correction needs to be, what the range correction needs to be. He's writing down things and running the radio to headquarters or the command post and getting them the information you have observed. And he's watching your back.

I was informed about a story a while back, and I don't exactly remember where it happened. A SWAT team went out on a planned raid, a drug raid on a rural house. The snipers were in position, and apparently the home owner snuck around behind the snipers and took them at gunpoint. Not having been there, I can't say exactly what the circumstances were, but your observer is supposed to be watching a full 360 degrees around your position. The sniper is not supposed to be doing anything but looking through that scope, ready to take the shot if he has to protect the team or protect someone's life. He isn't supposed to be concerned with local security. It was kind of interesting after we learned about that incident, because we included aspects of it with our training. The sniper has to be able to turn 180 degrees and engage someone if the spotter calls out the danger.

The observer has the hard duty; he has to do everything. All the sniper has to do is make sure the crosshairs are where they are supposed to be and pull the trigger. So the skill of your spotter is paramount for the success and safety of the team.

The relationship between a sniper and his observer is a very intimate one. He has to trust the shooter to make the shot, and the sniper has to trust him for just about everything else. In the training that I went through, the observer actually cranked the windage and elevation dope onto the rifle's scope. As the shooter, you never even touch the sight. The observer does his laser range finding or his mildot, whatever system you might use to make your range estimation. And then he reaches over and dials that into my sight. I have to trust that he knows what he's doing and he has to trust me to employ it correctly.

This makes the switch off between the sniper and the observer an even more important aspect of the team. You know each other's job. The main focus of the switch off is the physical aspect of not getting tired, not becoming complacent, or just not getting stiff from remaining in one position too long. But you have to really trust each other and communicate well. Whether it's hand signals or almost thinking the same thing at the same time, communications between the sniper and observer have to be sharp and quick.

There's an adage that if you get tired, you get lazy; if you get lazy, you get hurt. Having a partner helps you keep from getting tired. That was one valuable lesson I came away from the sniper school with, among others.

When we went to the sniper school, the guys who taught there were very well respected, longtime guys in the sniper community. They said that we should be proud of the term "sniper," that we weren't designated marksmen, or marksmen/observers, we were snipers.

"Sniper" has another connotation; it's a name that the public doesn't understand, like "suppressors." People say that the only need for a suppressor is for assassinating someone. They also think that the only need for a sniper is to assassinate someone. This couldn't be further from the truth, but the public has a long way to go in its education. At the sniper school, they told us to be proud of the term "sniper." It indicates a dedicated professional, someone who works hard at his craft. And the application of that craft is saving lives.

The reason I know well about the public's reaction to a suppressed weapon is that I carry one. At this point, my sniper team sergeant has a suppressed .308 bolt-action rifle. We're working on the rest of the team getting .308 suppressed bolt guns as our primary sniper weapons. Currently, I carry a suppressed AR-15 carbine, which is my patrol rifle as well as the rifle I take with me on deployments. When I deploy as a sniper, I carry my bolt gun in a backpack along with my spotting scope, happy sack, dope book, camouflage, water, and food—everything I would normally carry. But I deploy with my carbine going to the location. That's also the weapon I prefer to use when I take over the spotter's position. It has a more flexible application when I use it to protect both myself and the sniper.

Using a suppressed weapon just makes sense to me. It's tactically sound to use one since it lowers, suppresses, part of the firing signature of the weapon. For another thing, it protects my hearing when I practice with it, something I think a lot of. If we're out on a deployment where I get involved in a firing situation, I'm not going to have time to put in my earplugs or slip on my hearing muffs. I'll have to absorb the sound of my weapon firing with my unprotected ears. Using a suppressor to lessen that sound just makes good sense.

When a sniper team goes out on a deployment, we actually wear hearing protection under our helmets. While as a patrol deputy, if I have to roll out on a felony stop or something like that, all I have time for, if I even have that much time, is to grab my rifle. Otherwise, all I have is my pistol. Either one of those weapons, if they aren't suppressed, can cause some serious hearing damage, especially if fired indoors.

There's a Gemtech M496D suppressor mounted to the muzzle of my patrol carbine. That's a .223-caliber can that's quickly detachable from the weapon. As a sniper, using a suppressor is useful even when firing normal supersonic ammunition. With the bullet going faster than the speed of sound, even silenced, you're going to get a ballistic "crack" from the projectile going through the air. But the targets who are being engaged with the weapon can't tell where the shot is coming from. That's a part of being a sniper—not being seen or located.

The bulk of what we do as snipers is gather information and relay that information to the command post or the person in charge of the operation. That's the paramount duty we perform; not the most important duty, but it is the majority of what we do. I think the most important thing that we do is protect people's lives, whether it's the lives of team members or citizens.

As snipers, we have the equipment and training to better protect people's lives during a callout because we have the precise long-range capability of projecting force. The callout we went on the other night was a training callout, but we were using it as a tool, so we acted as if it was a real incident. The action was against a drug house, and the position that my sergeant and I got into—he's my partner—we were about sixty feet higher than the house and 144 yards out. From our position, we had a great view of the house. We could see three sides of the building and had a line of sight on the interior through a number of the windows.

We got to the site a good half hour before the team showed up. That gave us time to observe and note the tactical situation. In turn, that allowed us to give the team some ideas as to which way they could come into the house. The general plan was to try to approach the building with two teams. As observers, we were able to tell the team about a local creek, a barbed wire fence, and outbuildings. If we had to do some protection, take a shot to stop someone, we also had the capability to do that. That distance, 144 yards, is a pretty close, comfortable shot for us.

As far as the hardest aspect of doing the mission, what comes to mind as the hardest aspect is making sure that you have all the calculations made correctly, have your ballistics down right to make that precision shot right the very first time. There are a lot of variables in making a shot; you have elevation, distance, wind, lighting conditions, and the cold-bore shot—all of that comes into effect whether your target is moving or still. We shoot a lot of static targets, but we try also to shoot a number of moving targets—most people aren't sitting still during an incident. If they're on drugs, if they're agitated, if they're both, they aren't going to be standing there waiting for you to shoot them. So there's quite a bit to think about to make that first shot perfect.

If I'm taking a shot on a suspect to keep him from hurting his spouse, hostages, whatever, that shot needs to be made within an area marked off by a triangle between the eyes and the nose, depending on the angle of the head. That's the center section of the brain, and striking it will immediately cease the subject's ability to pull a trigger or use a weapon. So the perfect target is in the center of the suspect's head and about the size of a baseball.

The precision needed to hit such a target depends on so many factors. You can run seventy-two different scenarios in your brain, and the first callout you come up on won't be like any one of them. Hopefully, you will have done enough training and be confident enough in your own ability and your equipment's ability; plus you will have maintained your equipment in the shape so that it can make the shot. My understanding from what I've read is that the standard or the average law enforcement sniper engagement is somewhere around seventy yards. I've been deployed at forty yards, which is close to the point at which I'm not comfortable with the sniper rifle; that's a little too close.

The perfect target is the head of the suspect. But if you're too close, you can't see the hands. So the field of view of a precision long-range scope at forty yards is very limited. The danger doesn't come from the suspect's head; it comes from his hands. This is where the observer becomes even more important than usual. But I think I would feel more comfortable at forty yards with my patrol rifle and a reflex sight mounted on it, because it gives me an excellent field of view.

During practice, I've successfully engaged clay pigeons at five hundred yards, and I'm comfortable with that. The longest deployment I've ever been on was 144 yards.

So far I have never had to kill a human being, but I have had to consider it. It's a situation where I've had all this training and I think about the application of it. What am I going to do if . . .? I believe most people are the same, but as a professional, you want to be tested and know that you can do your job, complete the task that you've been given. If the situation comes up that I need to take a shot, take a life to save a life, or maybe several lives, I feel I can do that. On the other hand, I've talked to people that have had a real hard time with having had to do that. They have nightmares, lose their jobs, get out of law enforcement, because it's been so traumatic an action for them. I have also talked to, and read articles from, other people who have said that it wasn't a big deal. You had a job to do and you did it—that's what was required of you. And you saved someone's life by doing that. End of story, go on to the next chore.

Part of me hopes that I get tested. Part of me hopes that I don't. And I hope that isn't perceived badly, but I think that most guys who are on special teams think that way.

There was a situation out west some years back where a father was threatening the life of an infant. He was on a bridge and holding the baby out over the drop. There was broken concrete and rebar some feet below, and if dropped, the baby didn't have much of a chance of surviving. When the man pulled the baby back from the brink, a police sniper had the chance to save that child's life and he made the shot. I'm okay with that; he saved the baby's life. If the father, or the person—doesn't matter if they're the father or not—is threatening a baby's life, what happens if he lets go after you had the opportunity to take the shot and you didn't? He lets go and the baby dies, how are you going to live with the fact that you could have taken the shot, could have saved the baby's life, but you didn't? You hesitated; you didn't take the shot.

The man was unarmed in that incident, and that's what the public sees. As an officer, I see that the suspect had the ability and opportunity to take that baby's life. And he had given the indication that he was willing to do so. People can Monday morning quarterback that all day long. They have months and months to analyze the situation and say what they would have done. Statements may come out that the subject would have never hurt that child. How do these people know? Because they knew the individual, they've known him for years, they know that he wouldn't do anything to his baby, he loved that child.

Are they willing to take that chance on the child's life? If that shot hadn't been taken, and the suspect had dropped that child to its death, then the question from the public would have been why wasn't the shot taken?

A friend of mine teaches classes and he uses the first few minutes of the Bruce Willis movie *Hostage* as a teaching tool. I don't remember what the specific scenario was, but a woman and child were being held hostage in a house, and they had been there for several hours. My friend uses this as an example of the priorities of life. Willis's character is a police negotiator and he prevents the sniper from taking the shot. The chance for the shot passes, and the subject kills the hostages and himself. What should the sniper have done and why?

The new aspects of life, terrorism and such problems, haven't affected my life and work that much, except for the fact that not too long ago, the thought of using a suppressed weapon was scary to my department's administration. The news media coming up and asking the question why we were using silencers would be a difficult one for the average law enforcement officer to answer. With the new war on terrorism and the nightly news flow of military guys walking around with suppressors on their weapons along with all of the lights, lasers, and sights, the image has become more accepted by the public. They know that the soldiers need this specialized equipment to do their jobs better. Now our administration has come around to understand that suppressed weapons are a good idea.

It also leads back to the fact that the public has odd reactions. If you fire a shot in a neighborhood, all the people run out of their homes to see what's going on. That's the last thing you need, more people coming into a dangerous situation. With suppressed weapons, Mom and Pop are going to continue sitting on the couch and watching the news. They'll see what happened on the screen later, not run out and put themselves in danger.

The appearance of the police in full SWAT regalia draws enough attention all by itself. To educate the public as to just what it is we do, we'll go out to some of the high schools and put on demonstrations of rappelling, let the kids look at our guns and gear, stuff like that. A few years ago the term "jackbooted thugs" was being used to describe one of the federal agencies. The public was kind of in an uproar about that, but I don't think that perception exists anymore; at least I haven't seen it. My job has me in daily contact with the public, and most of what I get is positive reaction to us. They don't see us as the black ninja or jackbooted thugs. They seem pretty accepting of the use of SWAT and tactical teams.

Education is the key to this kind of acceptance. We had a county commissioner for a while who came out to the range one day. I let her shoot some suppressed weapons, something she had never had any experience with. After shooting them, she was very excited about us going to them to save lawsuits on hearing loss. We need the tools that we use because the bad guys can be better equipped. An example is that LA bank robbery with the criminals wearing armor and having fully automatic weapons. The police didn't have rifles; they had shotguns and pistols. It was the bad guys who had the rifles.

The public is more accepting of us having the tools we need to do the job. One of our new tools that is very well accepted and considered positive is the

A senior airman loads his M24 sniper rifle under the watchful eyes of an assisting air force captain during a sniper training and rifle advancement course at Andersen Air Force Base in Guam. ***U.S. Air Force***

taser. If you get in a scrap with a guy, you can solve the problem and end the fight with a taser, taking the guy into custody instead of the hospital, or you going to the hospital for that matter. The public's perception of that result is great. We did our job and needed to do what we did.

If you scramble and fight, rolling around on the ground, punching the guy, the public's perception of that isn't very good. The video will end up on the news somewhere, edited to fit the time slot, and the public will see that and just wonder why that poor guy was being beat up. Doesn't matter that he just assaulted a number of police officers and maybe innocent bystanders; that won't be on the film, at least it never seems to be.

The public wants us to have the tools to go in and take care of the problem right away. If the bad guy thinks he can get away, then he'll make a break for it. That's the old fight-or-flight aspect of the human psyche. If he looks out the window and we have the place surrounded, we have the equipment and look like we know just how to use it, the chances of him just giving up are better.

Whether it's a deputy or a SWAT team, not every tool works every time. So the more tools you have at your disposal, the better you are able to handle whatever situations may come up. If the taser doesn't work in one situation, then we'll get out the pepperball gun. If that doesn't work, we'll get gas grenades, or the long gun, the short-range guns, the ladder, less-lethal, whatever tool we need at that moment to do the job and rectify the problem.

LAW ENFORCEMENT SHOOTER

Officer Sheldon

As a patrolman in a midsized Midwest urban police department, I am assigned to a unit that combines our narcotics unit and our tactical team. The position I hold on the unit is the team leader for the six snipers in our team, and our day-to-day duties involve narcotics investigation. I've held this position for about five-and-a-half years now, about half of my career since becoming a patrolman in January 1995.

I grew up in Wichita, Kansas, having been born in San Diego and lived in North Carolina for a while. But I went through grade school in Kansas, so you can say that's where I spent my childhood, as a city boy who played in the country. Hunting wasn't really something I did until I got older; my family wasn't into firearms to speak of. One of my high school friends got me into shooting guns and bird hunting, skeet and shooting sports like that. That interest carried over into my college years, when I worked at a sporting goods store part-time while a student.

After I got out of college, I stayed on at the sporting goods store, working full-time and trying to decide what I wanted to do career-wise. That's how I ended up going into law enforcement, because of the people I met through the store.

An acquaintance of mine, an old friend really, had been hired on to the department about five years before I came on board. About the time that I was hired, he got onto the department's tac team (tactical team). He really enjoyed the work and always spoke highly of the guys he was working with and that the duties were interesting. As I got four or five years on the department,

I started looking for a change from working patrol, and one of the things that attracted me was the way my friend spoke of the tactical unit.

As I remember, it was on my third try that I finally got into the unit. When I first got there, we were still a full-time SWAT team; there were ten patrolmen, a couple of supervisors, and a commander. We worked the night shift, and everybody in the unit was tactical; that was our job. If we weren't training or serving warrants, we did directed patrol work, robbery surveillance, stuff like that.

Shortly after I got into the team, it was combined with a narcotics unit and theoretically doubled the size of the tactical team as a whole. But we also picked up the responsibility for narcotics investigations for the department, so our workload also increased quite a bit. Now we wore two hats. The reason for combining the units some of us thought was a finance issue for the command side. The way it had worked before, the narcotics guys ran the investigations, and when the case got to the point where they had a warrant to serve or an arrest to make, we would team up. They had what we would call some "SWAT-lite" training, and they would run perimeter for us on dynamic entries, where they would secure the area while we entered the target hard. They would also have some perimeter functions when we took on a barricaded subject. Part of their equipment included heavy armor and some Benelli shotguns at the time.

So we already worked together on tactical stuff anyway. The theory was that by combining us, the skills came together; the narc guys were already doing SWAT stuff, and in our non-SWAT time, the tac guys could augment the narcotics investigations. Very few of our warrants got served from a knock on the door from a plainsclothes officer, and our patrolmen did not serve warrants in uniform. Detectives might have a warrant for an arrest or search that they might go out and serve by themselves with a patrolman present, but the majority of the warrants that we served were narcotics warrants from our own cases or in conjunction with task forces, or "dangerous" warrants from our detective bureau. Dangerous warrants could include robbery suspects, homicide suspects, known violent offenders who had a history of assaulting officers, and things like that that would come to us from the bureau.

Going in heavy with a tac team on a warrant service in a situation like that helps prevent a situation from escalating. By showing the ability to apply force, we cannot only prevent many offenders from trying to fight their way out, we can protect the general public by not allowing anything to happen in the first place.

As far as my taking up the sniper position on the team, things just seemed to pan out that way. I had been a long-gun shooter by habit and was kind of into scoped rifles and accuracy. I was never really any kind of big deer hunter or anything like that. But I was an avid target shooter well before I got into the team. At the time that I joined the unit, there were only four snipers in the team and one of the slots was open. After I got into the unit and got my ears dry if you will and spent a little operational time in the organization, I was able to move into the opening on the sniper team.

Primarily, the job of a law enforcement sniper is to observe and gather intelligence. Just like any other law enforcement position, you can go twenty years without ever using lethal force, or even considering it. It's no different in the sniper's job. You spend hours and hours on barricade operation after operation where all you do is look, listen, and tell. At least that's all I've had to do so far.

We, the snipers, are the primary overwatch of the situation as it's ongoing. We have primary rights to the radio, the airtime. If a window or door opens, or if there's other significant movement, we're the ones who are supposed to report it, instead of having each perimeter guy call in. You get enough guys on the net trying to say "Side one, window three," and things get confusing fast, especially when every guy tries to report the same thing. Instead, the commanders on the scene wait on us to give the information. If for some reason we miss the action, then someone else will fill in behind us with the report.

So snipers do a lot of reporting, confirming details, information on openings—the door opens this way, the door is hinged on that side, it's a screen door, a security door, the windows are this high off the ground, the bathroom could be located here because of the vent stacks on the roof, this looks like a kitchen window—information like that can be vital in the control of the situation or planning of a dynamic entry.

On my first call out as a sniper, the situation moved so fast that I never even really got deployed. By the time I had gotten there, the suspect had already taken hostages and crashed through the perimeter with a vehicle. That chase extended itself over several law enforcement jurisdictions, the suspect shooting at police along the way. He was finally stopped, but I wasn't directly involved in that. For me, it was a case of being too little too late to get in on the chase, which really wasn't my job anyway.

My second deployment was only a couple of months later, during the winter and cold outside. It was a domestic scenario where a guy had shot his girlfriend, wounding her, and was holding several others as hostages for many hours. The situation was resolved without my having to take a shot. But when I finally got up, I was covered in frost and pretty cold. Looking at myself, I could see that the backs of my arms and along the backs of my legs were all covered with white—so much for the glamour of being a police sniper.

In my unit, the sniper designation is an additional skill set. We're all entry guys, although I moved from being an entry guy to becoming a sniper pretty quickly. Now we look for guys who have been on the team for at least a year before they can apply for a sniper position. As I said, specialties such as sniper are an additional skill; everybody who comes into the unit generally ends up with an additional skill, say, an equipment operator, such as the guy who drives our armored car. Anyone can drive it if they have to, but it's a big heavy vehicle and we have one or two guys who are good with heavy equipment, so they operate it best. We have grenadiers, guys who are armed with forty-millimeter launchers to deploy gas or shoot less lethal rounds. There are one or two guys who are really proficient with the ballistic shield, guys who are assigned as canine handlers, and guys who are assigned as snipers. So each job is an additional task on top of being a qualified general tactical officer.

The sniper position requires a little more dedication to the job. You've got one more weapon system to stay proficient with, maintain, and transport. The upside is that we get an additional training day over the rest of the guys, to practice on the range. In my opinion, it's the more even-keeled people who gravitate toward the sniper job, because it's a low-activity assignment usually. You get where you're going, after planning your route. You make your approach, then you end up lying, or sometimes sitting, down, and you are there until you're relieved.

That's as opposed to entry guys, who may be standing around, or moving about and completing tasks and all of that which takes place during a callout. The sniper position is more of a low-gear, long-haul assignment. It's a much more mentally demanding position, though the physical end isn't easy. You have to be physically able to handle holding a position for an extended length of time, and be able just to get into a position that may be on top of a roof. There are a few departments where the sniper ends up being the guy on the team who's too old or just doesn't want to do entry work anymore. His teammates say to just let him be a sniper so that he can lie around. In reality, the sniper needs to be fit for the same reasons as all of the other members of the team. He has to haul all of his gear, climb a ladder or go over a fence, or work all the way around the outside of a scene to get where he has to in order to set up.

Once on-site, a sniper has to go from transport mode and be able to get into position with the mental discipline to remain on task. That can be hard when all you have to do is lie there and look at a target from the prone position for a few hours before being relieved. There's no standing up and stretching. You can't go around the corner and look for a restroom. The commanders eat all of the sandwiches back at the command post. And you absolutely have to stay alert and on point in case the situation suddenly evolves and you have to take that shot in order to protect a human life.

There has to be that realization in your mind that one aspect of the sniper's job is the precise application of lethal force. You absolutely cannot do the job if you don't know that and agree with it. That mission is the whole purpose of the job; otherwise you would just be another observer. Coming to grips with that aspect of the job, possibly having to take someone's life, is something that you have to do before you decide to sign on the line. It's no different for me as a police officer who's a sniper than it is for a guy who's an entry officer or a detective or patrolman. The only difference is that if I have to do it, take a shot, I'm going to do it with a rifle as opposed to a guy who's going to do it with a handgun. You have to have made that decision, to be able to take that action, early on as a police officer. If you know in your mind that you may not be able to do that, then you should find a job outside of law enforcement long before you ever get to SWAT.

The rules of engagement for the application of lethal force are no different for a sniper than for any other police officer. Lethal force can be used when there is an imminent threat of danger to citizens, myself, or other officers—that's it; no other reasons can apply. Of course if I have a direct threat to myself, I'm going to defend myself just as any other officer would. If someone is a threat to another officer on the scene, or an innocent person, perhaps a hostage—whatever scenario you want to invent to put a person in jeopardy, it's probably been done somewhere—you may have to use lethal force to resolve the situation. The reason that's hard to deal with is in that situation where someone else has decided that the person involved is an imminent threat, and you are asked to take care of it. He's not an imminent threat to you or to anyone who's immediately around you. But the decision has been made that he is an imminent threat to someone, and you are tasked with ending the threat. That's hard, but you have to be able to do the job.

If you think about the way that a scenario unfolds, usually for a perimeter officer, a suspect comes busting out of a building and tries to escape the perimeter, and he can run face-to-face into that perimeter officer. If they have a gunfight, it's an instant, up-close, me-to-you situation. There's no time to think things out; the situation isn't preplanned, and the officer involved has to deal with it immediately.

The same thing can happen for an entry guy. You make an entry on a hostage taker or with a search warrant, and a lethal force scenario can evolve in a very brief time span. The difference with a sniper is that you may look at this person for two hours, watching him through your scope, absorbed in your own little "telescopic world." You see, you know how he moves; you know what brand of cigarettes he smokes, what his head does when he laughs, or how he holds the phone to his ear. I've never actually had to experience it, but you personalize that individual. He's no longer an inanimate object because you've seen him; you've seen his eyes, watched his mouth move when he talks, seen all the parts of how he moves, breathes, and thinks. And you could be called upon at any moment to intervene on what he's doing, or about to do, and result in his death, which is so often the by-product—that can affect some people. I don't know how to say anything more about that.

The public doesn't really understand what a police sniper is. The fact that the word has to be played with sometimes shows that. "Police marksman," "precision marksman," "marksman-observer," "tactical precision specialist," "police sniper"—many agencies dictate the name in order to diminish the negative reaction of the public to the job we have to do. We still use the word "sniper" where I come from; most snipers that I know want to be known by that name, and they don't want to be "airborne ballistic interventionalists," or any of those other crazy names that get used rather than just plainly saying who we are. "Sniper" is the title of a position of pride; absolutely it is. It comes from the historical use of the word. There are different stories on where the term came from; personally, I've never hunted snipes in India, or anywhere for that matter. But to hold the position of sniper, that takes dedication, skill, and commitment, and it earns you the respect of your peers. They trust you to help protect them, to hold their lives in your hands, and to use your skills to help protect the innocent.

The public's perception of a sniper, law enforcement or otherwise, is that we are killers; we use the lethal application of force to do our job in the ultimate way. I don't want to say that's right; although the thing that we do often ends in death, we're not killers. That's often the end result, but we are the protectors of other lives, innocent lives, lives that did not choose to be put in the situation where they have to face an imminent threat.

There are some mottoes used by snipers, all of them stolen from somewhere, I'm sure, but they apply to our job. Military snipers use a motto, "one shot—one kill, no remorse, I decide." Take some time and think about that. It cuts pretty close to the bone as far as what the nature of the job is. Some of the other ones include that used by the air force pararescue, the PJs, part of which is "that others may live." I take that one to heart. If you think about having to end one life to save another, that motto can mean a lot to you; it does to me.

The public, well, everybody thinks something different. Everybody expects something different from the officers of their jurisdiction. Some places love their SWAT teams; some places think that they're the paramilitary wing of government. I can't say that I know the perception of all the people that I encounter, what they think a SWAT sniper, a police sniper, is. But more than a few people have raised their eyebrows at my wife's Christmas parties if they don't know me, didn't meet me the year before, when they hear the answer to the question, "What do you do?"

"I'm a policeman."

"Oh really, what do you do?"

"I'm a sniper."

"Oh, wow."

That always seems to take the conversation down that odd fork in the road there.

The public's view of the sniper is based on a lot of misconceptions and just plain errors. A lot of people in general think that a sniper should shoot to wound or shoot to disarm (hit the weapon) rather than to take a life. Anyone who thinks that, I invite to go to their local library, pull their state statute books, and look up the use of lethal force. If they can find a statute where that action is legal under the law, contact me through this publisher.

As police officers and as a sniper, we're authorized to use lethal force to protect lives, be it our own or someone else's. If death is the result, then so be it. If we need to address an imminent threat, then we're going to address it as often and as rapidly as it needs to be, until the threat is no longer there. That's the job, and the end result is often the death of the perpetrator, but that's the way it is.

There's a neat new term been coined in recent years, "suicide by cop." Now that it's an identifiable statistic or category, something like that, people are keeping count of how often it happens. That makes the situation appear more prominent just because it has a name. How do you take a corpse and say, "Oh, it was suicide by cop"? You don't until well after the fact, when you're Monday morning quarterbacking the results. Looking back on a situation, people will analyze things and say, "He only did that to get you to kill him." I know there are people who will do that, but I don't think that the psyche of man has changed so much since the invention of law enforcement. The people who do such things create a situation where someone's life is at risk, a bystander's or an officer's. In such a situation, we act as we have to. There's no time to second-guess the perpetrator's motives. But now the public just has something they can call that kind of situation, and it reflects negatively on the department and the shooter. Nothing can be done to help that.

There's not the difference between the military and law enforcement sniper that there used to be. With our current operations in Afghanistan, Iraq, and around the world, with the war on terrorism, the missions are becoming more similar, along with the environments and the rules of engagement. The results demanded by higher command are the same: maximum precision in neutralizing the threat along with minimum bystander casualties.

The rise of terrorism has affected my job as a law enforcement officer and sniper. The biggest difference we've seen in my agency and in our area has been in the broadening in what is expected of us in response to terrorism. Weapons of mass destruction have resulted in some changes in the guidelines for imminent threat to the public. I think they've been broadened. Federal money is out there and it's flowing downstream. Our agency went from the underequipped, self-sustained kind of operation so many departments had as their small end of SWAT team deployments, to a much larger unit with

The wide use of airmobile assets has required that military snipers become used to firing from even moving platforms. Once only an action conducted by Special Operations snipers, here an army sniper sets up to fire on a target with his M25 sniper rifle while flying over water in a helicopter. His spotter, behind him and to his right, is observing the target area through binoculars while the sniper concentrates on his shot.
Private Collection

brand-new equipment including an armored vehicle that is way better than the air force hand-me-down that we'd been using for the last fifteen years. Hard to get parts for the old Peacekeeper vehicle now that it hasn't been made for twenty years. Command vehicles, EOD vehicles, robots, surveillance equipment—all sorts of gizmos have been added to our capabilities. We're relatively lucky at my agency on the equipment side in that we have brand-new submachine guns, very nice sniper rifles, and we're pretty well kitted as far as armor and communications goes.

I do wish we could spend as much money on training as we do on equipment. It seems to be kind of a head-scratcher when they hand you a brand-new $2,000 set of body armor and a new $2,700 subgun-flashlight-sling-spare-magazine loadout, and then, when you ask them for around a hundred bucks to attend a local SWAT school, you get told no because the budget for the SWAT team's training has already been allocated for the year.

Off the top of my head, there are two things I would like people to know about the sniper community and the law enforcement community. In the past, these organizations have been very insular—we don't talk about our problems, and we don't talk about our failures. We don't ask for help, and of course that's also carried over into the SWAT teams and tac units, as well as the sniper units. Those barricades are coming down. We're communicating with each other, learning each other's lessons. It's easy to point fingers after the fact, but it's hard to get together and talk about it.

"Hey, I'm so-and-so, we had this problem and this is what it cost us in lives, lessons, money, lawsuits, or whatever, and I want to share it with you guys so that you don't fall into the same hole."

Ten or twenty years ago, someone doing that would have been unheard of because you didn't talk about your failures. That situation is ending. Sniper organizations are now cropping up across the country to help end that isolation. The American Sniper Association is one of the organizations working to break down barriers and open communication. Snipersonline is the same deal, another excellent way to share information, and that sharing has moved mountains as far as bringing people with the sniper's mission up to speed and putting us all on the same page. I've traveled all around the country to attend other kinds of tac training, sniper schools, and stuff, and it's nice to know that now you can kind of walk in the door and know that a certain percentage of people in that class are on the same page that you are. That's a good feeling. It helps keep you from being that guy out on the left wing wondering what's going on.

Americansnipers.org is a nonprofit organization made up primarily of law enforcement snipers, that raises money to buy materials for military snipers who are deployed in Afghanistan and Iraq. The majority of members of Americansnipers.org, and when I say members—you don't join, you don't pay dues, you're invited by the other members of the fraternity. If they don't want you, you're not coming in. Right now, it's a group of about twelve guys who try to raise money so that we can buy equipment for the deployed troops. All of the members of as.org are members of snipersonline which is an e-mail list of active duty snipers, the majority of which are law enforcement snipers, though there are a few military snipers on board. The majority of as.org and sol members are also members of the American Sniper Association; it's no different than doctors, lawyers, or veterinarians.

Every profession in the world has different societies that they join, and snipers are no different. They're all ways to pool resources, collaborate on training, and really to exchange information. If I have a question about a certain procedure or certain protocol, a standard, a piece of equipment, or a problem that I have, I can go to any of my brothers online and ask them what I need to know, what happened, what broke, or where can I find a replacement, and just stand back and let it rain. I'll get answers from ballistics technicians at federal agencies, federal agents, rural sheriffs, metro and urban PDs, state law enforcement, and the military. They'll give you answers like "I found it here," "Don't buy that, it breaks," "Here's how our SOP works," "Our lawyers said this," "This happened in court," or "The guy got hurt because of this." Whatever the scenario might be, I can get an answer from the people who will know. The organizations raise the level of professionalism of everyone involved.

In reading this, I hope that the public can understand the level of commitment that it takes from an individual to accept the responsibility of (a) law enforcement, (b) SWAT, and (c) the sniper position. A lot of that boils down to stuff like equipment and training. One of the good examples is that the national average, the minimum number of hours put forth by the National Tactical Officers Association (NTOA) and the American Snipers Association (ASA) that a deployable sniper should train is no less than eight hours per month in his specialty—sniper stuff. Now, I'm also an entry guy, so I get an additional sixteen hours a month of entry training, general SWAT work. Then there's the eight hours of my specialty. That's the recommended minimum, eight hours a month; that's two hours a week.

How many hours a week do you spend caring for your dog? Kids? How many hours a week does your high school football team practice? Know anyone with a kid in ballet? Hockey? PlayStation? Two hours a week isn't much. When it comes time for somebody to save your life, or your family's lives, would you want the guy who trains two hours a week? Or would you want the guy who gets as much training as the kid who plays tuba in the high school band? That kid probably gets at least one hour a day for band class, plus marching band practice after school during the season. If I got that much training time, I would be in heaven.

31

LAW ENFORCEMENT SHOOTER

Officer Bourdo

For the purposes of this interview, I am speaking as a private individual and not in any way on behalf of my department. As a deputy sheriff in Kenosha, Wisconsin, I work in a department that's both rural and urban. Kenosha is the southeasternmost county in Wisconsin, right at the Illinois state line and Lake Michigan. There's a population of about 85,000 in the city of Kenosha, and a total of about 150,000 or better in the county as a whole. We primarily do patrol in the nonincorporated parts of the county. But the city is in the county, as we always say, so a good deal of our tactical work, our SWAT work, is in the city.

We are not facing too much in the way of crime moving northward from the Chicago area. The city police department in particular has done an absolutely terrific job over the last number of years in keeping gang issues squelched. There's always some degree of gang activity in an urban setting, but it's not like we have a huge gang problem. Even in what could be considered the worst parts of town, the graffiti issues are really minimal. So we're pretty lucky in that regard.

As of March 1, 2006, I will have been a deputy sheriff for sixteen years. Of that time, I've spent about fourteen of them working with the tactical unit. I started working with the team in 1992, in particular with the snipers. I trained with the rifle, trained with the team, came on board officially, and received my first bona fied sniper certification in 1994. And I absolutely prefer the name "sniper" to all the other euphemisms that are out there. It's what we do, it is who and what we are, and it is the most accurate description of the job that we perform.

There are many, many precision riflemen, or long rifle marksmen, or whatever the currently vogue politically correct term may be, but that only describes one facet of the mission that we perform. Better than 90 percent of what we do is not rifle work, it's information gathering, surveillance, and secreting ourselves into a position to perhaps do that surveillance. We provide an overwatch cover, much as the military does, for our assault or entry teams when they go into a target. That part of the job has specifically been an issue for the last year or so in the law enforcement community.

Growing up in Milwaukee County, in an outside suburb of Milwaukee, I've spent most of my life right here in Wisconsin. Being in an average rural setting, I was a kid who was in the woods a lot. Hunting, shooting, plinking—those sorts of things were something I did a lot. From my very first BB gun, I just seemed to have an interest in shooting tight, small little groups. I'd even put a BB in the ground and use that as my target. As the years progressed, I moved on to pellet guns, then .22, and so on.

I have no military experience, though. I was a teenager when Vietnam ended, and because of the nature of the family business and all of the time that that took, there wasn't a big push for me to go into the military. We didn't watch a lot of TV for those same reasons. Quite frankly, I just didn't know enough about the military life for it to have any appeal to me.

For many years, I had an interest in the law enforcement field. It wasn't until I was thirty years old that I joined the force. Prior to that, I had always held an appreciation for what it is that a police officer does, how things are done, and bad guys being caught and punished.

Our department has what is called the tactical response team. For me, volunteering to join that unit was just a furtherance of the job in law enforcement. My going in for the team wasn't at all because I thought the job would be fun or exciting. It was just that I saw it as a higher level of doing the job, certainly a more serious and challenging level.

The position of sniper in the tactical response team just seemed to be a natural progression from my interest in rifles. I did not necessarily need or want to be kicking in a door or that kind of thing. I do have a passion for the precision marksmanship aspect of the sniper's job, and having studied it a little bit even years earlier, just out of pure interest, it looked like something I wanted to do. From the very beginning of my time, I had been working with my supervisors, who were snipers themselves. They would tell me to get down onto the rifle

An air force sniper team settles in for a shot during a sniper competition. The men are concentrating on accuracy and speed, so they are operating without the need for camouflage. With the spotter to the rear of the sniper, he can look over the shooter while keeping away from the muzzle blast of the weapon and the ejecting brass. ***U.S. Air Force***

and give it a try. The first few times I put rounds downrange, resulting in a tight clover-leaf type group, my sergeant practically knelt down and kissed me on the back of my head, and said, "Thata boy!"

Firing that precision rifle just seemed to be a comfortable and completely natural thing for me to do; it quickly became a tremendous interest.

My formal sniper schooling started with basic law enforcement sniper training. The FBI has a facility immediately adjacent to the Great Lakes Naval Recruit Training Depot in North Chicago. Their facility is tied in some degree, but not directly, to the depot. It's right there on the shores of Lake Michigan, and they have a range going out to two hundred yards for pistol, rifle, whatever is needed, as well as a fitness area. My partner and I were sent to the school by our department. That's where my formal sniper training started.

First and foremost with regards to firearms during that training, it requires becoming intimately familiar with your rifle. It is a weapon. In some places now they insist on referring to it as a "firearm." Well, yes, it is that, too, but it is a weapon, one that happens to be a firearm, but above all it must be remembered as a weapon. The primary issue is that you have to be ultimately accurate with your weapon. The reason is simple enough. If you should be called upon to take a shot, it is because an innocent person or victim is in imminent danger of losing his or her life. In essence, you have to save that person's life by shooting the suspect, whoever that might be according to the circumstances.

Given the dynamics of a violent situation, that window of legal justification for the using of lethal force can be slammed shut without a moment's notice. Or it can remain unopened for many hours. You can go on a call and have to take up a position overlooking a situation and just observe for many hours. Or you can be in a situation as happened fairly recently I believe in Jacksonville, Florida. A man was holding a woman hostage. From what I have heard of the incident, he had a rope of some sort around her neck and was holding a shotgun to her head as he came out of the house. The tactical situation had developed quickly, and the sniper arrived on-scene after having had to run some four hundred yards with his rifle. He had to get down into a stable shooting position and, within moments of doing so, had to take that lifesaving shot. He took the shot properly, and it was a successful resolution to the situation. The felonious, murderous criminal was shot, and the innocent victim hostage had her life saved.

So the sniper's rifle work in that regard is ultimately important. Most marksmen go to the range and shoot tight little groups, which is what we desire to accomplish with our own weapons. They will not do that four-hundred-yard run and get into position, feeling their heart beating in their throat and forcing their breathing to come under hard control, before taking that lifesaving shot. That's one of the biggest differences between a marksman and a police sniper. And that is exactly what a police sniper has to be able to do.

The training for a sniper includes the physical aspects of the job. Movement is a large part of that in a number of ways. A human target, that potentially murderous, felonious criminal actor on the stage of the incident in front of you, is not just standing still like a paper target. Even while simply talking, human beings are normally moving their heads, turning, nodding, and so forth. That precision shot has to be taken against a target that is virtually always moving. Sometimes, the opportunity to properly engage that target may show itself for only brief moments during an incident. You need to be ready at that moment's notice, because there may not be another.

A lot of the sniper training is to be ready for that momentary situation. For my guys, almost all of our shooting during training will involve some sort of induced stress. Even verbally pumping somebody up with the description of a scenario that will ultimately arrive at a point of legal justification for the application of lethal force will get his mind thinking about the potential issues and the points of threat. This is to get the officer in a state of mental and slight physical stress. You feel some adrenaline go into your system; your heart rate might increase, your breathing change—all physiological changes that will affect a person's shooting ability.

A police sniper, a precision marksman, knows that he has to control the physical state of his body as best as he can; otherwise the crosshairs, the reticle of the sight, will be bouncing around on the target and things like that. Those factors increase the difficulty of our task that much more, so we must train to control them. Training is the time for us to do this because if you do not shoot the target properly on the range, it's not going to sue you for everything you own.

Sometimes that lawsuit possibility is a little bit overblown. No law enforcement sniper that I have ever known about would be on a call like that and not know what all of the ramifications are with regard to limitations, restrictions, and responsibilities. As far as I'm concerned, it is one of the most serious decisions that a person would make in his entire life, that decision that he has to take the life of that person who is putting others in lethal danger.

That decision has to be considered long before a sniper makes even his first callout. To do that, you talk about it with like-minded folks who have at least the same level of training as yourself and the honest-to-goodness understanding that taking a life is in fact what we possibly will have to do. When we really analyze it on a day-to-day basis for law enforcement work, that very situation could come up. There are many instances across the country, happening on what seems to be a weekly basis, where officers are in a situation where suddenly some violence erupts and there are shots fired—at them. These officers must have the capability and mental means to respond to that situation. In our end of things as snipers, we know that there is a very high likelihood of that highest, ultimate level of violence taking place in front of us; otherwise we just wouldn't be there.

Even in those situations where you have time to set up and concentrate on the shot, there is no indication of just how much time may pass before that final decision to take a shot will come up. You cannot allow stray thoughts to come in on your concentration; the lives of the innocent and your fellow officers may hang in the balance. Yet you must also remain relaxed, not tense up your muscles, in order to be as accurate as possible if you suddenly must take a shot.

Going back to my very first call, where I was on the rifle with my sergeant, it was at night. Roughly half of the documented law enforcement sniper shooting situations have been during the hours of darkness, and training for nighttime shooting should be mandatory for every law enforcement sniper. On this particular call, a man was barricaded in his house; I don't remember the reason, but it was a two-story residence. That man had already shot at the police, having shot at them with arrows, broad-head-tipped hunting arrows. I don't remember the circumstances that brought those first responding officers to the scene, but having hunting arrows fired at you is more than enough reason to bring in the tactical team.

We were called out, and as per our normal procedure, we created a perimeter around the residence, to control the situation and prevent the escape of the subject. Time frequently is on your side when it is a barricaded hostile person, because nobody else's life is in immediate danger. I was set up on the front of the house, my position being across the street at a neighboring residence, a nice brick structure so that we had some cover. We were out of immediate lethal range of the arrows, so we were able to begin our observations. As in 99.9 percent of the callouts, we watched and reported on the situation to the officers back at the command post. Some people call it the TOC, for tactical operations center; using that term depends on the agency's verbiage for such things.

All we were doing at the time was relaying what we could see the suspect do. We had a team on the backside of the house—they were actually up on the roof of a garage—to watch the areas that we couldn't see. At one point the suspect announced that he saw the team on top of the garage. Whether he did or not, I can't say for sure, but the after-action review caused me to have some doubts about the suspect's statement.

The other team was able to take up another position with the cooperation of the home owner to the rear of the suspect's location. They were quite happy to evacuate their house, and the sniper team set up at an upper bedroom window, moving some furniture about so that they were able to keep back from the window about five or six feet, staying in the shadows so that the suspect couldn't see them very easily, if at all.

It may have been just coincidental, but when the subject went to the back of his house, he would frequently be on the second floor, at a rear window shouting out and looking about. At the front of the house, he would be on the first-floor level, looking out the front windows, the living room window and what I believe was a dining room window. He would look out at the squads that were parked out there with their spotlights on the house. The officers had bailed out of those vehicles when the suspect started shooting at them; they didn't know what he was shooting at the time.

For about four hours, the subject would move around inside the house. He would go to the first-floor front window, then the second-floor rear, then back to the first-floor front and again to the second-floor rear. There are some textbook examples of those earlier stages during an incident when the subjects are scared or experiencing whatever feelings that they are going through. In those stages, the subjects will be moving around trying to see what's outside—exactly what this individual was doing.

His weapons changed with his location. At the front of the house, he was armed with a shotgun. When at the rear of the house on the second floor, he had exchanged his shotgun for a rifle. So we knew he had at least two potentially deadly weapons in addition to the bow and arrow he had already used. We could reasonably assume that he was a hunter of some sort. In addition, it was understandably considered that he was not in a completely stable mental state. Whatever had set him off in the first place wasn't known to me.

For the first couple of hours of the incident, I was constantly on the radio, as my sergeant was on the rifle and I was running the communications for our two-man team. My comms consisted of updates to the command post—suspect is a white male, he appears to be in the mid-thirties, has short brown hair with longer sideburns. He's at the front-door window; three seconds earlier he had been at the living room window. He very much appears to have a shotgun.

Then the subject would disappear from our view. A few moments later, team two would say: we have a white male suspect, appears to be in his thirties, he's got brown hair. Yes, this individual looks like he must be the same one, he has the long brown sideburns, and he has a rifle.

That caused us to initially question if he had a shotgun or a rifle. So the next time that he came back to the front window, I had my binoculars and spotting scope, and I verified what I had reported. It was a vented-rib shotgun, like a twelve-gauge, wood-stocked hunting shotgun, a bird gun, a Remington Wingmaster perhaps.

A few minutes later, he was again at the back of the house on the second floor. The team there reported that they were looking closely and could see that he had what appeared to be a rifle. In fact, it looked to be a lever-action rifle with a scope on it.

With those facts reported, we knew definitively that we had verified exactly what the subject was armed with when in different parts of the house. The rifle was of course the longer-range danger as it had significant lethal reach. The shotgun, out to a shorter range, also has a very lethal reach. So everybody else involved with the situation remained in the shadows and behind cover, that sort of thing.

After a number of hours, the subject made a threat. He had pointed the rifle out of a window toward this garage where he said he saw the other team.

Now, we knew that he hadn't seen them for certain, perhaps he just thought he had seen them in the first place, because the team had moved and were actually in the house.

The situation ultimately developed to where the suspect's house was gassed because he quit responding. I believe he got tired, because we'll just let suspects run themselves ragged inside of a location so long as nobody else is in immediate danger. If we think that a suspect is inflicting wounds on himself, then we may have to take another approach to the situation. Typically, when we have those kinds of threats and weapons are involved, we'll let them wear themselves down. A human cannot stay at a high degree of emotion, anger or excitement without naturally coming down at some point. So ultimately the house was gassed and the team made entry. They found him in the basement of the building, where he had buried himself underneath a bunch of old clothing, blankets, and I think a mattress, possibly as a way to help filter out all of the CS gas that had been deployed in the building.

Looking back, I think that incident ran for me from about ten o'clock at night to about six in the morning, about eight hours. This isn't a short length of time to be involved in an incident, and I really can't say offhand that it's average, but reflecting back on other situations, it seems that it may have been a fairly average amount of time to spend overlooking an incident location. Some situations will actually get resolved before the team even arrives. And there are some incidents that can take days to finally end. I would say that an eight-hour incident would be about the average time for such a situation to continue, without having any statistics in front of me.

The situation of an incident where a hostage is involved is different. On that particular call, a hostage incident, the potential horrendous outcome makes all of the difference in how it is handled. If there is somebody who is just tied up as a hostage someplace, and the perpetrator is going from window to window, we have to continue to try and develop information, intelligence, as to the condition of that hostage. That is usually done through negotiations with the hostage taker. The other issue is if the suspect has that hostage in one position, where we can see what he is doing. Again, we have to assess the level of threat that the hostage is facing and that the hostage taker is capable of delivering.

If the hostage is significantly injured, and we can reasonably assess that he or she may die if we don't do something, the team commander or sometimes the agency will dictate that it must be the chief or the sheriff that has to make a decision. And that decision is whether or not to authorize the use of deadly force otherwise not justified. Once the decision is handed down, then that is what will happen.

If the hostage isn't in that dire condition, where he or she may be slowly expiring, at any one moment the hostage still could be suddenly killed by the hostage taker. Then the situation becomes a little bit dicey for the snipers because, just as at normal threat levels, you have to make a determination based on what you are seeing, possibly influenced by what you are hearing. Being a sniper means that I work at some distance from the target so by and large, my decisions are based on what I see. If I see a hostage and the criminal has a weapon of some sort, whether it is a gun or a knife, and he's moving around, gesturing and so forth, then I'm not going to hear what's going on. So I'm going to rely on somebody else who may be located closer to the situation and may be able to hear something, perhaps through a negotiator.

If you have an open phone line with the hostage taker, the negotiator may then work through the operations center or command post, and have somebody get in touch with us over the radio to let us know what the circumstances are regarding what we are seeing. If we don't have any verbal communication in that regard, we may not know what the guy is up to, what the final intent of the hostage taker is. If all we see is that he's moving around and not directly threatening the hostage, where we still would not necessarily have the justification to use deadly force, we would remain alert, and simply observe and report back.

If the negotiator is on the phone with the person, and he's so angry that he says he is going to hang up and in three seconds kill the hostage, within those three seconds that negotiator better get us that information, because we won't know until the moment it happens, if we can see it happening, what's going on. If we have a visual on the situation and suddenly hear that information on the subject's intent, we're still going to want to verify it. Ultimately, if the snipers have to take action, it probably means that somebody is going to die.

If we're told that we have immediate authorization, that the perpetrator says he's going to kill the hostage in three seconds, even then, we still really need to double-check that information. Over the radio, we could say: 10–4, 10–9, repeat, verification target of opportunity authorization given (go, or whatever their verbiage is). So we still want to have that verification. That kind of situation was the topic of conversation on one of our networks recently, and we discussed it at some length.

Even if we are told yes, we may not know the reason why. Those giving the orders may not have time to give us the reason; they may just say "target of opportunity." What that means to us as snipers is that, at our earliest opportunity, we shoot the target suspect. We may not have time to get the reasons why, so we have to have implicit trust in our command that there is legal justification to complete that action, to apply deadly force and take that life.

That can be a pretty nerve-wracking situation to be in for that moment, looking through the sights at a target and trusting in someone else to make the decision for you to pull the trigger or not. I've been in a few situations over the years where my finger has been on the trigger and I was just ready to go, and the situation changed. It resolves or something else happens that removes the need for lethal force. The level of force in an incident can be up and down like a chart for the stock market as far as applying deadly force, not using it, applying, not applying. Things like that make a sniper's job interesting to say the least.

I think because we, that is human beings, are generally compassionate people, we are going to sense or feel that an incident involving a child, a younger person, or even a female versus a male, might be especially difficult. But the bottom line is that a human life is in danger, and the emotional aspect in pretty much any sense has to be eliminated from the sniper's assessment of the situation. We can't allow those feelings to fog any judgment of the situation. We have to look at the fact that an innocent victim/hostage/bystander is under the control of a perpetrator who may take that person's life.

If the situation has come to the point where the authorization to use deadly force has been given, that's because somebody else is in imminent threat of death, and there are very few options for the sniper. To try to wound the person who is in that very last moment, seconds, from potentially killing somebody, is an unknown. You have absolutely no idea what the perpetrator may or may not still be able to do after he receives a wounding shot. Whether it's that he can still move, pull a trigger, whatever, there have been many examples over the years of people continuing to act effectively after being shot. I have no specific hard facts immediately to hand, but between networks, publications, and police bulletins, there have been many instances where the action of only wounding the perpetrator has been tried, and people have still been killed. Then what, what do you do? Who's going to be held accountable because you've made this attempt that didn't work and things went bad. If there is a justification for the use of deadly force, when it gets right down to it, the sniper needs to shut off that light switch if you will. That perpetrator has to stop what he is doing immediately and irrevocably. There is only one way of doing that today, and that is why they call it "lethal" force.

Typically, as snipers we train for a central nervous system shot, and that is equated to being a shot to the head, specifically within the cranial vault. The vault is the hemispherical portion of the head that holds the brain. We know that a high-velocity rifle round will create such immediate and tremendous shock on impact with the central nervous system that it will immediately affect the brain, the brain stem, and the function of the body they are connected to. The body will immediately go into flaccid paralysis on impact; it will just slump to the ground.

The police sniper will have the most personal view possible of the results of his shot, through his telescopic sight. This is the kind of sight that will normally bother a person for a long time. For the sniper, it gets down to being mentally ready for the results of your actions. Not having actually shot anybody whose actions needed to be ended with such finality to save somebody's life, I still ultimately can't say how I would feel after the fact. Thinking about it many times over, discussing it, going to training sessions and talking to people and having presenters who will talk about these things will hopefully prepare us. Reading books on the topic—there are a number of books out there that address these questions very well—and learning of the experiences that other snipers have gone through—those are among the ways to also help prepare you to deal with that final eventuality.

You have to know that it's technically not "thou shalt not kill," it's "thou shalt not murder." In some instances, somebody must prevent that murder from happening. You must be at peace with yourself for being that somebody.

To prevent having to take a life, even in defending another life, some have said that the police sniper should shoot to disarm if he cannot shoot to wound. The target should be the weapon, not the perpetrator. This subject has been talked about many times in the sniper community. There are some publications and training videos available, by people "in the industry" as it were, that show the results of this kind of action. I agree 100 percent with their findings that tactical disarmament as it is called is not a very good option. Even though there may have been an instance or two around the country where a sniper successfully shot a weapon out of a perpetrator's control, there ultimately is no way on earth that you can say with certainty what is going to happen when you shoot that hard object out of somebody's hand. Plus, if the person is committing the act of killing somebody, tactical disarmament may not prevent that victim from being killed.

If the perpetrator hasn't or isn't committing the act of killing somebody, technically you don't have justification to use deadly force. But if a person is holding a weapon, where on that weapon do you target your shot to render it inoperable? Nobody can tell you how to do that for all of the weapons that are out there. There have been many attempts to re-create and test out those situations that I've seen and heard about through people that I know, where a loaded firearm is struck by rifle fire, it can still either be fired after the impact, or ammunition within the targeted weapon can detonate, creating a significant level of potential injury or death. If you are firing that shot to try and save somebody's life, then you obviously have a serious problem. You certainly may have a result that you didn't want, and you are still going to have to answer for it.

Tactical disarmament is an option that doesn't really exist. Some people who just don't know think that it's a viable option. My thought is that it is potentially an absolutely disastrous course of action. For myself, I would never try such thing.

As a police sniper, you have to know with an absolute certainty not only where the bullet you fired is going to go, but you also have to be aware of where it will end up after it strikes the target. That is another of the many topics that is shared among the law enforcement and military sniper networks. It is also one of the reasons for discussing different types of ammunition. Law enforcement snipers have used match-grade commercially manufactured ammunition for decades. Primarily, the reason for this use is because it is guaranteed, or you are certain, that it is going to have the accuracy you want and need.

The bullet of some match ammunition, and the military has recently addressed this, is technically a hollow-point bullet by the nature of its design and manufacture. The design is for accuracy; it is not designed for expanding on impact, such as a hollow-point hunting round would. In the military, using such a deforming projectile would be a Geneva Convention violation of some sort. Military lawyers and ballistics experts have established that the use of such a match-grade projectile is not an issue. So military snipers can and do use match-grade ammunition in combat whenever possible.

On the law enforcement side of the sniping house, putting that bullet through the cranial vault to get an immediately incapacitating, flaccid paralysis result is what we need to do. But that bullet, even though it is going through bone and tissue, is very likely not going to expand at all, and it will continue to go through the target and into the environment on the other side. The sniper has to be very cognizant of that. If a person is in the window of a building, and we're reasonably certain, based on all of the intelligence that can be gathered, that there is nobody in that structure behind him, we have a pretty high belief and expectation that if we have to take the shot, that bullet probably will just go into the physical structure of the building.

If that person is in the open someplace, we have to keep in mind just what may be behind him. If it is an open, rural setting, and there is nothing for the proverbial country mile, the risk from the bullet is relatively low. If the perpetrator is in an urban or community setting, lord only knows what could be behind him. That goes along with the tactical disarmament problem: What on earth is behind that object? The law enforcement sniping community is actively pursuing from the manufacturers of ammunition a technical solution or reduction of risk from overpenetration of a target. What that investigation is resulting in is ammunition being developed with a type of bullet construction that still offers the ultimate accuracy, but also offers a better terminal ballistics solution.

Terminal ballistics refers to what happens when the bullet hits its intended target. Those types of bullets which can meet at least some of the characteristics we are looking for typically are a plastic ballistic-tip projectile similar to what hunters would use for varmint hunting. That kind of projectile causes very fast, significant expansion, so that it will transfer its energy to the target and have a much smaller chance of overpenetration issues. The end result of

The advent of terrorism in the twenty-first century has forced an increase in sniper capabilities, even among law enforcement units. Here an officer practices with a Barrett .50-caliber rifle. The big semiautomatic weapon would normally be considered far too overpowered for police use. But in a few limited instances, the big projectile of the Barrett could be the only shot that will stop a hijacked armored truck or punch through to a barricaded terrorist before he can trigger a device. ***Private Collection***

this kind of thing will bluntly be very ugly. The situation is not pretty to begin with. The terminal performance of new ammunition will not change the final result of the application of lethal force. But it will minimize the possibility of injury to others. And I believe that even though the criminal isn't any deader, the view of such a thing could be a tremendous shock to the public. Unfortunately, that's something that we would have to deal with.

I think that in general, throughout the history of our nation, people have been called upon to do this kind of work, longer-than-normal-range types of shooting for accuracy. It started back in the Revolutionary War, grew during the Civil War years, and has become an integral part of our military arsenal. It is nothing more than that soldier, or that law enforcement officer, who has a fair to significantly higher amount of training coming forward to complete that task.

Whether the American public likes it or not, we need to have people who are willing and able to do that job because, no matter how you look at it, somebody may need to have his or her life saved. The term "sniper" in and of itself has always, unfortunately, been clouded in negative connotations. That's terribly unfortunate. The mass media will call any lunatic, gun-crazed person a sniper. Muhammad and Malvo, the Washington, D.C., gunmen weren't snipers. They might have had rifle training, but they were just criminals shooting at people using some sniper tactics. Using the name improperly gives military and law enforcement snipers a black eye. We are in fact professionals in the way that we conduct ourselves and the way that we perform our mission.

LAW ENFORCEMENT SHOOTER

Officer Klein

I've been a police officer in a northeastern U.S. department for going on twenty years now. As a member of my department's tactical services, I have held the position of sniper since 1990, after having joined the unit the year before. It was the kind of work that the tactical unit did that first drew me to it. It was something that I had always wanted to do, and I enjoyed the idea of being in a small unit that operated on serious calls. Not having to deal with the barking dogs, neighbor complaints, and things like that that you have to face while on patrol. Instead, the tactical unit dealt with things that were much more along the lines of serious crimes and real situations that would come up, not just what most police officers refer to as BS calls.

It would be very hard for me to say that one job or the other, the policeman on patrol or the tactical officer on a call, is the more dangerous of the two. People can argue it on both sides of the spectrum. I've heard people who are in regular patrols argue that their job is more dangerous since they walk into the unknown.

Frequently you hear of police officers being killed during supposedly routine traffic stops where the officer thinks he has pulled over somebody who has just committed a minor traffic violation. As things turn out, the individual is wanted for a much more serious crime and so he's going to resist the officer with unexpected deadly force.

On a tactical team, when we respond to something, it's expected that the individual has already reached a certain level of threat and he has to be dealt with by specialized personnel. Either job can be more dangerous than the other in a worst-case scenario. I think that the job of any police officer has an element of danger to it.

Growing up, I was a really just a suburban kid. I couldn't really say that I received some sort of childhood preparation for being in a tactical unit. My father did like to shoot, and as a consequence, I grew up around guns and I enjoyed shooting. If what I've done as an adult can be attributed to that background, okay. My father was in the service during World War II and my uncle and grandfather served as well. So we had a tradition in the family of service to the country and the people. For myself, I joined the marines and served my country for four years before becoming a police officer. I guess it was just the urge to serve that sent me to the law enforcement field as a career.

In the marines, I was originally trained as an 0351, which is an antitank assault man. Then I went to Okinawa and was trained as a recon marine, which gave me a different MOS. Then I had time to go to the Eastern Division shooting matches, where I competed in the rifle and pistol competitions. I received a primary marksmanship instructor MOS and taught others how to shoot. So I had a varied experience in the Marine Corps; it was good for me and fun. But I got out when I started a family.

The position I've held in the tactical unit is that of sniper. We call it that; some other agencies call the position by other terms—marksman, precision marksman, whatever. But the job is that of a police sniper and that's what my department calls it. We haven't had an issue come up where we've wanted to call the position something other than sniper, though some other departments, and friends of mine in those departments, have had to fight that battle for a name with their administrations. None of us are ashamed to call ourselves snipers; we're very proud of the jobs that we do. I think that the reason the name "sniper" has developed such a negative connotation is the way that the job is depicted in Hollywood and in the mass media.

People have learned to associate the word "sniper" with a negative image. Naturally, administrators in police departments, especially those who have never been trained as snipers or never had experience on the tactical teams, hear the word "sniper" and automatically put a negative meaning to it. Those of us who do the job like to be proud of what we do and to talk to people about just what it is that we do. The trouble is that a lot of times I've seen civilians react with some sort of weird expression on their face when they find out that I'm a sniper and I'm proud of the job that I do. They look at you as if they just learned that you're proud of being a psychopath or something like that. Those of us who do the job don't see it that way at all.

Constant practice is the rule especially in law enforcement sniping. Even on his own time, range practice for a police sniper is as necessary a thing as sleeping or eating; for some individuals, it holds a priority over even those normal demands. Here, an officer is sighting his personal weapon, purchased with his own funds. The rifle is a GA Precision "Rock" with a McMillan A5 stock. The sight is a Nightforce 3.5-to-15-power variable model with a fifty-millimeter objective lens. The objective lens at the front of the sight is fitted with a long sun hood to cut down on glare or reflections. ***Private Collection.***

The position of sniper is a difficult job, and it requires a level of dedication to be proficient at it and to stay current with the job skills and demands. All of the snipers that I'm associated with and I go to conferences with and such are spending their own money to improve their own training and skill levels and professional knowledge. It's an ongoing process; you have to stay current with the developments in ballistics, weapons, equipment, supplies, and training. You have to know about the incidents that have happened, why they went down the way they did, what the current laws are, and where things are going to change. It's a hard job and I'm proud of being able to do it. And I'm proud to say that I volunteered to do it and enjoy doing it, plus I think I do the job pretty well.

Most people, when they think of the sniper's job, imagine a guy who gets a rifle and goes out and shoots the bad guys. Being a police sniper just isn't doing things that way. Mainly, the job of sniper is being an information gatherer. He arrives at a scene, which for the purpose of this example, is a tactical response to as barricaded subject. Somebody has locked himself in his house and is threatening to kill himself or other people, or just shoot whoever shows up to deal with him. We've had many deployments like that.

The sniper shows up and gets into position. From there, you are able to give information back to the command post, to the officer, usually a lieutenant or higher, who's in charge of the tactical operation. You tell him what you can see, plain as that. We have those high-powered optics available to us, both the sights on the rifle and the spotting scopes our observers use, as well as binoculars. You lie still, observe, and report. That's 99.9 percent of the job that a police sniper does—simply observing and reporting. Kind of like hockey players, who spend the bulk of their time passing and skating and very little time actually shooting the puck.

So far, I've never had to pull the trigger on a live human being at all. I've never been placed in that position where it was necessary for me to shoot another person. I've come close, but I've never had to actually pull the trigger on somebody. You prepare for that ultimate eventuality by training and you develop a certain mind-set. You know that you are there to do a certain job, that you are there to save people's lives; whether it's the lives of your fellow officers or the lives of the innocent public that have been put in danger by this individual, or, if the individual has taken a hostage or hostages, the lives of those hostages.

That's what your mind-set and focus has to be. If in order to save an innocent life, you have to immediately stop the actions of a bad guy, a perpetrator, then you have to do what the job calls for. If using that level of deadly physical force in order to stop someone from killing another person becomes necessary, you have to be able to do that. That knowledge and commitment, that you will do what is necessary, has to be in your head before you go out on the job. If it isn't, if you won't do what is needed, then the time to find that out is not when it's time to pull the trigger. You have to ask yourself that question ahead of time, when you're training or even contemplating accepting the position of sniper. If you answer with anything but a solid "Yes, I can do this," then you need to consider another line of work. It doesn't mean that you aren't a good person, that you aren't a good guy. I know plenty of good guys who don't want to be snipers, and I don't blame them. It is a difficult thing to look through a rifle scope and see somebody moving around and talking, watch him for a period of time, and then maybe have to pull the trigger on him. But you have to be able to do that. If you can't, you cannot be a police sniper.

I have known several snipers who have had personal difficulty after having had to pull the trigger. I think the worst thing that can happen to a person in that situation is to feel that he's been cut loose from his agency and his peers.

The most important thing in my opinion is the support that should be given to any officer that's involved in a shooting, whether he's a sniper or a regular patrol officer. You have to feel that you have the support of your peers, that you have your agency behind you and backing you up, not cutting you loose because maybe your actions don't seem popular in the mainstream media or your actions have a question raised against them. It could be in a press conference that the question is raised as to whether or not this officer should have shot in that situation, but that agency should never come out and just be politically correct and cut their officer loose and not support him. They have to stand with him and get that officer the help he will need, get him to talk to his peers about the situation as soon as he can to help him get through what has happened. It's natural to have a reaction to the killing of another human being; it's natural to be upset, to have trouble sleeping and things like that. The officer has to know that he's not all alone. You need to know that you have support, that your peers support what you had to do, because you did it to save their lives or the lives of innocent persons.

From what I have seen, most police snipers do not get any sort of mental preparation training for the job they may have to do. Some departments are very progressive, and they send their officers to a seminar where you might listen to somebody such as Lieutenant Colonel Grossman speak on that aspect of the job. You might listen to other officers tell you about their experiences with having to shoot someone and tell you about the entire incident. You learn from that; you learn from their experiences with the aftermath of their shootings. There are departments that after a shooting seek to quickly get an officer right into some sort of peer support group with other officers who have been involved in shootings. I think that helps those officers who can have that sort of support group available to them. Other officers come from departments where they don't have that support set in place. Worse still, they don't have a screening process set up ahead of time to ask the officer what he would do if he was placed in the position where he had to shoot to save a life. Would he do what he had to do?

Plainly, there are officers who would not do that, not take a life. I don't fault them for it. I don't think they are less of a police officer for that stand; I don't think less of them at all. An officer who has the ability to look at himself honestly and objectively and say that he doesn't have what it takes to do that, to take a life, can still be of value to his agency in another aspect of the job. I applaud that officer's honesty; I think he's a great guy because he didn't put himself or his fellow team members in a position of finding that out in a moment where it would matter the most. He was honest with himself, assessed himself, found that he didn't have the temperament or ability to do the job, and decided he would be better off working someplace else.

In order to become a police officer, you go through a tremendous amount of screening. In my case, you had polygraph exams; you had background checks where they talked to your former employers, your family members, friends, people who had known you for any length of time, coworkers, and folks like that who knew you long before you became a police officer. Also, there are psychological tests administered, interviews are conducted with a psychiatrist, and all aspects of your persona are examined. Then, after you have served in your department for a period of time and you apply for duty with the tactical team, you're screened again. A lot of the time it's almost the same kind of screening you went through in order to originally become a police officer.

Again, all of the psychological screening and all of that is done to see if you have the kind of temperament to do the job, or if there's some aspect of your personality that might indicate that you aren't suited for that kind of work. Personally, I've never met a police officer who's a tactical team member and has some kind of mental disability where he seems to enjoy the deadly physical force aspect of the job. Peer pressure would probably eliminate such an individual from remaining on a team, even if he managed to get through all of the filtering. Tactical units are tight-knit organizations; they have to be to do the kind of work that they do. The guys work together a lot and depend on each other to be there for each other when they are needed. If somebody's exhibiting signs of aberrant behavior, if he's not right for the job, there is definitely going to be more than a little bit of pressure, not just on that officer but also on his supervisors, to do something about the situation.

The sniper is not the ultimate application of deadly force. I would say that any application of deadly force is the ultimate force to be used, whether it's done by a regular patrol officer or a sniper. Much is made about the sniper because of the negative connotation of his job. A sniper is looked at as somebody who ambushes somebody else; it's not a fair fight and you didn't face him down in the street and give him as much chance to kill you as you have to kill him.

Dealing with rifle shots to the cranial vault—a high-powered rifle shot to a human being's head is most likely in almost all cases—is going to cause that person to die. That's a realistic way of looking at things. If there is a way of instantly ceasing a subject's actions, and guaranteeing that he will not be able to function, and at the same time will not endanger your fellow team members, or the possible hostage that is there, or members of the public, long enough for you to take him into custody without killing him, then I wish that somebody would come out with it so that we could use it. Unfortunately right now, the only way to resolve that type of incident, such as a hostage taken by an armed subject, or someone who has a weapon of some type and is actively firing at police officers, is to use deadly physical force. The best application of this force, to keep the police officers and members of the public safe, and the safest to employ by a tactical unit, is the police sniper. He must use his skills to stop that individual.

The sniper may not be the best application of force, but he is one of the better options. His employment doesn't endanger others, because so few people have to be directly involved with the sniper's shot. If you use an entry team, that team has to get close to the target, which generally means they have to expose themselves to that individual. That increases the possibility of one of the entry team officers being hurt, seriously injured, or killed.

In the case of a hostage rescue, the perpetrator could have the hostage standing in close proximity to him; usually they will do that because they want the hostage to be a shield against the officers that they know are outside and are possibly looking to use deadly physical force against them in order to resolve the incident. The hostage taker, if he's not exceptionally stupid, is moving, trying to make it hard for the police to finalize his location. In one case that I know of, a fellow officer was within a very short distance of the hostage taker, who was constantly moving around behind the hostage. He was moving so much that the entry team did not think it was safe for them to conduct an entry; if they tried to, they might accidentally hit the hostage. The officer took his rifle and, using a support, was able to resolve the incident through the use of deadly physical force without hurting the hostage at all. That's the real point for a police sniper; he has to know that when he's going to use deadly physical force, he will only hit the perpetrator and not injure the hostage. That's one of the differences between the sniper and the entry team; the team's weapons are designed for use in close-quarter combat (CQB) applications. They are not to be used in the same way that a police sniper uses his weapon, which can place a single shot in a very precise location over long range.

For the sniper to end an incident before it can escalate into harm for innocent persons or fellow officers, he has to be able to place his shot to take out the central nervous system of the individual involved. This produces an instantaneous stop, so that the individual cannot continue the hostile actions. The only way to do that is to insure that you have a hit in what we refer to as the cranial vault area of a person's head. As long as you put that crosshair on the center of mass in an individual's cranial vault ahead, when that high-powered rifle round strikes it, it will stop someone's actions either instantaneously or nearly so, to the point where they could not even pull the trigger of a gun they were holding to a hostage's head.

Shooting to wound as an option for a sniper has been brought up in the public's eye. I'm aware of an incident where perpetrators took hostages on a bus, and the tactical teams set up to contain the situation. One by one, the hostages were let go and the deranged armed individuals who had seized the people and the bus decided to march on the police and force them to shoot them in order to stop them. The popular term for this action is "suicide by cop." The officers were tasked with shooting the individuals in the shoulders of their weapon-holding arms. In the video I've seen of the incident, one individual just switched hands once he was shot and continued his actions, which were to kill his companion and then himself.

A wound, even a serious one, does not guarantee that you will take a person out of the action to the point where he won't be able to hurt somebody else or a fellow officer. It's an unrealistic expectation to have a sniper do something like that. And there's another completely unrealistic expectation of the public, that a sniper could just shoot the weapon out of a perpetrator's hands, or that he could shoot the perp in the hand, or the arm. If you have ever watched somebody through a rifle scope, you would know that the movement of an individual is so great, and can be so quick, that it is almost impossible for the sniper to keep his sights on the target. The chances are great that the sniper would end up throwing his shot someplace where he didn't intend for it to go—something that is completely unacceptable to a police sniper. An innocent hostage could be endangered even more by the sniper missing his target. And if that round continues downrange, it could intersect some innocent person who's not even part of the incident. Then there's the possibility of hitting the hostage, and not necessarily with the bullet fired by the sniper.

If the hostage is being held at gunpoint, it is a fairly sure thing that the weapon will be in close proximity to the hostage—held up to either the head or the body. If the hostage taker suddenly moves his hand, the sniper could inadvertently hit the hostage with his shot. Or the weapon could discharge or even explode if hit by the sniper's projectile, wounding the hostage with a secondary missile. Shooting a weapon is Hollywood; it's the movies, and this is reality.

"Tactical disarmament" was a popular term that came about as a result of an incident or two that were resolved by a SWAT sniper shooting a weapon out of an individual's hand. The most famous one that I'm aware of has a video of the incident that shows an individual sitting on a chair while he's armed with a revolver. He threatens to shoot himself several times, and officers surround him and try to talk him into surrendering his weapon. The man refuses to give up his weapon and continues to sit in the chair that is in the middle of a residential street. The sniper's superiors authorize him to try and shoot the weapon out of the individual's hand. The individual drops his hand down between his legs, the revolver now dangling below the seat of the chair, relatively immobile. The SWAT sniper takes the shot, successfully engaging the revolver, smashing the weapon and rendering it inoperative. The individual remains seated in his chair, more than a little shocked that his gun has just been shot out of his hand. Then the entry team officers swarm him, taking him into custody without him being injured.

Everyone saw this incident; it made national news for days. The public thought it was a great thing and said why couldn't the police do that more often. My answer to that is that the shot was almost unique. The individual, the weapon he held, and the shot made by the sniper—all lined up perfectly on that day for there to be a successful outcome of the incident. As far as what would happen if I shot a weapon, I would have no guarantee that hitting that weapon would render it dysfunctional. I've seen instances where weapons have been shot and remained perfectly functional because they were not struck in exactly the right spot. Also, shooting a weapon creates a problem just by the physics of the situation. You are shooting a hard metal object with a high-powered, high-velocity rifle round. Where is that projectile going to go after it strikes the metal of the weapon? Is it going to glance off and into the neighborhood and possibly kill some innocent person? Where is it going to go? There is no way to guarantee where that projectile is going to end up after it strikes that hard metal object we call a weapon.

Another question is would the impact of the projectile possibly cause the weapon to discharge anyway? If you have the case of an individual holding a weapon to either his head or somebody else's head, you may cause the weapon to discharge, thereby killing that person. How are you going to justify that later on during the investigation that's sure to follow? To me, "tactical disarmament" is an unrealistic expectation based on an almost unique incident that was successfully resolved using the technique. It is very dangerous to do in my opinion.

No one should expect a police sniper to engage in tactical disarmament, and no police sniper should ever validate this technique and say that it's a good thing to do—it isn't. In that one incident, it worked out, but it may not work out in the next incident. And we have no way of knowing which way it will go. With that kind of uncertainly, I would say you cannot engage that technique. You have no way of knowing where that projectile is going to go, what that weapon's reaction is going to be, and all you will most certainly be accomplishing is endangering people even more than they have to be.

Other things have affected the job of the police and the police sniper more than the single isolated incident of a unique shot. The advent of terrorism in this country has made everybody more vigilant, more aware of what could be going on around them. After 9/11, all of the incidents that seemed relatively innocent before, or strange things that would have been shrugged off, now make people think twice. They call law enforcement and report things that they see. The police sniper now has to think about the fact that the nature of his target may have changed. Before, it was common that you would be facing an individual who probably was not formally trained with a weapon, as a military person would be. There have been a few cases where formally trained people have gone "bad" so to speak and police officers have had to deal with them. But generally an officer has to deal with someone who is mentally ill, or under the influence of drugs or alcohol, and is upset over something. Those situations are peacefully resolved 99 percent of the time. The individual is taken into custody without serious incident, and that's the type of individual we're all used to dealing with.

With the advent of the terrorist, you're now looking at a possible individual who's very dedicated, has had some sort of formal training, and is not afraid to die for his cause. In that kind of incident, your normal procedures of negotiation and things like that are probably not going to work the way that they do with the average person. The usual perpetrator has at worst a mental disorder that has caused him to perform the actions that drew the attentions of law enforcement. Maybe he's had a particularly upsetting action take place in his life, possibly he was fired from his job or whatever. Those kinds of things can often be resolved through negotiation and talk with the individual, getting him to surrender without a major incident. But that isn't the situation you'll have when you are dealing with somebody who's a terrorist.

When law enforcement runs into somebody who's possibly trained, possibly going to try and engage a SWAT officer and kill him, we have to be prepared for that eventuality. We have to be more careful with what we do. During an incident, we have to be even more cautious that we remain out of the public eye, that the person who's involved in the incident does not know how we are deploying or where. These people know how to watch the local news media just as well as we do. In the back of our minds, we have to retain the thought that this might be the incident where we meet a terrorist. We have to be careful for ourselves and the public that we serve.

There's also the factor of terrorists' favorite weapon apparently being the improvised explosive device. We have to be aware of the possibility that they may be manufacturing a device in a location that we're looking at. They may be seeking to escape that location while armed with an explosive device. We have to operate on the information that we have on an individual, as well as what we can see and observe on our own. Decisions have to be made based on that information, so we have to see as much as can be observed either during an incident or at a location being watched.

If a sniper sees something that would obviously inform him that an individual may be in possession of an explosive device, or possibly wearing an explosive vest, that intel has to be relayed back to the command post as quickly as possible. They would have to take measures to insure that no one else is endangered by a situation like that.

I would like to say that the military and police snipers have always been looked at as very separate and distinctly different jobs. Generally, that's true, but in detail, it isn't true. The police sniper is different because of the precision nature of the shot that he has to make. Nowadays, what's happening is that the snipers who are being deployed overseas are finding themselves in an urban or suburban environment. There, they are dealing with distances and a job nature that is a lot like what a police sniper does.

Observation is always a big part of the job, even with military snipers. A lot of them will go a long time without taking a shot. They will observe activity and report on it, remaining hidden so that the enemy forces will have no idea that they are being watched. Once you take a shot, they know right away that a sniper is in the area. You will either have to leave or take action based on what they do in response to that shot.

A lot of the time, the greatest asset that a military commander has in the field is the sniper who can get into a position where he can observe the enemy for a length of time without ever letting them know that they are being watched. The police sniper does the same sort of thing. He gets into a house, perhaps across the street from the target house, or maybe he slips into a neighboring yard, and he sets up a hide. He observes and reports what he sees so that a proper course of action can be decided on. Generally, the time frames for a police incident are much shorter than for a comparable military mission.

The nature of the job is the same—saving lives. The military sniper saves lives when he goes out and engages the enemy, eliminating their leadership and leaving them lost and confused, unable to operate effectively. The police sniper also saves people's lives by what he does. The execution of the two jobs might be a little different—the military sniper deals mainly with distance shooting and stuff—but with the environment they are finding themselves in more and more today, military snipers are now operating in the domain of police snipers, at distances that we find comfortable and shooting where there are a lot of people around. In an incident where there are a lot of non-combatants milling about, you have to use what military snipers like to call themselves—the cheapest smart bomb in the military.

There are no civilian casualties; only bad guys are getting hit. And I think it's a great thing that these guys are saving innocent people's lives by what they are doing, engaging the bad guys in a precise manner. In that way, our jobs are intersecting and meshing together more and more.

In a hostage rescue operation, a police sniper is looking almost exclusively at making a cranial vault shot. That keeps his possible engagement distance relatively short, much less than a military sniper. If a military sniper sees an enemy, he can usually hit him. There have been military sniper shots taken at unbelievable distances, twelve hundred, fourteen hundred meters, or much longer. Military snipers have successfully engaged enemy combatants at distances that can be comfortably measured in kilometers. A police sniper doesn't have that option. He cannot afford to miss a target and he cannot accept just wounding a target. In the military it is acceptable to shoot at the center of mass on an enemy target and strike him in the arm or leg. Actually, in some cases in combat, it is preferable for a military sniper to wound his target rather than simply kill him outright. That way, the enemy combatant has to be helped off the field of battle by additional enemy personnel, personnel who won't be available for the fighting for at least the time it takes them to assist the wounded individual. So by wounding one guy, the sniper has taken three people out of the fight.

The police sniper has to take a shot where he is certain that the target's actions will stop instantaneously. The only way to do that is with a cranial vault shot. So naturally the target range is much shorter than those of his military counterpart. That's why a police sniper has to train a lot, so that he is confident in his abilities at whatever range he may have to operate. The longest true police sniper shot that I'm aware of was a cranial vault shot at 187 yards. That doesn't seem like all that far when compared to the military guys, but to take a cranial vault shot at almost two hundred yards—and humans move all of the time—is an accomplishment. Try to hold an unmounted telescopic rifle sight on a person during a TV show. The camera is always on the person but they do move around the set. Try and move when they move—that's hard enough.

To accomplish that task, make that shot, we have to train very hard and very long hours. We have to keep our margin of error to an absolute minimum. To do that, the sniper has to learn just what his personal comfort zone is against a range of targets and distances. The average police sniper shot is fifty-one yards in the United States. We train to engage moving targets at longer ranges, and those that are behind intermediate barriers such as glass, doors, or windows. And we train to develop confidence in our rifles, abilities, ammunition, and fellow officers. All of that comes together when we go into an incident and know exactly what we can do to stop those who wish to harm others.

THE SHOOTERS—
MILITARY

MILITARY SHOOTER (MARINE)

Travis Mitchell

Montana is the state where I hail from, born and raised there. We lived on a river in a little neighborhood with about thirty homes. My dad was a butcher in a small town called Evergreen. Growing up, I did a lot of hunting and fishing, never had beef until I was probably about seventeen; everything we got, we had killed ourselves. So you could call me an outdoorsman, I guess.

I probably went out hunting for the first time when I was about eight, done a lot of it since then. When I was seventeen, I enlisted into the United States Marine Corps. I still had a year to go since I went in under the delayed entry program. So three months after I graduated high school, on September 7, 1989, I went into active duty.

As far as which military to go into I really didn't know what I was doing. When you look at the marines, you know it's something that you're going to have to be one of the best to stay in, and they look sharp. Everyone who thinks of the marines knows that they are the best at what they can do. That was something that I wanted to be involved in.

When I came into the military, I was very fortunate when originally I was supposed to be an 0311, a basic marine infantry rifleman. Once I went to my school of infantry and passed, I went on to my unit, which was the 1st Battalion, 9th Marines; back in Vietnam they were known as the Walking Dead. It was the only battalion in the marines where their flag was actually captured by the enemy.

Once I was at my new unit, they had a bunch of us called into a room where some sergeants came in to speak to us. Basically, we were asked who

Two sniper students having been dropped off just prior to beginning a stalk during training. The men are both wearing ghillie suits that they have made themselves, part of the sniper training curriculum being how to produce the very efficient camouflage item. ***U.S. Army***

wanted to be a sniper. I didn't know much about the military at the time and asked them just what being a sniper meant. From what I was told, it mostly involved shooting guns a long ways. Sounded interesting enough, and I volunteered to try out for it.

Usually, they don't take someone right out of infantry school for the sniper training course. Normally, they want a marine to have had two or three years in the Corps and have been overseas and have at least a little more experience than I had prior to allowing him to volunteer for sniper training. But they let me try out, and I attended a three-week indoctrination program for snipers.

Along with a couple of other guys, I made it through the indoctrination course successfully and went on to a sniper platoon prior to attending sniper school. Being a sniper turned into my job the whole time I was in the Corps. I never did serve in the infantry; instead, I was in a sniper platoon. I may not have been a sniper all the time, but I spent my time in a sniper platoon my whole eleven and a half years in.

Eventually, I did find out why a group of us were asked to volunteer so soon after joining the service. They knew they were short bodies and needed people. When you're in a sniper platoon, most people only stay in the military four to five years and then they get out. So they thought if they got someone young, brand-new in the service, they could get him to do at least two overseas deployments with the platoon rather than the normal one. There was a problem with training up a sniper, spending all of this money and time on him, sending him through training, and then having him just get out of the service. So by using us, it was an experiment in economy.

Going through basic, I wasn't what you would call a stand-out marksman. I shot expert on the rifle range, but nothing super-high score-wise. All of my hunting experience had been with optics, telescopic sights, so going to iron sights in basic wasn't a big deal, but it was a transition. Shooting with iron sights is much harder than using optics. But I had shot expert and was in good physical shape back then.

Going through sniper school, I probably had it easier than most people because before you went to sniper school back then—I don't know if they do it as much now—you had to go to a pre-sniper school, a regimental course. That was three weeks long. What was different for me was that I had been in a sniper platoon almost three years before I even went to sniper school. They had even sent me to U.S. Army Ranger School before I went to marine sniper school. At Ranger school, I passed the course my first time through. Immediately following that, I went on to jump school. Finally, after coming back from jump school, I went on to sniper school.

Since I had been in the platoon for so long, I already knew a lot of the material we were being taught. It was hard physically, but as far as all of the knowledge and studying, I already knew much of it. The only holdup in sending me to the sniper school in the first place was finding a slot for me. There were other people with higher priorities than I had, sergeants, corporals and such, and we already had a couple of snipers in the platoon before we deployed overseas. So an open slot just didn't come up for me for a long time. Ranger school was ten weeks, and jump school another four weeks on top of that, so there went fourteen weeks out of the year gone.

The sniper school back then was very difficult. There were different phases of training that you went through—marksmanship, stalking, mission planning, and land navigation. Land nav, as we called it, didn't really cut a lot of people at the Camp Pendleton sniper school where I attended. It was a lot easier than the school at Quantico, where you had to do a lot of dead reckoning using a compass. Most people get dropped from sniper school because of failing either stalking, shooting, or mission planning. Those are the three hardest things taught in the course.

Back then, the snipers operated as two-man teams; now they are fielding four-man teams. A two-man team can have a lance corporal like me telling a battalion commander where to move the troops according to what I can see out ahead of everyone. That's a lot of responsibility. Additionally, as snipers, we knew how to call for fire, drop in artillery, close-air support, or naval gunfire on an enemy position that we could see from our sniper's hide. During some of our mission planning, we would do what was called a TEWT, tactical employment without troops. That's something that the officers learn how to do when they go through basic training. A TEWT is about a hundred pages long and you write it out from the go point and list everything you're going to do on that mission. Most officers have to do that when they go to their school. As a sniper, you are taught how to do that when you are just a private first class or lance corporal. That way, everyone up above knows just what you are planning to do out there in the field.

Learning that kind of stuff is hard and involves staying up long hours writing it up. Every day at sniper school, you had two or three hours or even more of homework to do. We took two or three tests every day, both written and practical. Camouflage, concealment, definitions that you had to memorize verbatim, a lot of subjects that had to be learned—just everything that encompasses performing the mission of a sniper.

The schoolwork and training wasn't that bad for me since I had that prior experience in a sniper platoon. Nothing was really harder than any other subject. Stalking was fairly easy for me; I was almost the high stalker as far as the score goes. My partner actually broke the school record for high stalker, getting into position quickly, smoothly, and without being detected. As the high stalker, you had basically received so many points on your previous stalks that you could practically just fall off the truck and still pass. You didn't have to go out on any more stalks. If you did, you were trying to build more points, continue being successful conducting a stalk. Most people had to get one hundred points to get through the stalking portion of the training. My partner and I already had 128 points when we still had three or four stalks left in the training course.

Things were so different for me since I had been in that platoon for three years. I already knew the basics of the sniper's craft and some of the advanced things as well. A lot of guys on their first stalk don't know what they're looking for, how to find their window of opportunity to approach the target, make sure that there are obstacles between their avenue of approach and the observer's post during the stalk, things like that. Just trying to get into the position and think about what you have to do in order to take the shot with a blank, making sure that the walker doesn't walk in on you, so that you can take another shot

A Navy SEAL sniper sights his McMillan M88 .50-caliber sniper rifle. The large muzzle blast from the weapon would kick up a large amount of dust and debris, forcing the sniper, and his spotter to his right rear, to leave the area after firing. ***USSOCOM***

and the observers can't ID you—these were things I already knew and was able to think about while most of the rest of the students were still learning about them.

When you're on a stalk at sniper school, basically what happens is that they drop you off about a thousand yards from instructors who are acting as targets. They're going to try and spot you as you move to within two hundred yards of their position. The instructors know what area each of the fourteen students in the class is going to be coming in from. And they're looking over that line of approach with binoculars. You set up in your FFP (final firing position), set the proper range dope on your rifle, and shoot—"boom," one blank shot. Then one of the walkers comes over to near your spot and asks who shot.

After you've identified yourself to the walker, he comes around to within ten feet of where you are hiding in your FFP. Now the observer knows that the walker is within ten feet of a sniper's position. He'll look over that area very carefully, trying to spot where the sniper is hiding. If he sees where you aren't quite blending in, he'll direct the walker over to where he thinks you are. When the observer thinks that the walker is right on top of you, he'll say that there's a sniper at the walker's feet. If there is, then he gets you and you only receive something like six points for the stalk.

If the walker is where the observer has moved him and the sniper isn't at his feet, then the observer will have the sniper take another shot. If you've done everything right, the observer won't see any sign of your shot, no blast or flash, no movement of the vegetation. If he still can't ID you after the second shot, then you receive a perfect score for the stalk. If he sees the blast, and you really can't do a lot about that since you're so close to his position, then you're caught again. If your veil or something sticks out from the cover, the muzzle blast will draw his attention to it. Even if you aren't detected, the walker will lean in to check the dope on your rifle. If that dope isn't correct, the windage and elevation on the scope isn't set right, you won't get as many points for the stalk.

A lot of people just wonder how you can do this. How can you expect a student sniper to move to within two hundred yards of an instructor and take a shot? And they know where you are to begin with. It's all about concealing yourself correctly and learning how to move. You learn how to set up your FFP so that you can back off from it a little. That way you don't have a problem with the muzzle blast from your weapon giving you away. During a stalk, I've actually taken my shoes off while lying there and stuck grass in them to make kind of an umbrella out of them. Then I backed up and shot through that to cover my muzzle.

Back when I went through training, they didn't have you extract from a stalk after you fired your shot. Things have changed a lot in training, and now they have the students slip out after they've fired their blanks. It's just as important for a sniper to be able to get away unseen as it is for him to be able to slip to within range of a target. In real-life missions, you don't worry as much about the extraction because you won't fire within two hundred yards of your target, more like six hundred yards or more. At that range, you just take your shot and then leave. Distance is your friend.

Shooting for a sniper is actually a luxury. You are there for your eyes and ears and gathering all of the information that you can. That's what a lot of people get hung up on. As a sniper, you're supposed to shoot. No, you are the eyes of the commander, you are observing. If you shoot, you have to pull out, get away from your hide. That's why it's a luxury to shoot when you're a sniper in the field.

The information you can gather, that's what the battalion commander and all of the people around him want from you. You're also supporting the units around you with your weapon. But if I'm on a hilltop, why would I want to fire my rifle when I can use my radio and call for fire? You have to think outside of the box as a sniper, but these kids who take up the job want to go out there and fire their weapon. What they have to remember is that they have a team there with them. The sniper has to take care of everyone plus get them back out of there. Stuff like that is what you have to take into consideration when you plan and conduct a sniper mission. That's why a lot of the sniper guys will prefer working with someone who's older, rather than a youngster who still thinks he has to shoot all of the time.

The major portion of the mission of a marine scout-sniper is centered on the scouting aspect of the designation. And most any sniper will say that, or at least they should.

My first deployment with the sniper platoon was right after the Gulf War had been going on. We went over there to conduct some training. I wasn't even a sniper yet. When you go to the sniper platoon, they call you a pig, a personally instructed gunman. When you graduate sniper school, then things change. Now you're called a hog, a hunter of gunmen. I don't know if they do that anymore because of how the military is now on that sort of name calling. But I was a pig for about three years.

During that first deployment, the training was pretty easy as far as I was concerned. We didn't do anything all that extravagant on that float because it was right after the war and things had wound down quite a bit. It was okay; a lot of traveling around followed, going to Okinawa and training there, setting up missions where we did observations. People would come in, and we would watch what they did, who they were, what they had, and report back on what we saw. For that training, we were the R and S, reconnaissance and surveillance, team. Not the most exciting kind of training to do. As a matter of fact, that deployment was a little boring.

Probably the best time I deployed with all of the guys was when we went to Somalia. We arrived only seven days after the Blackhawk down incident had taken place, and things were still stirred up. Command pulled eight of us off of the boat, and we basically worked for JSOC (Joint Special Operations Command) as a detachment of the Marine Corps assigned to the Special Forces unit out there. We had our own air-conditioned tent and everything was set up great. The support we received was great, too. This was nothing like what we were used to in the Marine Corps.

For our part of the mission, we set up observation posts. One OP was on the top of the U.S. Embassy, while another one, code named K7, was on the outside of the embassy. K7 was seven stories up in the air, with the Pakistani

UN contingent. We had a good secure perimeter with the UN guys nearby. We were not going to go out into Somalia and do an op. It was plain to see that we were two white guys, and all we would have accomplished was to get killed. Instead, every sniper had a secured position along the perimeter occupied by UN forces. If something did happen, we could get people inside of the perimeter and fight off any attackers.

One of the things that made that a really good time for me was the experience it gave me. You learned a lot about the stuff that had been studied so hard back in sniper school. The only problem was that you found out that some of what you had learned in school didn't work that way in real life.

Observing was one of these things. Using binoculars didn't work, at least for us in Somalia. The targets came up so fast that you didn't have time to switch over to a rifle. You had to do your observations from behind your weapon, set and prepared to fire. That was something that we learned from experience. Normally, using binoculars, binos as we called them, is much easier for general observation because they have a much wider field of view. They are only about eight-power magnification, not that much less than the ten-power Unertl scopes we had on our M40A1 rifles.

The targets we had moving through our field of view were out and visible instead of behind cover or concealment on those city streets for only a very short time. Anyone with a crew served weapon, we engaged; we didn't have to ask permission, we just fired. Those Somalis knew what they were doing as far as moving around their city went. They would cross at intersections where they were only exposed in the open for about two or three seconds. I think the whole time that we were there, we only engaged about eight or nine people between all eight of us.

It was a very interesting time at the very least. As a sniper, you spend your whole time training, training, training. Then, all of a sudden in 1993 we were using our skills in real-life situations. That's what a lot of these new guys are doing now in Iraq and Afghanistan, putting their training to work. You don't want to spend your whole time in the military doing nothing but training. If you are a warrior, you want to go to battle and put your skills to the ultimate test.

I had my test in Somalia. The first shot that was taken with our platoon, I was with my partner Corporal Schmitt. We were on top of the U.S. Embassy and he was cleaning his .50-caliber. You had to clean your weapon every day over there because the high humidity caused rust. While he was working, I was observing another area. We had 350 yards of open field in front of us and then a brick wall. On the other side of the wall was a hospital. Then we had a road off to our left and K7 was to our right, where another team was observing.

Schmitt told me that he was going to look through his optics since he had just cleaned his lenses. All of a sudden, I just heard a loud "bang." He had just shot without telling me anything or giving any warning. Later Schmitt finally told me that he had seen a guy with an RPG and didn't have time to give me a warning before he pulled the trigger on his Barrett .50-caliber. Moving my weapon over, I indexed on where Schmitt had taken his shot and saw a guy just lying there with the RPG next to him. There were a bunch of people around the body. Another Somali just came up and grabbed the RPG and ran with it. He was too fast for either of us to engage him. Everything had just happened so fast that we barely had time to react. I had thought we were getting shot at when that big Barrett went off. The muzzle blast from that huge weapon came back and just smacked me; I didn't know what the hell was going on.

That was the first incident where Corporal Schmitt took a shot. Later on, about three or four weeks after that first incident, we kept getting intelligence on the situation in Somalia from the Fifth Special Forces guys. The story was simple enough: There were guys walking around with the black body armor they had taken off the Delta guys killed in the Blackhawk down incident. If we saw the guys wearing the body armor, they were targets of opportunity. We didn't have to ask permission to engage, or wait to be fired on. Bluntly, we should just shoot the SOBs.

Roger that, we all thought, and none of us had a problem with it.

The Somalis had also taken weapons from the Ranger, Delta, and Blackhawk guys. There were a couple of M60 machine guns taken, a sniper rifle, and some other stuff, from the bodies and the downed birds. The intel people wanted us to watch for anyone who might be running around with that gear.

A lot of the OP works that we did involved watching the locals go about what looked like their daily business. They didn't realize how well we could see them sometimes, and we watched them hiding weapons in the grain that was coming into the city. Weapons were the big deal in Somalia; they were all over the place. When the 5th Marines first went into Somalia and Mogadishu, they secured parts of the city and took away what weapons they could find. When the UN came in, everyone was saying that they couldn't protect themselves from the warlord Aideed, so they handed weapons back out to the locals; only they identified the weapons by putting tags on them.

It doesn't take a genius to figure out that you could put a tag on any weapon that you wanted to, and it would look like it was okay. So the Somalis were hiding guns everywhere. Some of these places were the kinds of places you see on TV with the Save the Children Foundation. I don't know if the militia members were telling the locals that they had better help hide the weapons or they would kill them or what. But we saw a lot of guns moving about.

All of the intel that the army says we supplied them with, 70 percent of that information came directly from the sniper teams. It was the way we could observe what was going on that made us so valuable to the whole effort. With our information, covert people could go in to talk to the people and learn what was happening for the civilians in the area. That was a job for the CIA and those kinds of folks as far as we were concerned.

One day while we were talking with the team at K7, we watched a car come up. It was a little white Toyota Corolla. The driver got out and had an M60 machine gun with him. This was one of the weapons we had been specifically told to look out for. As I watched, I could see him pass it to somebody, but there was a wall in the way that blocked my field of view. Calling over to Schmitt, I pointed out what was going on. Before we could really do anything, the guy passed the weapon back to the driver and he took off. Now, I was wondering just what in the heck they were doing down there.

There was nothing to do but wait and see if the car came back. We wrote the incident down in our log books and reported what we had seen up the chain of command. After we had returned to viewing the area, that car came back around. Again, the guy in the car handled the M60, but we couldn't get a clean shot at him. Once more, the Toyota drove off and we kept watch. Nearly two hours later, the car came back; only this time, the guy carrying the M60 moved into plain sight.

The Somali took the M60 and walked clear around the car, from the driver's side over to the passenger side. This was enough, and I told Schmitt that I was taking the shot. There was another guy with an AK nearby, acting as a bodyguard by the looks of things. Schmitt said that he would take the AK holder out. Lining up my sights, I could see the Somali with the M60 clearly in the crosshairs of my scope. I squeezed the trigger and felt the rifle buck in recoil as the shot roared out. Next to me, another shot rang out a few seconds later as Schmitt pulled the trigger on his guy. Down by the Toyota, the Somali with the AK had started pointing up toward our position as the bullet from Schmitt's rifle reached him.

There wasn't time to do anything but react to the situation as it was unfolding. I jacked another round into the chamber of my rifle as I scanned the area. There was a guy in the backseat of the Toyota starting to point a weapon up toward us. I brought my rifle to bear and shot him through the back of the window. I knew I had hit the guy in the Toyota when other Somalis came up and dragged him from the vehicle. The bright red blood stood out clearly against the dust in the street.

It wasn't thirty minutes later that there were people demonstrating down the street from us. We had to call a react force to come and get us out, and that was about the end of the incident, which was pretty good for us. Right after I had shot for the second time, Schmitt and I were on our guns and I was having a hard time. It was maybe two minutes after everything had happened when I turned to my partner and told him that I couldn't even shoot anymore. It turned out that he didn't think he could shoot either.

When I tried to get on my optics, sight through the scope on my rifle, my eyes wouldn't focus or come to bear. We were both so jacked up on adrenaline that we would have been easy meat for the Somalis right then. I've hunted my whole life, and this was a lot different than any hunting I've ever been through.

For me, it was a neat experience. After I got back from another deployment after Somalia, I went to the Special Operation Training Group (SOTG) as the head urban sniper instructor. But that incident on the roof in Somalia stuck with me. You hear all of these guys talk about how they can shoot people and get right back on the gun. Well, I probably could have shot if I'd had to, but it would have been very difficult. Both Schmitt and I were so amped at that time—I'd never shot a person, and it was a very different situation—that we had a very hard time settling back down. I didn't feel bad about the shot or anything like that. That Somali had been one of the militia members we had been specifically told to keep an eye out for. There was no mistaking the weapon he had been carrying. But shooting a man is way different than hunting an animal.

Politics didn't matter to us on that deployment; we were supporting other soldiers. And we had been told to shoot anyone with an M60 machine gun, which was a pretty damned clear order. There was only one place that weapon could have come from. And we had been asked what we would do if we did spot someone with part of the gear from the Blackhawk or the D-boys. What if it had been an eight-year-old carrying the machine gun? That was a specific question the intel guy asked us. I said that I would probably try to fire at the machine gun to disable it, rather than fire at the boy.

The problem was that every time we got an engagement on an armed target in Somalia, the rifle would drop, but someone else would pick it up. There was a question among ourselves if we were doing any good, and the answer we decided on was yes, we were. If nothing else, they weren't bringing those machine guns around when we were shooting. And they weren't shooting at the embassy; they weren't launching mortar rounds at it. The Somali militiamen knew that we were up there, and that if we saw them setting something up, they weren't going to be around much longer.

We had Specter gunships on call at night, all the way through to the morning. There were sixty-millimeter mortars set up on a position to fire illumination rounds if we called for them. Back then the night vision we had available to us just wasn't as capable as what's being used now. We could only see to aim our weapons accurately out to about 150 or 200 yards with what we had. Now there's night vision available that lets a sniper see out to six or seven hundred yards or more and see great. There are lasers now, IR illuminator sights that allow you to see out into the dark. They're actually using IR sights now that let you see the boat tail of the bullet pass through the air every time one is fired at night. The technology has come a long way and is still advancing.

I didn't go to the Quantico sniper school when I became an instructor. Instead, I was sent to the Special Operations Training Group (SOTG). Every marine expeditionary force (MEF) has a train-up before they go on a MEUSOC (marine expeditionary unit, Special Operations capable) deployment. Force Recon has snipers, and they all go through training that's put on by the SOTG. We put them through training and then the units go through exercises to practice that training. They are considered SOC qualified after they have done all of the exercises successfully.

I went to SOTG in March of 1995 and stayed there until I finally left the marines in 2000. The biggest key task at SOTG sniper training was making surgical shots. It's of great use during hostage rescue operations. If there's a hostage involved in the scenario, then you have to take a head shot on the hostage taker. That's what we trained for. The students had to shoot from one to three hundred yards and hold their accuracy to one minute of angle—that's a maximum of three inches at three hundred yards. Probably the biggest thing that I tried to teach my students was that they wouldn't do that kind of shooting very much. So if they were going to take a shot, fire at the center of mass, a chest shot, whenever possible.

What most people don't realize is that shooting at paper targets is easy. But when you are shooting at a person, he is normally very animated. A person is almost constantly moving, both his body and especially his head. I had people miss shots on paper at one hundred yards just because they were firing on a count; they had only a very short window in which to finally pull the trigger. That caused them to jerk the trigger because they would be kind of guessing on when the instructor was going to call out the certain number that they had to fire on.

Certain other things happen in an urban area that the snipers had to get used to. The marines don't use bipods for shooting, so the students had to make their own support for their weapons when they established a hide. Support is key in accurate shooting. The better your support is, the better a shot that you are going to be. I've seen bipods fall off right when the student was ready to take the shot. SIMRAD night-vision devices have even come off weapons right after the student took the shot. All kinds of things can happen, and I've seen a lot of them.

Judging by the gear issued it seems as if the military doesn't support the sniper community as well as they should. Here, you have the most cost-effective weapon system in the battle area, and the marines are still operating with old Unertl scopes. There's a new rifle being issued, but instead of going forward in time, the committee went backward. It weighs close to nineteen pounds all-up; that's the equivalent of damned near a machine gun. Yes, it's accurate, but the caliber is still .308 (7.62-millimeter), so it's only going to shoot so far and still be effective. Why not go to a bigger caliber like so many other countries' snipers are

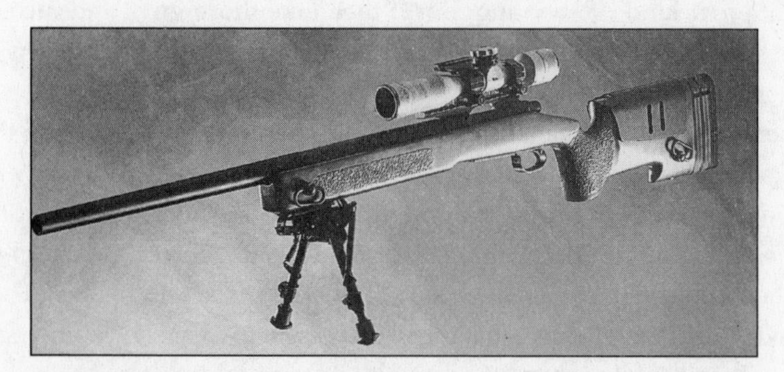

The M40A3 sniper rifle as presently fielded by the Marine Corps. The development of this precision bolt-action weapon was begun in 1996 and the final design adopted in late 2000. Full production of the M40A3 was initiated in February 2001. Further minor changes have been included in the weapon as experience has been gained with it in the field in Iraq and Afghanistan. The basic action of the rifle is the Remington short-action Model 700 receiver mated to a twenty-four-inch Schneider match-grade stainless steel #7 barrel with five lands and grooves with a one-in-twelve-inch twist. The barreled action is epoxy-bedded into a McMillan A4 stock, adjustable for comb height and length of pull. The telescopic sight is upgraded and refurbished by U.S. Optics from a Marine Corps Unertl ten-power sight, with a mildot reticle and bullet-drop compensator optimized for use with the 7.62-millimeter 173-grain M118 Special Ball ammunition. ***USMC***

doing? A .338 would be a longer-range projectile, and an accurate semiautomatic weapon would aid in the firepower that could be needed to break contact with the enemy. For a super-long-range round, the .50-caliber weapon is still the best thing available right now.

No one I know still goes with the manual operation, bolt-action rifle for the sniper anymore. Modern design should be able to come up with an accurate semiautomatic weapon that is the equivalent of a bolt-action gun. The problem is in part the fact that politics are involved in the adoption of, and bids put in for, a new weapon. When they finally introduced the new rifle, while I was still in the service, I laughed. They had been working on that gun for years, and it had nothing new over those from fifteen years ago. They had a new base and rings to hold the scope and a different stock, but it was old technology at best. With today's technology, we should be able to have a ten-to-twelve-pound sniper rifle with top-of-the-line accuracy. In the year 2000, why did we go to a nineteen-pound gun?

One of the problems that the marines have had with new rifles is that not all of the sniper people get together at any one time. If they had an annual, or biannual, meeting every year, they could talk and compare ideas. Snipers have a difficulty in working with match shooters, competitive marksmen, because it's a different kind of shooting. But if they take those guys and incorporate them at the schoolhouse, sit them down with the snipers, then something would get accomplished.

There are competitive marksmen in this country who are among the best shooters in the world. They know how to call wind way better than any marine sniper does. When you just go through the sniper school, you learn how to read the wind, call its effects on the ballistics of your shot, but you don't get to go to the range and train as much as a competitive marksman does. That way they know instinctively just what the wind will do to their shot, something that comes only with experience.

In SEAL sniper school, they bring in what you could call professional wind callers to help train the Team snipers. They brought in a woman a few years back; she was something like seventeen years old and had won the Wimbledon Cup. She taught these SEALs how to read the wind. They bring in match shooters to teach the operators to call the wind, which to me is one way to make the school much more effective.

To my mind, calling the wind is the hardest skill to teach a shooter. I can teach anyone how to shoot in a fairly short time. Although to get really good at it, you have to practice and learn all of the tricks. Ranging is also a hard skill to

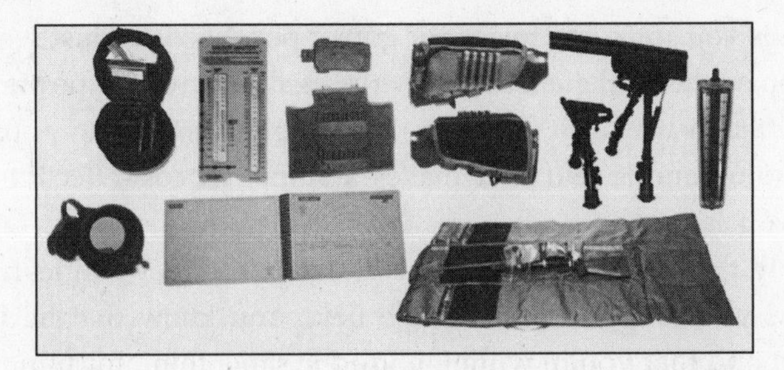

The M24 sniper accessory kit assembled by the Army Project Manager, Soldier Weapons Office. The kit is of off-the-shelf items that army snipers had been purchasing for themselves, or had had sent to them in the field by sniper supporters back home. Across the top from left to right is the Otis compact cleaning kit, the mildot Master, a slide-rule type calculator for quickly determining range and other data, an ammunition pouch that retains twenty rounds of ammunition in protective loops, a cheek-piece stock pack that gives the shooter a nonslip surface for his face while holding spare ammunition and additional small supplies on the opposite side, an improved adjustable Harris bipod, and a small, light plastic wind meter.

Across the bottom of the picture from left to right is a polarized light filter that can be slipped over the lens of a telescopic sight, an improved sniper's logbook, and a drag bag that can be unfolded to give a protected, padded rest for a sniper lying prone. *U.S. Army*

learn, but with all of the lasers and technology they have now, knowing the exact range to a target is not any harder than pushing a button. Back then, when we didn't have a lot of that stuff, you had to learn the mathematical formulas to use the mildot reticules on the rifle scopes and binos to determine range to target.

The mildots are still a great tool to use; the batteries can't run down on them. But it does come to a point where you have to say screw all the math and just know where your gun is shooting in all conditions. You know your holdovers at a long range by firing them on a range. The guys back in Vietnam knew at what range they had to aim at the top of a guy's head to hit him in the chest. So that's what it comes down to, just shooting your weapon often enough to learn.

In our sniper platoon, we might have shot once every month, when you should shoot once or twice a week. Sometimes, it's just difficult to get the ammo that you need. One time when I was in, we had all of four hundred rounds for a three-month period. I can shoot four hundred rounds myself. Support just wasn't there. And that comes down to the table of organization, how many rounds you are going to get per gun or per shooter. It just seems like the best weapon on the battlefield in both services, the army and the Marine Corps, isn't supported as well as it should be. That's pretty sad; it only costs 46 cents for one of our rounds and that makes a sniper as cost-effective a weapon system as you can get.

Probably the best thing about being a sniper is that you learn that your partner and you are everything in the field. You know that he's willing to die for you, and that you're willing to do the same thing for him. You can be married and all of this other stuff, but you'll never be as close to anyone as you will to your partner. When you finally transition to the civilian world, you realize that most of the people out there just aren't the same. It takes you a while to realize that it's a different world, that it isn't the military. It's just the way people are.

That's probably the biggest thing I learned in the military—the value of teamwork, and how much to depend on that person to the right or left of you. You and your partner are everything. Now it's four-man teams, and I don't really agree with it, but with the radios, batteries, and everything else that you are supposed to drag along with you now, the larger team makes sense. And if you get caught by the enemy, that extra firepower is a nice thing to have on your side. But that's the point of being a sniper—not to get caught. With two people, it's a tight pair; now with four men, I might as well be in a recon team. With a two-man team, as long as you're doing everything by the book and remaining stealthy, you shouldn't get caught. If you do get caught, then you had better have air support on call and get the hell out of there. It just seems that two men are easier to hide.

I've never really met many of the snipers who have come before me, the Vietnam and World War II guys. I'm not one of those people who go up on the sniper Web sites and register. I'm not into all of that. I just did my job in the Corps. That's not to say that all of that kind of stuff isn't good; that just isn't me. So I've never met a lot of older snipers, but I have a lot of respect for them. The stuff they did back then was a lot harder than what we're doing now, because of the technology if nothing else.

I can't even imagine what it was like to be a sniper in World War II. Now, with all of the smart bombs and such, we can put support right up close to a sniper team in trouble. Back then, you were really on your own. We have lost sniper teams in today's conflicts, but I think it was because the training didn't cover what they got themselves into. Or if they did train properly, they were just caught by surprise. That's just the nature of war. You can only train so

Wearing his ghillie suit on an open lawn with his rifle uncovered, this air force sniper practices tactics during a run-through exercise. ***U.S. Air Force***

much until it becomes boring and people start not to take it seriously enough. That's when it becomes dangerous.

As a past military man and sniper, I do have problems with the media today and their naming every criminal with a rifle a sniper. It's all about perception, I guess, one of those things where the newspeople need to be educated. That's a place where I don't think law enforcement or the military has done to the best that they could. The same problem exists in the military as far as people knowing who and what a sniper really is. When I would go to battalion, I would always try to brief the officers in regards to just what a sniper was and how you would deploy him. A lot of the officers then had no idea of how to use a sniper in the field. At best, they thought he was some kind of point man who would go out and lead their unit through an area.

Snipers aren't point men; they have big rifles and know how to use them. You put them out on gaps or overlooking avenues of approach. You let them leapfrog ahead of an advancing unit and conduct an overwatch. These were some of the capabilities that we had to brief officers on. A sniper can call for fire; he can act as a forward observer for artillery or bring close air support in on a target. That's all material that we are taught in sniper school.

The press needs to be educated. When they call someone a sniper, they need to be told that he isn't a sniper; he's just a damned gunman.

ial group to operate with. They worked hard at their job but knew how to h at things.

About sniping there are always going to be differences of opinion from rent services. The British thought that we had one of the strongest sniping rams going on. In my opinion, they have a great one. However, when we there, the operational differences did show. For example, the British ers would wrap their rifle barrels with ghillie materials for camouflage. We taught that doing such a thing affects the harmonics of the rifle, and it will shoot as well as it will with a plain, unadorned barrel. There were always t differences in how the snipers operated in the field, things like how you ld conceal yourself, how operations would be conducted. For the most we tried not to get into too much of a debate, because we were all trying ork together at the time and get the job completed.

There didn't seem to be any negative feelings about the use of, or being, a er from the Dutch or the British troops. The British especially had their heavily invested sniper programs and had incorporated such men into military for years. The Dutch, who I was impressed with at the time, had sights on their weapons, optical sights that really increased your ability ck up a target and hit it quickly with a service rifle. The U.S. forces did get nd to issuing such sight to all of the troops—they're very common over in now—but at that time, in the Marine Corps we were still working with the sights on our regular M16 rifles. Every single Dutch Royal Marine had an and telescopic sight and was able to view a lot farther out there than we l with our standard weapons. It was kind of disgruntling that we could hat the Dutch were outfitted with a lot more gear than we were at the time. ot going to knock our government policy in outfitting and supporting the ers in Bosnia as far as gear went, but nowadays, I'm looking at what the have and I'm thankful that things did step up as far as the general issue gh-speed gear goes.

he main advantage we gave as snipers operating over there was to provide ity, especially to the bounding or assault elements on the force. Most of me while we were patrolling, if we made contact, then it was "advance to ct," which is an offensive operation intended to make contact or reestab- ontact with an enemy force. That's as compared to running defensive ons or conducting regular patrols through an area such as is going on now q. Since we were operating in a largely mountainous area, we held over- positions. Both the snipers and machine gunners would be dropped off osition overlooking the patrol route. The machine guns could lay down a

MILITARY SHOOTER (MARINE)

Jason Bombaci

While serving in the Marine Corps during the latter half of the 1990s, I was a reconnaissance sniper with the 2nd Recon Battalion. After going through the Amphibious Reconnaissance School, I went over to the Camp Lejeune Scout-Sniper Course and graduated with the additional MOS of 8541. As well, I went through the urban sniper course twice. That's a three-week course which is part of the workup for a normal Marine Corps six-month MEU (marine expeditionary unit) overseas deployment.

This was a long way for a suburban kid who grew up in Long Island, New York. Growing up, I didn't do anything really in the way of hunting. Out in my area, the most you were going to do growing up was stalking with a pellet gun and things like that. My mother was a firearms instructor for the probation department, so I had a lot of pistol time from about age fourteen to eighteen, but not much time firing a rifle or anything along those lines.

Like a lot of young men, I didn't want to move right on into college after high school. Instead, I wanted to do something with my life, not just start working in a pizza parlor or a deli. I made a choice at one point, not a spur-of-the-moment thing—I thought it through, and decided that I wanted to join the service. So I ended up signing up for my four with the Marine Corps.

The marine scout-sniper course is some pretty hard-core training, even among a bunch of hard-charging young marines. Personally, I felt that the physical standards of the reconnaissance aspect of my training were a little more demanding than the scout-sniper course required. In scout-sniper training, the instructors focused on working with a smarter marine. You're dealing

His face heavily covered in camouflage paint, this sniper silently steals ahead during a stalk. The hood of his ghillie suit is pushed back so that his face is more visible for the photograph. Local foliage woven into the hood of his ghillie gives the dull cloth a hint of brighter green. *USSOCOM*

more with the long-range precision interdiction of a target as a scout-sniper. As an urban sniper operating inside a built-up area, you have to be fast with the ballistics you run in your head, as well as quick in the acquisition of your target. In an urban environment, a target may be exposed for only a short time, what with all of the cover and concealment offered in the average city or town. Out on the battlefield, you may have a target that remains exposed for a much longer time, just because of the local environment.

For myself, I did sort of a different mission than the normal scout-sniper does. The primary personnel who need to have 8541 qualifications are the scout-snipers who operate at the battalion level. For us, there were the 1st, 2nd, and 3rd Reconnaissance Battalions. Each one of them would have three teams with scout-snipers within it, each made up of a sniper and his observer. Those three teams, depending on their mission, determined during the predeployment workup, would have sniper-qualified personnel in order to give the recon asset another ability to employ in the field. That way they can pick up on targets of opportunity while the recon teams are out there operating behind enemy lines.

As snipers, we're the eyes and ears of the commander.
more than we shoot; at least that was the case when I spent
duty. For the guys who are seeing action now, even though
provide eyes and ears, target interdiction is a lot more active
past, just because we are at war.

While I was in the service, I didn't see any active comba
As far as sniper-related missions, I had a few of those where
the field while we were over in Bosnia. Our mission then wa
lages and run patrols and such in the area. We patrolled wit
at the time, and it helped give us the ability to provide ove
areas in our sectors.

A sniper mission in that kind of area was very different fr
prepared for. Bosnia wasn't what we had trained up to expect
prior to a deployment, you are doing all kinds of patrols, in our
Camp Lejeune, and you also conduct patrols as part of exercis
the United States. But when we were over in Bosnia, because
threat, it was hard to drop off a reconnaissance or other team a
to patrol from Point A to Point B. There was this little probl
Betty antipersonnel mines and other booby traps all over
encounter them while on the move.

So the commanders tried to keep us on the hardtop r
missions. The group I was with was attached to the Dutch Roy
time. With our mixed unit, we operated in the British sector wh
Bosnia.

There was a noticeable difference in working with a for
than within a U.S. marine group. The Dutch were a very squa
of guys. There was no difficulty in communications; they spoke
ent languages it made things easy for us. It was great interfa
The Dutch Royal Marines had a different mentality than we h
the United States. Probably the simplest way to say it is that
little more "cowboyish" in our operations, while the Dutch
professional. When they sign on, they'll do a normal two-year
if they sign on again, it's for life, a career.

That's a pretty strong commitment as compared to here
States. Here, there always seems to be a question about wheth
going to sign on for another tour, whether you're going to stay
what. For the Dutch, they were all like a brotherhood, just as
Corps, but they were always looking at the long term. The Britis

suppressive fire if necessary while the snipers could strategically interdict threats as the assault force moved forward. What snipers do is bring precision and range to the party. In either an assault or withdrawal, from our overwatch position we could help protect our units on the ground by taking on targets that were either out of their range, or that they couldn't see from their positions.

In spite of the hostile feelings throughout Bosnia, I was never put in a position to have to pull the trigger on a live target. I know that personally I absolutely missed that challenge to my skills. Since then, I've talked to other snipers who have been in the same situation, and they, too, said that they wished they had gotten that active mission, that one where you have to go out and see if you are as good as the other guy. After being out and talking to others who have served in the sniper community, you realize the fact that, as in my situation, you go and train hard for six years, constantly going out on missions, and not being able to use your skills. The only real comparison that I can give to it is from the point of view of someone who trained for his whole life to compete in the Olympics, only to come in fourth place in the trials. You have to watch everybody else who's in the game and just wonder what would have happened if you had gotten that chance to compete.

Afterward, you can become an instructor, as I have, and pass on the skills you worked hard to learn. But there will come a time where people will ask you if you saw anything. And you have to answer no. Not the most fun thing you can do, but I've done well enough.

Currently, I am the training coordinator and deputy of operations for SPA SIMRAD, the U.S. manufacturer and distributor of SIMRAD night weapons sights. In the past, SIMRAD had changed hands a number of times. In 2000, the company I work for picked up the rights to manufacture the SIMRAD night vision systems. The only training that was actually supported with the product then was the operator's manual, and it was basically up to the schoolhouse instructors for the services to come up with their own training plan for the employment of the system. They answered their own questions about how things worked and then went on to pass their information on to the students.

When I was brought aboard as the training coordinator for the company, I was tasked with developing a full training outline that could be issued as a PowerPoint presentation. I went to all of the bases and spoke to the instructors, getting feedback from them. Then I took that feedback and incorporated it into our training course, as well as improving the product itself to include exactly what the operators wanted to get out of it. We added options such as Picatinny rails, what are known as 1913 specification mounting systems, quicker acquisition for the system, and basically supporting the product. In addition, I was able to use my own skill to test and evaluate new products that we're trying to put out on the market. We are trying to do things such as increasing the resolution of the sights to increase the ranges that snipers will be able to successfully engage targets at night.

Our primary product is a family of night vision sights that can be added to the normal optical system of a user's weapon. The SIMRAD sight mounts on top of the telescopic sight already on a weapon. By amplifying the available light in a situation and magnifying it thousands of times, the SIMRAD sight projects that now visible image down through the objective lens of the telescopic sight. The user can see his target through his normal sights and be able to efficiently engage it. No rezeroing has to be done to the weapon, and the SIMRAD sight can be easily mounted or dismounted, even switched between weapons if necessary. It has proven to be a very popular system with some of our end users who are out there right now using them in the field.

Night vision brings the obvious ability for snipers to conduct their missions after dark. It also allows them to positively identify their target at night. The worst-case scenario would be if someone was operating alongside friendly troops at night and didn't know the troops were there. They could see just a muzzle flash and think it was rounds coming in toward them. Men would return fire, engaging what they thought was the enemy, without positively identifying the weapon or the actions of the personnel they were about to interdict. This is how some fratricide incidents, friendly fire, have taken place in combat. The availability and easy use of a sight such as ours helps to prevent such incidents from ever happening.

Law enforcement has an obvious use for night vision equipment such as we make. Currently, we have over 650 law enforcement agency customers throughout the United States. Those customers include NYPD, LAPD, Miami-Dade, the Houston Police Department, and some of the bigger government agencies. Even though the distances that police snipers are expected to engage at are shorter than the normal range of a military sniper, typically less than one hundred yards, 99.8 percent of a law enforcement sniper's mission is observing and reporting, gathering information and turning it over to the SWAT team commander or to the crisis negotiator. That gives the people who need it more intelligence on the situation, so that the attempt can be made to deescalate it. That would mean the sniper doesn't have to take the shot. Any SWAT team commander or police chief would rather his sniper not take a shot. Law enforcement is there to save lives and not to take them.

One of the key differences between the law enforcement and military sniper concerns the operational ranges each works at, the distances at which a target can be acquired and engaged. Most law enforcement snipers, I believe, shoot at an average range of seventy-five yards. That is a much shorter range than military snipers would work at. Working at that range in a combat zone could quickly cause a military sniper to be detected and engaged by the enemy, no matter how good his concealment was. Police snipers do not usually have to worry about their targets being able to call in artillery fire on their hides.

Now that the war has been going on, and will continue until the conflict is over, sniping has played a huge key role in the fighting. Basically, the modern sniper is changing the face of the battlefield. He is stopping insurgents with precision fire, as compared to dropping in 155 shells all over the place. That helps protect the civilian populace of a war zone. And one thing we have always wanted to do is reduce the number of civilian casualties in a conflict. That is the primary goal of our military in Iraq, to engage insurgents and not bring fire down on anyone else. The ability, especially with protests taking place, to have surgical shooting available to a local commander can be key to the safe control of the situation.

What has to happen is that, for one thing, support for the military as far as funding and for another, support for the sniper community has to stay there. After World War I and World War II, the troops and officers in general always looked at the snipers as some kind of cowards since they hid in the shadows and shot people from distances. This was in direct comparison to the obviously brave men who stormed the beaches of Normandy or Tarawa, Sicily or Iwo Jima. The general military community back then had the wrong idea about sniping; they looked on it as a vile necessity of war, not a valuable skill that needed to be kept up within the military forces. The sniper community as a whole has stuck together, more now than ever before, and we're making sure that it is very well known that the sniping programs in the military can't go away again, not like they did after World War I and World War II.

One of the key issues was always funding, and that translated into proper gear. Training and skills are kept up by the teachers. We've always taken it up to the next level; the schoolhouse instructors stay out until well past midnight on a lot of the courses, doing night fires and training, and training, and training. In the past, we always had overseas countries that had a lot better equipment and support for their sniping communities than we had for ours in this country. That's changed a lot, and the United States sniping community is

The end result of the M88 PIP (product improvement program). This is the Naval Special Warfare (SEAL) .50-caliber Special Applications Sniper Rifle (SASR). The weapon is a repeating bolt-action rifle with an attached five-round magazine, seen extending out from the bottom of the receiver in front of the trigger guard. The spiral lines cut in the bolt help clear the weapon of dirt and debris when the bolt is moved. At the wrist of the stock is a joint where the stock can be removed to reduce the overall length of the weapon.

In spite of the M88 PIP being a repeating weapon, most SEAL snipers have reported that they only load the big weapon with a single round. The huge muzzle blast of the .50 cartridge, even when fired through the big muzzle break on the front of the barrel, kicks up a lot of material, enough to give away the sniper's position when fired from a hide. Used inside buildings and other cover, the weapon can be fired repeatedly without giving away the sniper's position except by the sound of firing. ***Kevin Dockery***

leading the world in a number of areas. That's not to say that the foreign units can't work well with us and bring their own skills to the game.

The Canadian military sent snipers who were able to blend right into the operations we were conducting in Afghanistan. They were able to operate on their own as separate sniper elements without any major restrictions put on them. Those Canadians pulled off some major accomplishments in the fighting over there. During Operation Anaconda, a Canadian sniper using a .50-caliber rifle took down an al-Qaeda fighter at a reported 2,430 meters.

These are the kinds of skills that we've developed within the sniping community in this country. They are hard-won and show the level of professionalism and commitment that it takes to be a sniper today. It's a community that I'm proud to be a member of.

35

MILITARY SHOOTER (ARMY)

Staff Sergeant O———

Currently, I'm a full-time member of the Army National Guard. I grew up in south central Pennsylvania. My whole life it seems my mother wanted me to be a priest, so I spent a lot of my childhood in Catholic school. I had other plans for my life, and immediately after I graduated high school, I joined the army.

Growing up in the city, I had never fired a weapon before going into the army. Enlisting as an infantryman, my first duty station was at Fort Drum, New York, with the 10th Mountain Division. For my first two years in the infantry, I was a SAW (squad automatic weapon) gunner. Somebody told me that the battalion reconnaissance platoon was doing tryouts. I had seen snipers in the field working from the recon platoon, and it looked interesting to me. Having read a lot about military history, sniping had always piqued my interest. After trying out, I did make it into the platoon and worked my way up into the sniper section, and it kind of stuck with me; I loved it.

Working independently as a sniper had real appeal to me. I was the kind of person who could work with little or no supervision, and so I could do well in a sniper position. I didn't end up going to sniper school with the 10th Mountain Division. I had PCS'd (permanent change of station) to Berlin, Germany, for duty with the Berlin Brigade and was sent to sniper school from there. Established in 1987, the army sniper school I went to in 1993 was in Fort Benning, Georgia.

It was a different school then as compared to now. The course was only three weeks long back then; now it's five weeks. The sniper manual, FM 23–10, had just been written then, so they were focusing on your basic, traditional sniper skills. Of course stalking was a big part of that, and we were still in that mind-set back then. There were a lot of green ops, working in the woods and brush lands.

Propped up across the roof of his Humvee while in Iraq, SSGT O—— sights in on a distant target with his Barrett M107 .50-caliber sniper rifle. The round from this weapon can punch through walls of light vehicles in order to reach a target on the other side. ***Private Collection***

Back then, the light infantry concept was fairly new. You've got a division of light fighters now who basically carry on their backs everything they need to fight with. Counterinsurgency-type warfare is what they do, but not in the way we're doing it now, more along the way it was done in Vietnam. Tactics-wise, we were still fighting our past wars, and that's how the snipers were being employed within their parent units. Put the big green pile on your back and sneak in. Nobody thought about the problem of how the sniper team was going to defend itself if they made contact with the enemy.

The procedure then, if the sniper team was detected, was that the sniper broke contact with a bolt-gun and the observer broke contact with his M16A2 and M203 combo. This was because we had no way to carry the M24 on our backs with a select-fire weapon in our hands to fight with. You really had to sneak around, snoop-and-poop, and avoid being observed. Being compromised was the worst-case scenario, because you would be operating about ten kilometers in front of the battalion. There was a lot of stuff that was great in theory, but when you really looked at it under a magnifying glass and in practice, you ended up really hoping that you didn't have to do that, because you just weren't sure that the procedures were going to work.

No one had thought of an alternative way of allowing the sniper to fight and operate back then, certainly not the way we are now. In training then, urban stuff was touched on, probably all of a paragraph's worth of class lecture.

An air force security sniper team practices their skills during a training scenario at Spangdahlem Air Base in Germany. *U.S. Air Force*

Otherwise, it was green ops and marksmanship. One week was spent on the range, one week doing stalks, and the last week doing field training exercise (FTX) type stuff. Then you graduated and were sent back to your unit.

When I went through training, we were still using M21s and M24s mixed together. I know that my unit had both the semiautomatic M21s, ART-II scope-mounted National match versions of the M14 rifle, as well as M24s, the militarized bolt-action Remington 700 rifle, so some of what I learned in training from using both weapons moved right with me back to my unit, where I could employ it in the field.

The hardest part of sniper training for me was the stalks. It takes a big imagination to do well at a stalk. The guys that are most successful at stalking can detach themselves from their bodies, their personal environment, and picture how they would look to somebody at a distance from them. I think it's important to be able to do that, look at the big picture of your surroundings, the area that you are moving through, and how you fit into that landscape. So many people that I've seen fail the school for stalks just can't do that kind of detachment. They just can't get outside of their own minds. Instead, they just look around at their immediate surroundings and don't envision what they look like in that specific environment.

When traveling, I spend a lot of time when I'm a passenger in a car looking out windows at the countryside going by. And I envision myself in that area, thinking about how I would be more brown in some areas, more green in others, how I would make sure to use such-and-such a plant in my personal camouflage. That's the kind of mentality you've got to have in order to operate properly as a military sniper. If you're not going to do it full-time, make it a passion, then you're just not really going to be successful at it.

There are different variations of stalks during training, both in each school and among the services. The Marine Corps does their stalks a little bit differently than we do in the army. The typical sniper school stalk that I did while in training involved starting with your ghillie suit on. There's a start point for the exercise that may be a road, and the vegetation may vary. Some stalks are fairly easy because it's thick vegetation, easy to move through. Other stalks can be a little bit difficult because the vegetation is scarce. It's very easy to track somebody through high grass, which is easily trampled down by your passage. The difficulty of a stalk can also depend on the time of day. If it's early morning, you may leave a dew trail, what's called a snail trail, behind you from where you smooshed down the grass during your crawl and wiped the dew from the vegetation.

When you get to your start point for a stalk, you are usually given a certain amount of time, ten or fifteen minutes, to prepare your individual camouflage. This is more of a buddy thing rather than just an individual effort, kind of depending on the guy next to you to check you out. He'll tell you if you look like the ground, or look like your surroundings. Then you start.

The instructors have dropped you off and given you your orders. You have to prep yourselves, set up your individual camouflage, and make sure your ghillie suit is straight, and add the natural vegetation, and then begin stalking. You can use any number of the sniper movement techniques, from walking erect if the vegetation is thick and you need to make up time, to walking stooped over, crawling on your hands and knees, crawling by pulling yourself forward on your elbows, or using the sniper's low crawl. In the sniper's low crawl, everything is in the dirt and you're pushing yourself along with just your feet and fingertips, pushing and pulling yourself with an absolute minimum of movements.

You have a certain amount of time in which to make the stalk from the start line to what you have been prebriefed is your final firing position, or FFP. You can't go any farther than your FFP, and once you arrive there, you identify your target.

The final goal of the stalk, besides not being spotted, is to fire two blank rounds at the target and identify either a letter or number that is being held up by the target after the fact. If you have fired both rounds successfully, and have identified the number successfully, all without being caught by the target that you're firing the blank rounds at, then one of the instructors will check the dope on the rifle, the elevation and windage, to make sure that it is indeed what it should be for the situation you are in. And you are checked to make sure that you are in a stable firing position, that you could have made the shot accurately. Then you are a successful "go" at that stalk, and you receive your full twenty points.

Keep in mind that the entire time of the stalk, there are instructors moving as walkers out there among the students. Their job isn't to spot you, but to act as guides for the person (spotter) who is trying to identify just where you are. The way it works is that if the spotter thinks that he has seen a sniper, then he will yell "Freeze" at the top of his lungs. Each of the walkers will echo that command. Then you have to literally freeze in position immediately. If you are in the middle of climbing over a tree stump when the target yells freeze, then you have to stop in mid-motion and try and hold that position. You have to hold it well because one of the ways to get kicked off the course, or get a zero for the stalk, is to do what they call "pancaking out."

When you pancake out, you slowly flatten yourself out down into the ground. Get caught doing that and it's not good for you as a student.

When he yells "Freeze," the spotter will use the other instructors in the field, the walkers, to attempt to locate you. The spotter will call a walker on the radio and tell him to take, say, a left face and move ten paces forward to the big tree. At the big tree, once the walker gets to where the spotter thinks he sees a sniper, the spotter will say, "Okay, sniper at your feet."

If there is indeed a sniper at the walker's feet, then he's busted. If there isn't, then the walker says, "Negative, there is no sniper at my feet." And the order is called out to continue the stalk.

It's difficult because you are being watched by the men who taught you how to move, hide, and create your camouflage. I don't think that stalking is hit on during training as much as it used to be. Now, I believe that Fort Benning has added a whole section to the sniper training curriculum on urban sniping and urban warfare. That's sniping as is being done in Iraq now, shooting and moving among buildings, houses, streets, and structures. That's a completely different kind of sniping than I was originally taught.

One place where the army and marines differ in their sniper training is in the withdrawal after a successful shot has been made during a stalk. The Marine Corps required you to stalk out of the objective area after you'd taken your shot. While in the army, at least back in 1993, you got to the point where you successfully made your two shots and you were a go, you received your twenty points and were happy. The problem is the combat focus of the

A SEAL sniper during a public demonstration in the 1990s. By this time, the SEALs had had a mature sniper training program in place since the early 1980s. They had a number of weapons developed specifically for their use and applications. This sniper is laying the hood of his ghillie suit down over the action and sight of his rifle to protect them from flying dust. The heavily padded underside of his ghillie suit is visible. The thick, tough, material protects the wearer as he crawls along the ground during a stalk. *Kevin Dockery*

job means that there's a tactical withdrawal from that shooting position. Everybody who has been in the field knows that.

Anyone can get into a target area undetected because nobody knows that you are there. Once you have fired, you have made your presence known and it becomes a whole lot more difficult to find your way out. This is an aspect of the mission you have to learn through experience and later training.

Once the snipers get back to their home stations, that's when the training tempo picks up a little bit and you start rehearsing stuff like a tactical withdrawal. Your training then includes an opposing force, the "opfor," who's out looking for you. I would guess that the Fort Benning course was too limited back then. It was just a three-week course, and there was only so much they could cram into those twenty-one days. I'm the biggest critic as well as being the biggest fan of army snipers. The army, in general, has a long way to go. It needs to catch up with the Marine Corps mentality of what snipers are and what they should be.

In the army, the primary duty of a sniper is no different than that of any other sniper. His primary mission is to deliver precision rifle fire on key selected individuals. When a sniper graduates training, he is able to achieve 90 percent first-round hits on a target at six hundred meters. The secondary mission of the sniper is to gather battlefield intelligence. However, up until just recently, most army snipers were dual-hatted in that the sniper was a soldier in the battalion reconnaissance platoon ("Oh, by the way, we need a sniper for this operation and that's going to be you."), as opposed to an individual dedicated solely to the art of sniping.

That's two different mentalities about the employment of snipers. I can only speak about the light infantry side of the house, as that's all I've ever done; I've never been in a mechanized unit. There, you've got a light infantry reconnaissance guy whose only job is to stay hidden. If he fires his weapon, then something went bad on his operation. It's ingrained into the heads of recon men not to shoot—"don't shoot, don't shoot, don't shoot" is the mantra of recon. When the enemy is walking by, stay low, keep quiet, and don't shoot unless he sees you.

Then you are going to take that same individual who's oh, by the way, assigned as a sniper, and send him off on a sniper mission when it becomes time for such. Then you find that the two thoughts sometimes will conflict. When a sniper is on his mission, if it is purely a sniper mission, his position will be to reduce that target. If he is primarily a recon soldier, then his urge to hide and not draw any attention by firing will directly conflict with his sniper mission.

The amount of training needed to sustain the skills required to be a successful sniper is a whole hell of a lot more than what most light infantry battalions are willing to dedicate. Light infantry reconnaissance squads consist of five men. The minute you take two of them away to conduct sniper-specific training, that light infantry reconnaissance squad is mission-ineffective as a recon unit.

The army never caught on that maybe they should make a fourth squad in that reconnaissance platoon that's made up strictly of snipers. Some units did it on their own, but it never was official MTOE (modified table of organization and equipment), at least it wasn't in any light infantry units that I served in.

So you found some units that would just put the sniper training on the back burner. These kids would be lucky if they fired their M24 rifles once a year, let along quarterly, monthly, or what's required to maintain skill levels. And they would be lucky if they remembered anything that had been taught to them at sniper school, unless they did it on their own, conducted their own training and practice sessions.

Recently, with the reorganization of the army into the brigade combat teams, and the modularity system, all the snipers have been taken out of the reconnaissance platoons. Those snipers have been formed into a ten-man sniper section, which is what we've been requesting for a long time and is what was needed to properly employ and train these highly skilled people. The army went from two-man sniper teams to three-man teams. Now instead of just the sniper and observer, the teams have that third man for security, which is a good thing as well. That third man brings a little bit more firepower into the equation and can allow the sniper and observer to maintain their attention on the target while remaining relatively safe from unwanted attention.

What you have now is three three-man sniper teams in the section, along with a sniper employment officer, the section leader. The sniper employment officer's sole job while in garrison is to see to the proper training of those snipers. In the field, he's a liaison between the battalion staff and the snipers out in harm's way. Or he maintains the link between the logistics officers and the snipers in the field. His job is to stay in that TOC (tactical operations center) while he has snipers in the field and make sure that they are supported the way that they need to be. He insures that the battalion staff understands what their conditions are out in the field, and when they make decisions that may affect these guys, that they're the right decisions.

I did a lot of that kind of work when I was deployed to Iraq. Not only was I working as a sniper section leader, I was also the sniper employment officer for the battalion. It was a major help to both the battalion and the snipers in the

In the open desert, an army sniper team sets up for a shot. The sniper has his M24 rifle supported across his rucksack while his spotter scans the target area. The mechanical arrangement at the rear of the M24 sniper weapon system (SWS) stock is for adjusting the length of pull to better fit individual shooters. *U.S. Army*

field to have a guy like me advising command as to just what they had available to them from their sniper section. You can have a battle captain who sometimes is a first lieutenant and has no experience with snipers at all. He has no idea what they are capable of, what they are doing out there in the woods. If you don't have somebody standing there, looking over his shoulder and making sure that he doesn't make a stupid mistake, he can do things like forget that he has a sniper team operating in grid square such-and-such. This fact is really important to remember when you have an Apache flying over and he sees thermal images of several men lying along an MSR (military service road). When that pilot calls back and asks if there are any friendlies in that grid square, it's best for that sniper team that the lieutenant says yes. When he doesn't, that's when we have fratricide, death by friendly fire.

And the problem comes back to that lieutenant not being used to operating that way; he's not used to seeing that kind of thing, a single three-man sniper element out there operating apparently all alone.

The U.S. Army Sniper School just recently started a sniper employment officer course. I believe it's going to be a two-week course for the battalion staff, reconnaissance platoon leaders for the units that have them, and some infantry officers. This is something that is way overdue in the army. The Marine Corps has been doing it for years and years, while the army kept putting the idea on the back burner.

The average officer just doesn't have any real image of what a sniper is or what he is capable of. Sometimes, they don't really know anything more than what they see in movies. The employment of a sniper asset is just not something that's taught in the infantry officer basic course. Using a sniper is not something that they have had to deal with a lot. Up until just recently, the battalion snipers were a battalion asset. If the battalion commander or the battalion operations officer wasn't involved, then the snipers just were not being used.

When the sniper teams are taken and flexed down to support an infantry company, if you have a company commander who thinks outside of the box, is willing to give you the latitude to do whatever it is that you need to do, then the sniper is fine. But sometimes, you get assigned to a company, and immediately after that, you get told that you're going to be attached to, say, the first platoon. Now you're dealing with someone who could be a brand-new second lieutenant right out of the infantry officers basic course, and he's looking at you—and in the army, he's looking at Specialist E4 Smith and his PFC observer—and saying that he needs your team to do this, something that has nothing whatever to do with the sniper's mission or skills.

The specialist who knows better is going to tell him, "Sir, that's not what I do."

Usually, that conversation will be short, with the lieutenant telling the specialist that he needs him to do this thing, whatever it may be. That's the situation where the sniper employment officer comes in handy, because that specialist can now get on the radio and explain just what the story is. Snipers are limited-availability, high-value assets; you don't just stick them out on duty as warm bodies. The liaison part of the job starts working at that point when you contact that officer and tell him just what he has available to him.

I've been in that position so many times, having to tell a young officer what he has been given and what the capabilities are for his mission. It's frustrating; it really is, to just keep having to say the same things over and over. If you can educate these young officers, it makes things much better for them, their mission, and the snipers working for them.

I was very fortunate during my last deployment in that I had a battalion commander who was a big game hunter. He flew to Africa to hunt, was really into ballistics, and understood the missions that snipers were trying to perform. So he allowed me the opportunity prior to deployment to have a leader's professional development seminar within the battalion. It was an eight-hour brief on sniper employment for every E7 and above in the battalion. That was helpful. I had prepared handouts, a PowerPoint presentation, and pretty much gave them rules to follow. And they were pretty open-ended rules. To be frank, in my opinion, the only thing that limits the use of snipers is a lack of imagination on the part of their leadership. There's so much that a sniper team can do, and there are principles to their employment that you shouldn't break. Anything within those principles is fair game.

Among those principles is communications. A leader should never, ever, leave a sniper team without reliable communications to their support. If communications are lost with a team, and you don't have a good no-commo plan, then it's time to go out and get those snipers back. If you're not capable of providing them with comms whatsoever, then you are not to send them out there. They are not expendable. Some commanders are just lackadaisical about using such assets. They feel they can just figure out how to employ them later, once they have reached the operational area. I tell then no, that we need to figure out those problems now.

Selection is another principle, probably one of the weakest points in employing snipers. A lot of commanders don't understand the sniper selection process, but when you explain it to them, they do get it. The basic idea is that you're going to put two individuals out there by themselves to operate on their own in a hostile environment. Isn't it worthwhile to make sure that the right kinds of individuals have been chosen to carry out that kind of mission? You want these individuals to be able to work within the commander's intent unsupervised. Otherwise, you don't know if these guys are just going to go out there, get behind a tree, and take a nap, calling in every hour and saying that nothing is going on.

A commander wants to make sure that the guy chosen for the sniper position is the kind of person who is going to do the mission and give it what it takes. The army sniper manual gives you a list of characteristics and traits to look for. That's not a sure guarantee than the person chosen by a commander is going to make a good sniper, but it is a start. And I see too many commanders who just go by the marksmanship scores. If a guy is a good shot, then they think he would make an awesome sniper. There's the saying that "Every sniper is a marksman, but not every marksman is a sniper." I believe that rule 100 percent. I've seen guys who were great shots but just couldn't hack it in the sniper position.

The job is something I can speak about with direct experience since I did go into the field as a sniper in Iraq. My section was pretty understrength when we were deployed in Iraq, due to injuries or guys rotating home for leave, all the normal things that can happen in a war zone. So I found myself doing a whole lot of sniper missions while over there, just sharing the workload really.

Maybe it was subconscious, but when I went out in the field on an active sniper mission in Iraq, I didn't think about anything but the mission at hand. Once you stop and start dwelling on the fact that you are out there basically by yourself, just you and your observer, all of the bad thoughts can sneak in on your mind. What would happen if someone crawled up on the sniper hide? And all of the worst-case scenarios start going through your mind. That's just not good. When you do start thinking like that, you have to just pound it right out of your head; that's what I had to do.

Start thinking about something else rather than the "what if?" kind of thing. Is my dope (ballistic information) right for the kind of weapon/ammunition that I'm using? What's the terrain look like around the hide? You start focusing on the mission again and you're fine.

What does mess with your head is when you're back at the FOB (forward operating base) and you start reading reports about other sniper teams that have been compromised, massacred, killed, or just wounded. Or you read about the civilian contractors that have been captured or kidnapped and later beheaded, stuff like that. Then the other thoughts go through your mind; you think about the fact that you just had a daughter. But then you have to crunch down and put it out of your mind, just focus on the mission, concentrate on the task at hand.

There have been a couple, three times that I have had to pull the trigger on an enemy in my sights. The first time was the Friday before Easter Sunday, Good Friday, and there was an awful lot of enemy activity in our area of operations (AO). Actually, there was a lot of enemy activity in the theater at that time. It was April 2004, and there was an Iraqi ING (Iraqi National Guard) FOB in our battalion area of operations. They had been mortared fairly heavily the day before, and the rumor was that insurgents were going to try and capture this FOB for the weapons that were there. The Iraqis had a number of their ING soldiers not show up for work after the mortar attack. Our battalion commander wanted to show solidarity and support for our allies. He ordered that a patrol be mounted up that consisted of a section from one of our battalion's eighty-one-millimeter mortar platoons, a platoon from our Bravo company, our battalion reconnaissance platoon, and my sniper section.

We convoyed down to the FOB and pulled into the front gate. We weren't in the FOB more than five minutes before we started receiving heavy mortar fire in and around the position. It was getting to the point where the incoming

Utilizing an AN/PAS-13 thermal weapons sight, this sniper examines a distant target while pulling duty in Iraq. This heavy version of the TWS was developed for use with the M107 Barrett .50-caliber sniper rifle as shown here. This type of sight uses thermal imaging rather than visible light to identify a target and can see through smoke and other obscurants on the battlefield. It has a range of 2,200 meters to match the capabilities of the M107 rifle. A spare ten-round magazine is sitting on the platform just in front of the magazine loaded into the weapon. ***Private Collection***

fire was starting to become pretty effective, when my own instincts took charge. My first thought as a sniper was that I had to get into a position where I could see what was going on. So I grabbed the M107, the .50-caliber Barrett sniper rifle, and sprinted across the compound for the top of the stairs leading to the rooftop of a building inside the FOB.

While heading into position, I wasn't thinking about how I might have to squeeze the trigger and shoot somebody. Instead, I was thinking that I had to get up there and see just what was going on. Once I got up on the roof, I started looking around to pick my position. I had the battalion forward support officer and the B Company commander with me, and they pointed out what looked to be an Iraqi taxicab at what looked to me to be about six hundred meters range. The cab was just sitting stationary on a canal road. That canal road offered a very good vantage point for an observer.

Getting into position, I set the M107 up and cranked the magnification on the telescopic sight up to its maximum, fourteen-power, and glassed the cab to see if there was anything I could identify inside the car. Sure enough, there was somebody in the taxi with a pair of binoculars looking over at our positions in the Iraqi FOB.

I didn't need anyone to tell me to shoot; I just saw a danger and decided to see what would happen if I took it under fire. Taking my helmet off, I got behind the gun and chambered a round. Taking the weapon off safe, I prepared to fire. My aiming point was the driver's compartment of the cab, where I could see the man with the binoculars sitting and watching us. Squeezing the trigger, I fired off my first round.

The first thought that went through my head as I rode out the recoil of the big rifle was "Ouch!"

In the excitement of the situation, I had forgotten to put hearing protection in my ears. If you have ever been around an M107 when it fires, you know that the noise can be incredible. It felt like someone had driven a tenpenny nail into my ears.

Even though I was paying attention to my shot and handling of the weapon, I thought, "Oh my goodness, that hurt."

What I saw through the scope when my eyes stopped watering was that my first shot had been ineffective. My elevation was set on five hundred meters, and I had hit the car, but the point of impact was low, on the floorboard. Then came the astonishing part. Looking through my scope, I could see that the guy in the cab had no clue that someone had just fired a .50-caliber round into his vehicle. He just sat there clueless while directing mortar fire.

The applications for the big Barrett M107 sniper rifle are apparent as this picture looks along the left side of the weapon far out into the desert. Any enemy force appearing from one of the bunkers or other cover could be brought under immediate fire by the sniper. ***Private Collection***

This was a great opportunity, so I put on my hearing protection, reset my sights, and put my next round where it needed to be. Thirty seconds later, the incoming mortar rounds stopped impacting. That was my first experience in doing the ultimate job of a sniper.

There were no later observations on what I did. To be frank, I hadn't even thought much about that situation until my unit and I had been back from Iraq for three months. Then, I was presented with an Army Commendation Medal with a V-device for valor. I hadn't known that the forward support officer had nominated me for a Bronze Star for valor; it had been downgraded to the award that I received. You look at things like that and wonder just what it is. I guess I was instrumental in stopping the mortar attack. But up until that point, I just hadn't thought about it much; it was just what I had to do at the moment.

The following evening after that Good Friday, we received random mortar fire that wasn't really effective; it wasn't landing anywhere near us. I hadn't really had the opportunity to let my guard down much from the day before. It seemed like every day after that one it was constant back-to-back missions for us, and I had just forgotten about that incident. The next mission came up, and I forgot about the one before and just concentrated on the task at hand. Do that one time after another, and you don't really dwell on the earlier incidents; you just don't have the time for it. Until now, I really haven't talked about that day to many people at all.

That first shot didn't really have any effect on my second firing mission, except that I remembered to put my hearing protection on. Other than that, and some of the technical stuff—making sure that the weapon was definitely ready—I had no reflections on my earlier shot. Knowing that in the confusion and everything, your own tactics, techniques, and procedures have to be honed a little bit better, I had no regrets or bad thoughts about my job. I just felt that I had to be better prepared technically and tactically to do my mission.

There had been the driving thought to go out and perform my mission as a sniper, to see if I had gotten it "right." I was fortunate in a way in that I was able to handpick my sniper section. In selecting my personnel, I looked for the kind of guy who was the "I want to get out there because I have been practicing and practicing and I want to see what I can do" person that I was. This was who I wanted to be with. And it shouldn't be confused with the kind of person who just wants to go out and kill people. No.

People join the military because they want to serve their country. Guys come to the sniper section because it's something that interests them. No one becomes a sniper because he enjoys shooting at paper targets. It's an art, and a science, and the skills you learn you want to ply in defense of your country.

Yeah, sure, these guys wanted to go out on a hot operation. They were like "Sergeant O, you've been training us pretty hard. Let's see what the hell we're made of."

Everybody was curious about what he could do and if he really was good enough for the job. I was no different.

I don't think we thought about our overall mission in Iraq at all, at least not while we were there. My deployment to Iraq was with a National Guard unit. The unit was made up of primarily civilians, at least they were such 98 percent of the time, and they were being taken from their civilian jobs and their families, to be sent over to Iraq. The big picture that the soldiers in the battalion were thinking, and there were a lot of them who were not terribly enthusiastic about going over there for a year, was: "I can't get out of this, so I make the best of it." Yes, we all missed our families; my wife was pregnant when I left. But we knew that we had a job to do.

The mentality of the guys in my section was not to stop and dwell on the bad aspects about where we were going. We didn't want to feel sorry for ourselves, whether we wanted to be over there or not. If we did act that way, then something was going to suffer. Focus was going to go down. And I didn't want the guy next to me to be out in the field thinking about something other than what was going on in our immediate surroundings. Everybody understood that and came to just about the same conclusion. I have to tell you that I was pretty fortunate with the guys I had with me; they were a level bunch of troops.

After we had been there for a while, I found myself being tempted to question our presence over there. Were the Iraqi people in a shitty situation? Yes. Did we do something to make it better? I certainly hope so. It wasn't any worse than it was before we got there, and it was a lot better in a number of ways. That in itself was good.

Do I think that we made an impact on the global war on terror? Absolutely! I've seen the bodies that were at the other end of that bullet, and they were bad guys. They were terrorists. True, there's a criminal element in Iraq that we're dealing with as well, but there were terrorists out there.

The way I looked at the situation, and the way my guys looked at it, was that we would rather be in Iraq, doing what had to be done, and doing it over there, than have to worry about dealing with these same terrorists in our hometowns. Or anywhere back here in the States. That more than enough justified our being over there and doing what we were doing.

For the most part, I wasn't concerned with the politics of it all. The enemies that we were facing were cowards—smart, not to be underestimated, but cowards in general. Using women and children in war, whether as shields or pawns or lying for you, is not in my opinion worthy of a noble cause. That's not being a warrior at all. The guy who would put it out on the line with an AK or whatever, fire an RPG at a convoy—that was at least a worthy opponent. But the guy who would hide behind women and children was a worthless coward. And I was more than happy to show him how much of a coward I thought he was.

But the enemy, they are not to be underestimated. They are smart people. Anybody who's over there in Iraq and looks at the Muslim or Arab culture as backward or unsophisticated is an idiot. These guys are geniuses at the craft of the improvised explosive device. They will surprise you whenever you let your guard down.

Our battalion intelligence cell was putting out information that we didn't have to worry about the enemy having night vision capability. Okay, that was half-right. We didn't have to worry about the enemy as a whole being night vision capable. However, to assume that the enemy had no night vision capability was a bad, bad idea.

How many times had U.S. units been overrun and their equipment captured, including night-vision devices? How many convoys had been attacked from ambush and their night vision captured? How hard is it to order a $400 piece of night vision over the Internet? Or get a piece of Soviet or Chechen gear brought across the border? It's not outside of the realm of possibility, and every time that I went into the field, I assumed that the enemy had night vision capability.

Prior to the war on terror, light infantry units would focus on noise, light, and litter discipline while in the field. One of the rules was that you couldn't have a white light on out there at night. Then the introduction of all of the infrared (IR) devices, pointers, lasers, passive and active night vision came along. Nobody really realized that now we had to have IR discipline as well. You would have units in dismounted ambushes actively marking their positions throughout the night for aviation assets with strobe lights—infrared, but a flashing strobe light. I thought that was the most idiotic thing to do in the world. You have to assume that the Iraqis have some form of night vision available to them, and there was an American ambush, right over there, the one with the flashing strobe light. That's just one example of why you don't underestimate your enemy. And we have found night vision devices on some of them. Stuff was turned in at our FOB by Iraqis who found it out in the field and just wanted to come turn it in.

Finally, I think that equipping and preparing the sniper teams is still something that has to be worked on. Organizations like the americansnipers.org, the fact that they have to exist is just wrong. The job of the military sniper is intense enough where it warrants its own job specialty, its own separate military occupation specialty. The Marine Corps does that; the army does not. The job is more than just a kid with a rifle who shoots straight. There are a lot of things that have to be taken into account in order to get the sniper's mission done.

To be an effective sniper, you have to be a subject matter expert in ballistics; the environmental impacts and their effects on a projectile in flight. That's not something that you sustain by training once a year. You have to be a subject matter expert in land navigation because there are just two of you, and who knows where you're going to be for an operation. And you have to get there and make sure that you are in the right spot. You have to be a subject matter expert in calling for close air support. Again, you are going to be on your own, you won't have an Air Force TAC guy tagging along behind you like an infantry unit does in the field. You have to know that stuff. You have to be a subject matter expert in field medicine because if one of your team gets hurt and you are so many kilometers away from help, you have to stabilize that guy until you can get him back to support. All of that stuff, those subject fields, whether they are directly or indirectly related to the sniper field, is all material that's important.

Intelligence gathering is more than just getting on the radio and saying that there's three of them at this grid coordinate. Knowing what you are looking at requires familiarity with order of battle, equipment, personnel, and the local area and its inhabitants. This is all stuff that has to be constantly worked on in order to remain sharp and up-to-date.

Being a subject matter expert in optics, whether it is day optics or night optics, is important because you have to know how your equipment works in order to properly employ it. This knowledge is a lot more than knowing that you turn something on and it turns green and it lets you see at night. You need to know the concept behind that, why and how does it work? That helps you know the strengths and weaknesses of the system. And it lets you understand if the other guy can see that kind of active IR source with the kind of equipment that he may have available to him. From what direction can he see it? All of this is important stuff that just needs to be focused on more and more.

The equipment used is an area that the army really needs to catch up on. They're issuing us a laser range finder that's huge. We're being issued spotting scopes that just aren't cutting it in competition with the other materials that are available on the market today. They aren't adaptable to night vision devices; they aren't as durable, not as compact. The civilian marksmanship and hunting community is somewhere that cutting edge equipment is being developed in some fields. I understand that the military has different needs than civilians, but it's important that new gear be integrated into the systems the army already has in widespread use. The technology is out there.

One of the things that frustrated me about being in Iraq was that this was a global war on terror, and they make specific use of the word "war." Okay, when I hear the word "war" and the United States being at war, I think back to our grandparents in World War II. Back then, the country mobilized. You had that soldier who asked Rumsfeld why did he have to dig in the garbage to find armor to get his Humvee across the border. Okay, so it was a setup question by a sneaky reporter looking to embarrass Rumsfeld. But we're at war. Why can't we shut down a civilian Humvee plant line and just mass produce armored Humvees if that's what's needed to be successful at fighting this war?

It's not that I'm suggesting that we go back to rationing like we did during World War II. But at some point, you have to commit to the action you've undertaken. And I didn't feel like we were committed to what we were trying to do, because there were still a little bit of half-assed efforts going on. And that was a little frustrating.

But we worked with what we had, and I think we did a good job. We're still doing a good job.

36

MILITARY SHOOTER (ARMY)

Sergeant First Class Dillard J. Johnson

At this time in my life, I'm a twenty-year veteran of the U.S. Army with a military occupation specialty (MOS) of 19 Delta (Cavalry Scout). My present duty station is as a platoon sergeant in the 3rd Infantry Division, Charlie Troop 3–7 Cav, stationed at Fort Stewart, Georgia. We've recently returned to the United States from a deployment to Iraq.

Growing up in the hills of Kentucky and the Ohio River Valley, I lived down around the Green River area, in a little place called Island. This was where I hunted and fished as a young man. Squirrel hunting, turkey and deer hunting—I did them all in between working on the farm. In high school, I played football, so I was just an all-American kid, I guess—doing all of the things you're supposed to do while growing up.

I don't know if the way I grew up had any influence on my adult career; I had just always wanted to be in the military. During high school I attended ROTC and stuff like that, and I always was a big Audie Murphy fan. So I just wanted to come into the army when I was able to.

After I did join the army, I've been in the reconnaissance scouts most of my career, basically going out and finding the bad guys and telling everyone else where they are so the troops can go up and engage them or whatever has to be done. To add to my skills, I attended a unit scout-sniper course when it was offered. That's where the sniper instructors come in from Fort Benning, Georgia, and they give a fairly quick in-house sniper course. The training includes the basics such as manipulation of the sights, minutes of angle, how to hit your targets, and stuff like that. There's some movement techniques taught in the course, how to track your target, adjust your fire, and similar skills. This isn't the full sniper course as it is taught at Benning, but it's a pretty good set of instructions.

The scout-sniper course really just gives you the meat of the army's formal sniper course. They give the information to you real quick, and it's a course taught to most units preparing to deploy into combat. The information on how to hit on a target makes it a really good course; it's just not as lengthy and detailed as the sniper course itself. Most military schools, and schools in general from what I've seen, when they have a long curriculum that you have to go through, they throw in a lot of fluff and filler. This is stuff that you don't really need at my level. When a unit is heading into combat, they need to know how to kill a guy; how to make sure they can get on their target, hold their weapon, and ensure that they can hit the target and get a first-round kill. That's what the scout-sniper course was all about.

Actually, for the most part my own shooting skills come just from growing up with guns and being around them. My dad taught me how to hunt, and I was always working with weapons. I was never really a competitive marksman of any kind; my use of guns was as a sporting marksman for the most part, hunting and things like that.

The training I received was good, but for the most part I only used a little bit of it. As a scout platoon sergeant, my job is to make sure that the platoons get all of the right information during their PCIs (pre-combat inspections), the primary checks and inspections we do before we go out and do the missions. I have to make sure that the guys are disciplined and getting the equipment and everything taken care of. On top of that, I go and help directly with the missions. As my platoon didn't have an officer, I was acting as the platoon leader as well as the platoon sergeant. Among doing that stuff, I also was the guy who planned the missions and insured that they were executed. When they were completed, I made sure that the information that was generated got to the commander. So I also went out on the ground and made sure that the mission happened and that my men were taken care of.

Most platoon sergeants stay back in the rear with the gear and a lot of other stuff. They act as support and bring up supplies. But with the new battlefield such as we have in Iraq, it's hard to do that. Plus, I was brought up through my military career learning from some really great NCOs and leaders that you lead from the front.

To fill some sniper positions or whatever else had to be done, it didn't matter to me if the guys had been to the formal schools or not. I took the guys who were the best shots, and those were the people I chose to be the marksmen for the unit.

I never had a reason to regret that way of doing things, none at all; it stood up well for me. If a man could shoot, then he could shoot, that was it. Once a rifle scope is zeroed, it doesn't matter who did it, the sight picture is the same for anyone who knows how to use it. All you need to do is get a guy who can shoot and hand him the weapon. He'll be the sniper for that day. So, as far as having a designated marksman in my platoon, I actually had Sergeant Williams, who had gone through the designated marksman course before we went over there. Like the scout-sniper course, it was a pretty good chunk of training. It was taught by what we called small arms master gunners that the army has now, as well as some trained marksmen-instructors from Fort Benning. They had come in and taught our guys how to fire accurately at long ranges with the M16A4. So Sergeant Williams was the one guy in my platoon who had been formally trained to be able to act as a designated marksman. However, I also had several very good shooters in the platoon.

We were able to get an M14 rifle as well as a Barrett .50-caliber from the units that left the area as we arrived. I had been able to get a Bushnell telescopic sight donated to me from the Bushnell Outdoor Products company, and that's what I put on the old 7.62-millimeter M14 we had. That became the sniper rifle of choice for me; it just felt comfortable and was easy to hit with using that Bushnell Elite 4200 scope. That was a good piece of equipment to use, a real nice scope with a wide field of view, and it made it possible to see the targets clearly at a good distance. It was hard to miss using that rig. As long as you have a rifle that will reach out there and is properly zeroed in, you do your part with the proper breathing and trigger control and you will hit the target.

There was really no specific advantage that the platoon received from having that kind of weapon available to us for the most part. Most of the stuff you have to deal with is really close in, like IEDs—improvised explosive devices. You don't know who this is or who that is as far as the population goes.

That weapon system was an advantage when I had a sniper kill at long range on December 14, 2005. We were working near Salman Pak along the Tigris River when we started taking fire. There were actually two snipers on the far side of the river who both had SVD sniper rifles during the incident. Their fire was ranging pretty close to us and already had several kills on both Iraqi and U.S. soldiers when we got into the firefight with them. With the laser range finder I had available, I determined the range to their position to be exactly 852 meters. To my advantage, I just had a little better weapon than they

A sniper peers through the scope of his rifle during a mission in Mosul, Iraq, as part of Operation Iraqi Freedom. His weapon is an upgraded M21 sniper rifle fitted with an extensive Picatinny rail system for attaching various accessories. A Harris Engineering bipod is under the front of the stock, and the weapon is mounted with a high-quality optical sight. The power of the 7.62×51-millimeter round fired by the M21 gives this sniper an extended effective range over the 5.56-millimeter M16A4 or M4A1 carbines his fellow soldiers are carrying. The semiautomatic action of the M21 rifle also offers greater firepower and a faster follow-up shot than the standard, but more accurate, M24 sniper rifle. ***U.S. Army***

did, along with better optics. I don't think that I was all that better a shot than the other guy; I just think that I had better equipment than he did. That made me able to hit him at a greater distance than he could successfully engage me.

With my M14 and Bushnell scope, I made my first shot at 852 meters. My observer, Sergeant Kennedy, who had never fired a Barrett .50-caliber before, was acting as my spotter because my other spotter was on leave at that time. Sergeant Kennedy was another of the good shots in my platoon, so I stuck him on the Barrett. Once I told him the range and had confirmed it with my shot, he was able to click the Leupold scope on the Barrett up to exactly the right setting and was able to hit his guy with the first shot as well.

I had no difficulties with making the shot; there wasn't anything personal about it. I've been in the army long enough and have several combat tours to give me the experience needed to do my job right. Going through the first part of Operation Iraqi Freedom, I was in the lead vehicle for my unit and ended up going into a massive firefight with that one. The enemy suffered a lot of casualties during the engagement. I've also done several close kills during my time in combat, one of them where I had to kill a guy with a knife; that's about as up close and personal as you can get. So pulling the trigger on somebody across a river with a sniper rifle was no more different than if I was shooting a deer or something else.

The only thing that I had to worry about at that time was finding the target and getting on the guy, because he was ranging in on me. Rounds hit within six inches of me and several of my men. So he was close to hitting us; we just hadn't found him yet. The only thing that I was concerned about was taking out the threat to the rest of my guys.

In general, I think the enemies we face over there now are cowards. They want to hide behind explosives and stuff like that. They don't want to come out and fight because every time they do, they die. So maybe they aren't cowards, maybe they're smart. But they are not the caliber of soldier that we are. There are very few of them that actually stand up and fight. Most of them just fire a few rounds off and then run away. I do respect the guy who was on the other side of the river that day. He stood there and exchanged rounds with us. They had not received accurate return fire when they had engaged the Iraqi Army forces in that same area, even though the Iraqis had shot back with heavy machine guns and other weapons.

So that pair of shooters on the far side of the Tigris had stood their ground, at least until we showed up and put accurate rounds in on them. Basically, they lost the fight. But in no way do I consider that man to be my equivalent on the other side. He wasn't a soldier, not in any way whatsoever. The people he was shooting at were soldiers. The guys in the Iraqi Third Public Order Battalion (POB) of the present Iraqi Army, those are soldiers. Those are guys out there doing the right thing for their own country, and they are in uniform, not hiding until they can shoot and run. Those Iraqi soldiers are out there every day facing the threat. That guy with the SVD that day, he wasn't a soldier and I don't look at him that way. Those individuals who I fought and killed on my way up into Baghdad during the first part of the war, those were soldiers and brothers in arms. There was sympathy for those Iraqis during the battles, when we engaged them and when we treated the wounded afterward.

Even though we were in different armies, we are comrades in a way because we shared the same spirit; we loved our country and what we were doing. That doesn't compare to those snipers who were on the other side of the river; they were just insurgents at best, cowards, who sneak out and try to hide, shoot at people and then run away. That's what that guy and his partner had been doing.

In the real military, having a sniper or a designated marksman, someone like that, changes the whole battle focus that you have inside of your element. It makes you so much more flexible for your missions. If you were an armored battalion, instead of just being able to do a route reconnaissance where you would drive up and down the route to make sure that it was secure, a sniper gives you the ability to go out and set up covert observation posts and sniper positions. In those, you can wait and look for the enemy. That way, you don't have to control the route with a visible armor presence. You can actually move your armored vehicles to another location and have the sniper position set up to provide precise cover fire and information gathering.

Using that technique, we were able to capture an individual emplacing an IED and not just kill him. That was only possible because we had identified him from the sniper position and tracked him. Watching his movements, we were able to put him in a position where we had him solidly covered. Then we put in our security elements and captured the guy. By doing that, we were able to gather a lot of information on where he was getting his IEDs. Following up on that intel, we went out and conducted raids on the insurgents to gather more intelligence on how the stuff was getting into the area. That proves the value of the kind of tool a sniper is. You go from being a defensive force all of the time to being an offensive force out there that can cover the area. And if the sniper needs to take the shot, to take out the guy who's emplacing the IED, then he can.

Or, as Sergeant Williams, my designated marksman, did, he can take out the individual's car when he stops to emplace an IED. That left the guy stranded so that we were able to come by and pick him up. That way, we were able to get information from him that allowed our Iraqi allies to go out and arrest the guys. That's an important part of the actions of the democracy that we're trying to instill over there. Instead of just killing the guy or taking him prisoner as a military combatant, he gets to go through the Iraqi court system. Now he's charged and goes to jail, just like any other criminal. This is how the Iraqis can see for themselves that we're not just over there killing people indiscriminately; we're doing the right thing, getting the individuals planting the bombs off the streets, putting them away in the Iraqi system.

With the ACOGs (advanced combat optical gun sight) that we have on the M16A4 rifles now, along with the EOtech sights, the lasers, and everything else, along with the marksmanship training that we go through in the military, you have a lot of control of the situation as far as firepower goes. The sniper gives you an advantage in the unit as far as range and precision goes; I just

A pair of soldiers sight in on the enemy with two significantly different weapons. Lying prone is a sniper armed with an M24 SWS painted in desert camouflage colors. The large sight he is peering through is the AN/PVS-10 sniper night sight first issued in 1998. The sight can be adjusted for use during the day or, with light amplification, at night. Looking through the ACOG sight on his M4A1 carbine, the soldier to the rear has a suppressor mounted on the barrel of his weapon. Strapped on top of the front of his carbine is an AN/PEQ-2A laser designator. The sniper is also carrying an M4A1 carbine for close-in combat. The compact weapon is slung diagonally across his back. *U.S. Army*

don't know if it's a deciding factor as far as fighting a battle goes. In my opinion, a sniper is a specialty job, and he is used to take out specialty targets, targets of high value, or he's set up to watch areas or targets of particular interest.

Basically, as a designated marksman, I acted as a sniper for my platoon and Sergeant Williams was also a designated marksman. For myself, I carried an M4 carbine most of the time with my sniper rifle in the trunk of my vehicle getting banged around. Williams also carried an M4 most of the time with the upper receiver for his A4 rifle—the long barrel assembly that could be mounted on his M4 receiver—left stored back in the arms room. We just didn't need it all of the time due to the normal range of the shots that we were taking. Working in Baghdad was a MOUT (military operations in urban terrain) situation most of the time, and we had a wide variety of operations we had to conduct.

There was no sniper per se assigned to our platoon. Even the infantry platoons that have the guy in them who's a sniper, such as the 101st Airborne units we worked with, the guy doesn't necessarily get to operate in his trained specialty. The first raid we took the airborne troops out on, where we were showing them our area of operations, they had kids breaching doors and going in armed with M14 rifles. That's a big weapon, and the same thing that we were using as a sniper weapon, with a scope mounted on it. What you need to do is remain flexible and able to conduct a wide range of missions.

Snipers are great shots. They've been trained to hit their targets in all types of weather, wind, and other hard conditions. The army is very good about teaching that sort of thing. But to actually be able to shoot somebody, to nail a bad guy at range, that's a difficult thing to do. Bad guys don't give you the opportunity to kill them at a great distance. He may have his rifle on his shoulder, or he could be doing something else with a weapon nearby. The way that Iraq is now, if you shoot a guy at eight hundred meters because he has a rifle over his shoulder, you don't know if you just killed an Iraqi soldier, police officer, or just a local who was out watching his own property.

You have to be very careful with the targets that you pick to engage. That's why if I have an individual I see with a weapon, and he's engaging someone else, I have no problem firing at him and taking him out. I used my sniper rifle to fire at the individual who was shooting at us because I couldn't bring my Bradleys up to engage him with their weapons. Instead, I used the weapon that would reach out and hit that target at that location. That's why the sniper rifle was used rather than calling in helicopters. He would have heard the helicopters coming in and just disappeared, run away to shoot at us or our allies at another time. And the Bradleys with their 25-millimeter cannon and machine guns couldn't get into the proper position to engage him.

The sniper's job over in Iraq right now is really the scout's job. He's watching the area and letting his people know what's going on around them. If he does see a viable threat, then he can take the shot that eliminates that danger. But 90 percent of his job as far as Iraq goes—I can't speak on Afghanistan in wars past—is observation. In Iraq and the battle I just came home from, it is the sniper out there in his hide, his observation post, who's making the difference in gathering intelligence rather than just killing the enemy.

I don't really see a difference between law enforcement snipers and what we do in the military. When you get right down to it, they're doing the same job that we are. They're observing a situation, and they are prepared to take out the threat if necessary. No sniper that I've talked to—including myself, and I have twenty-eight kills in Iraq; from a Bradley my first time over there, I had confirmed kills well past the hundreds—none of us want to shoot somebody just for shits and giggles, just to say I killed somebody. You want to make sure that the guy you kill is the threat that's out there and who's going to try and kill you if he gets the chance. Military snipers, we watch and maneuver people to go out there and eliminate the threat. If they can't get him, or he gets away, then we have to take the shot. That's not what we want to do; we get too much information from a capture to waste the chance.

Law enforcement snipers I see as doing the same job. There are no police officers out there on SWAT teams who want to make a name for themselves as a shooter. They're up there with their sniper rifles and looking at somebody who could be somebody's father, son, brother, or husband, and that individual may have just had a bad day. Those snipers are watching that individual to make sure that the threat doesn't escalate. The last thing that they want to have to do is put the guy down, because when you go home at night, that's what you sleep with. And you have to be able to make peace with yourself. So I don't see any difference between a military and a police sniper; they're both professionals who are trained to fire their weapons only to take out a danger at a distance, or to just observe a threat. A sniper is more of a security blanket for the "what if?" problems, able to do the things that sometimes must be done, rather than someone who has to shoot to end a situation.

I think a lot of people look at snipers as killers, just somebody who goes out and is trigger happy. You hear a lot of remarks about guys with deer rifles, hunters and such, where people say that they just want to be snipers. I think the term "sniper" has been brought up to be not really a dirty word, but to denote somebody as a killer. That isn't the case at all. A professional marksman, that's what a sniper is.

It takes a heavy dedication to the craft to be a sniper. To be a sniper is to be a high caliber of shooter. I don't consider myself to be a sniper; I was able to hit a guy at a great distance with a good weapon and a very good optical system, along with some great training. But it takes a lot of real dedication, time, and work to be a true sniper. Getting your breathing right, firing one shot, going downrange and marking exactly where that one shot hit, then coming back and doing it all again, that's very painstaking work. It's not like taking your pistol out and emptying fifteen rounds and then going to look at your target. Or even when we're firing our M16s or M4s and going downrange after firing our nine rounds and looking to see the different patterns of where the bullets hit. To be a sniper, you shoot one round, and then every other shot you fire is treated as a new round. You go back and adjust what you did. You look at your breathing, your holding position, the time of day, the wind, the sun angle, even the type of shade that was out there.

Every shot you fire as a sniper, you mark down and record as an individual shot. Each time you pull the trigger is a carefully considered event. A good sniper may spend forty-five minutes on the range and only shoot two rounds.

He would spend the bulk of that time making sure all the conditions are right for his shot, that the bullet will go exactly where he intends it to. And if it doesn't, he will spend more time figuring out why the shot went astray. That's a lot of dedication to go with a job, including loving your rifle, caring for it, and making sure it's being properly taken care of. And you have to make sure that you're up on the different elements of the situation of the day, checking the ballistics type, making sure that you fire the same lot number of ammunition that you practiced with.

It's not just picking up a rifle and going out and hitting something. Anyone can shoot a Coke machine at a hundred meters with a rifle. Anyone can kill a deer at a hundred meters, or fifty yards, using a rifle with a big scope on it. But it's really, really hard to hit somebody at five hundred plus meters with the wind blowing, at night, shooting across a river or across a field, where you're going to have two or three different temperature changes. A sniper can take his weapon and accurately hit his target in conditions like that.

The big thing that I would like everyone to know about what's going on in Iraq is that everything that you see on the news is a facade. The reporters don't leave the Green Zone; they stay in there where it's safe. You see your news footage where a vehicle is burning or something else is going on, and you don't see the photographer or reporter with it. You just see the news footage, the action on the screen. Usually, that footage came from some soldier that was out there, and they ask him for anything he has that they can use. Or they send an Iraqi cameraman out there. As far as the reporters go, you always see them in the same building, behind a mosque, in one of the most heavily guarded places in the world. There's an Iraqi Army division watching the place, along with an American tank battalion. There are several freelance support agencies in the Green Zone along with the Iraqi police brigade. It's a very fortified position and nothing gets in there.

Yeah, you hear about stuff on the outside, but nothing gets in where the reporters are staying. The people need to understand that it's terrible that they watch the news and learn that three marines have died. The part that they miss was that yes, those three marines died, but 150 insurgents were also killed. If we don't kill these insurgents in their backyard, then they're going to come to our backyard. I'm just one of the many soldiers who don't want to have to worry about these insurgents, these terrorists, coming over here to Las Vegas, or Disney World, or LAX and committing their acts on American soil. We don't need another 9/11 in the United States because we quit something when we were only halfway done. The American forces are winning in Iraq. The reason that there continue to be casualties among the U.S. troops over there is because we are always shifting our forces to the trouble spots.

Once we get a spot secure, we give it over to the Iraqi Army, or the Iraqi police take over controlling it. There are no more issues out of that area. Then we move on to another area where there's a large insurgent push or something else going on. It's not that we're over there and then suddenly get someplace clear and everything is hunky-dory and that unit stays there. They don't do that. That unit moves on to the next hot spot. And we're continually putting out the hot spots.

The American people need to support the soldiers and to support what they're doing. And they need to realize that it's not the Iraqis that we're fighting. Four out of five of the insurgents that we kill over there are either from Saudi Arabia, Jordan, Syria, or Iran. Those are the places that the terrorists are coming from, someplace other than Iraq. The people of Iraq are there for us, they are really with us. But are they going to go out into the streets and start standing up for themselves? It's very hard for those people to do that. They just got done with living through thirty years of a brutal dictatorship. You protested or stood up for yourself and you were taken out and killed.

It also has to be remembered that during the Gulf War, we made a lot of promises to the Iraqi people and then left. Saddam Hussein's regime came out and killed a lot of people who had been on our side. So there's a lot of distrust over there. The Iraqi population is just not sure about what the American people are going to do. It's hard to stand up and start fighting the people who have taken over your neighborhood, to try and make the place better for yourself and your family, if the people who say that they're going to support you may be gone in a few days, weeks, or months. The people who have to live there are sort of waiting things out to see what happens. Democracy has taken a hold over there, and the people are enjoying freedom, and we are making a big difference for them. But if we leave, the difference won't be there.

I've seen a difference in the military since I first enlisted; what I don't see is a difference in the caliber of the soldiers themselves. The battle that we fought when we invaded Iraq was basically the World War II battle. It was like Kasserine Pass in North Africa, a very bloody conflict. The U.S. marines were bogged down in some of the street-to-street fighting. But all of those young men were people who loved their country and had a job to do, so they went out and did it. And they did the same thing that had been done by their fathers and grandfathers in World War II, Korea, and Vietnam—they fought for the guy next to them.

We go up to now, when we're trying to fight this big insurgent battle, which is basically Vietnam. The only difference is that our firebases aren't out in the jungle; they're in little towns—what we call FOBs, or forward operating bases. Now, instead of the big battles taking place in the jungle on the dirt trails, they're happening on the highway. The weapons have gotten a little smarter, and the emplacement of them has gotten smarter as well, but there's no change in the troops. The American soldier is over there doing a job.

This is the same thing that Sergeant Sammy Davis, the Medal of Honor recipient, did in Vietnam, and Sergeant Paul Smith did over in Iraq. Paul Smith got up behind a .50-caliber machine gun and was defending his guys until he was wounded several times. Finally he died, and he's being considered for a Medal of Honor for his actions. This is the same caliber of soldier who's done what had to be done for over two hundred years now. There's no change in the American Army, or the way that the average American thinks about things. We're over there for the other country, and we're bringing freedom to them. Just like we did in Germany after World War II when we handed it back over to them when the war was over and the people there were freed. We trained the German military after the war, too, got them back up and running in order to defend their own country. Even some of the German Army officers picked up their military and ran with it after the war was over.

It's the same thing we did in Japan and the same thing we did in Korea with the South Koreans after that war ended. That's exactly the same thing that we're doing in Iraq today; we're giving them democracy and letting them learn how to run their own affairs again. And democracy is a very scary thing, because it's contagious. There are a lot of governments around the world, monarchies and things like that, especially in the Middle East, where all of the power is concentrated in the hands of a very few, who are very afraid of their people getting a taste of democracy. And the American soldier is the one who's going to make a difference. Once democracy has gotten hold, once it has its feet on the ground, that's when you're going to see a change.

There are a lot of people back here saying that the government should get our soldiers out of Iraq. And there are some soldiers who want to come home, and I understand that. It's very hard, and it takes a lot out of you to go out there and have somebody shoot at you, or to get blown up. I had eight IEDs in one day, so I understand the anxiety that goes through a man's mind when he doesn't want to go out there again. A lot of people feel that way. But the main thing that the American people need to understand is that there are people who want to continue with this fight to the end. We're not warmongers. Nobody hates a war like a soldier does, because we have to fight the war.

I've lost very close friends and comrades over there, and I've had to pick myself up by the bootstraps, motivate my platoon, and go back out only hours after these men died—and return to the same positions where these individuals were killed. Why? Because that's our job, that's what we do to defend the American people. And if the American people think that the best thing that they can do for us is to pull us out of Iraq before we can finish the job, then the only thing they're going to do is demoralize us. We started a job and we've lost many brave men. We've lost brothers and fathers and sons and daughters over there, individuals who have given the ultimate sacrifice of their lives so that we could spread democracy over there and free those people.

Soldiers don't need to look at the TV and see all of the negative stuff that's on there. They need to look at the real news. If they look at what's on TV, they'll see how the stations chastise the president and the military, and then see what happens when one of the media's own gets hurt. It was unfortunate that a photographer and reporter were injured over there. But the same day that those two media people were injured there were three American soldiers who were severely injured. All they got on the news was barely a little bit of attention, a fragment of ticker tape passing by on the bottom of the screen. They were out there doing their job as compared to the reporter, who I am sorry got injured, but he left a heavily armored American Humvee and got into one of the worst civilian vehicles over there, one that could be easily punctured by 7.62-millimeter AK-47 fire. And he rode on top of that vehicle in an exposed position with very little body armor on. He was pretty much asking to get his ass blown up, and that's what happened.

That's in comparison to the American soldiers who are over there doing their job and getting blown up in the process. And there are the marines who are doing exactly the same thing and facing the same dangers. Those are the people who need to be on the news every day. There are very few news agencies that are covering what's really happening every day over there. This book is going to go out and people are going to read it, and the news isn't going to touch the same subjects, and the mainstream public isn't going to touch it. The people who read this are going to be the kind of people who actually care about the soldiers and what goes on. Then they're going to talk to some of their friends and the story is going to get around. But until somebody actually stands up and says, "Look, you fools, this is what's going on!" nobody is going to care.

A piece of expended brass flies through the air as this marine scout-sniper works the bolt of his M40A3 rifle during range shooting in the desert. The fingertips of his gloves are cut away to allow him more sensitivity when firing his weapon.

These are the men who watch over their fellow soldiers on the battlefield. Their skills and their rifles make them the ultimate in precision "smart" weapons for America's war against global terrorism. ***USMC***

No one cared about the Vietnam vet until we had the big win for Operation Desert Storm. Vietnam was the forgotten war, or one the public tried to forget. The soldiers are coming home now and things are different. I walk around in my uniform here at the Shot Show in Las Vegas, and we have really great men like Sergeant Sammy Davis walking around with his Medal of Honor on, and few people acknowledge what he has around his neck. They have no idea of the sacrifices men like that have made for their country; the man is totally disabled, and the government expects him to live on less than $1800 a month. He can't work, and that's all of the compensation that somebody who has saved American lives while fighting on foreign soil can expect to receive. We give Freedom Medals away to other countries that cost upward of $20,000 or $30,000, and the American soldiers who get wounded in combat receive a Purple Heart that you can buy for $15.00.

It's just a sad state of mind in how the public looks on everything that a soldier does. I'm not a hero. I have a Silver Star, I've been wounded four times, and I'm not a hero. I'm doing my job and taking care of my soldiers so that my wife and kids can live in freedom. I want to fight them in their yard, not in mine.

Snipersonline UA through its www.AmericanSnipers.org original "Adopt-a-Sniper" program has several members and sniper friends actively engaged in the battle against terrorism. Our friends are only the tip of the iceberg when it comes to the number of snipers deployed around the world right now. These snipers are deployed in the theaters of combat in Afghanistan and Iraq as well as other locations overseas. Many of these are full-time law enforcement officers who have been activated for regular duty action.

Our memberships, all full-time law enforcement officers and swat snipers, have chosen to help support our fellow snipers, and brothers in arms, by helping supply them with items that they need to get the job done better as well as things to make their lives easier while on deployment. This has been everything from basic hygiene-type items to tactical products such as new optics, rifle accessories, and cleaning equipment, to spotting scopes and laser-range finders to mini binoculars, batteries, and rucksacks.

We invite you to directly support the effort of those who are going in harm's way on behalf of all of us. Visit our website, www.American Snipers.org and make a contribution so that we may continue to supply our brothers with the mission essential equipment they cannot get through normal channels. The equipment not only helps them to do their tasks, but also helps all U.S. military personnel to heighten their chances of survival in this war on terrorism.

Drawing and website information courtesy of AmericanSnipers.org

BIBLIOGRAPHY

BOOKS

Appleman, Roy E. et al., *United States Army in World War II The War in the Pacific—Okinawa: The Last Battle*, Historical Division, Department of the Army, Washington, D.C., 1948.

Ashdown, Charles Henry, *British & Foreign Arms and Armor*, T.C. & E.C Jack, London, 1909.

Bartlett, Derrick D., *Snipercraft: The Art of the Police Sniper*, Precision Shooting Inc., Manchester, Connecticut, 1999.

Benton, William, pub, *Encyclopedia Britannica*, Encylopaedia Britannica Inc., Chicago, 1965.

Bilson, Frank, *Crossbows*, Hippocrene Books, New York, 1974.

Blair, Claude, gen. ed., *Pollard's History of Firearms*, MacMillan Publishing Company, New York, 1983.

Brookesmith, Peter, *Sniper: Training, Techniques and Weapons*, St. Martin's Press, New York, 2000.

Brown, G. I., *The Big Bang: A History of Explosives*, Sutton Publishing Limited, Gloucestershire, 1998.

Burkett, Matt, *Practical Shooting Manual*, Hot Shots, 1995.

Butler, David F., *United States Firearms; The First Century 1776–1875*, Winchester Press, New York, 1971.

Cacutt, Len, ed. *Combat*, Chartwell Books Inc., Secaucus, New Jersey, 1988.

Canfield, Bruce N., *U.S. Infantry Weapons of World War II*, Andrew Mowbray Publishers, Lincoln, Rhode Island, 1994.

Cerasini, Marc, *The Future Of War: The Face of 21st-Century Warfare*, Alpha Books, Indianapolis, Indiana, 2003.

Chandler, Roy F., and Norman A. Chandler, *Carlos Hathcock White Feather: USMC Scout Sniper*, Iron Brigade Armory Publishing, Jacksonville, North Carolina, 1997.

———, *Death from Afar: Marine Corps Sniping Volume I*, Iron Brigade Armory Publishing, Jacksonville, North Carolina, 1992.

———, *Death from Afar: Marine Corps Sniping Volume II*, Iron Brigade Armory Publishing, Jacksonville, North Carolina, 1993.

———, *Death from Afar: Marine Corps Sniping Volume III*, Iron Brigade Armory Publishing, Jacksonville, North Carolina, 1994.

———, *Death from Afar: Marine Corps Sniping Volume IV*, Iron Brigade Armory Publishing, Jacksonville, North Carolina, 1996.

———, *Death from Afar: Marine Corps Sniping Volume V*, Iron Brigade Armory Publishing, Jacksonville, North Carolina, 1998.

Chapel, Charles Edward, *Guns of the Old West*, Coward-McCann, Inc., New York, 1961.

Coughlin, Gunnery Sgt. Jack, USMC, and Capt. Casey Kuhlman, USMCR, with Donald A. Davis, *Shooter: The Autobiography of the Top-Ranked Marine Sniper*, St. Martin's Press, New York, 2005.

Crosby, Alfred W., *Throwing Fire; Projectile Technology Through History*, Cambridge University Press, Cambridge, 2002.

Crouch, Howard R., *U.S. Small Arms of World War 2*, SCS Publications, Falls Church, Virginia, 1984.

Culbertson, John J., *13 Cent Killers: The 5th Marine Snipers in Vietnam*, Presidio Press, New York, 2003.

Diagram Group, *Weapons; an international encyclopedia from 5,000 BC to 2000 AD*, St. Martin's Press, New York, 1980.

Diez, Octavio, *Armament and Technology: Assault and Precision Weapons*, LEMA Publications, Barcelona, 2000.

———, *Armament and Technology: Support Weapons and Combat Equipment*, LEMA Publications, Barcelona, 2000.

———, *Special Units: Assault Police*, LEMA Publications, Barcelona, 2000.

———, *Special Units: Commandos*, LEMA Publications, Barcelona, 2000.

———, *Special Units: Special Police Task Forces*, LEMA Publications, Barcelona, 2000.

Ezell, Edward C., *The Illustrated History of the Vietnam War Volume 15: Personal Firepower*, Bantam Books, New York, 1988.

Faust, Patricia L. ed., *Historical Times Illustrated Encyclopedia of the Civil War*, Harper & Row, New York, 1986.

Flack, Jeremy, *Sunburst Military Series: Rifles and Pistols*, Promotional Reprint Company, London, 1995.

Flayderman, Norm, *Flayderman's Guide to Antique Amercan Fireams . . . and their values*, Follett Publishing Company, Chicago, 1977.

Gilbert, Adrian, *Sniper: The World of Combat Sniping, the Skills, the Weapons, the Experiences*, St. Martin's Press, New York, 1994.

———, *Stalk and Kill: The Sniper Experience*, St. Martin's Press, New York, 1997.

———, *The Encyclopedia of Warfare*, The Lyons Press, Guilford, Connecticut, 2002.

Hallahan, William H., *Misfire: The History of How America's Small Arms Have Failed Our Military*, Charles Scribner's Sons, New York, 1994.

Haskew, Michael E., *The Snipers at War*, St. Martin's Press, New York, 2005.

Held, Robert, *The Age of Firearms; A Pictorial History*, The Gun Digest Co., Northfield, Illinois, 1957, 1970.

Henderson, Charles, *Marine Sniper: 93 Confirmed Kills*, Berkley Books, New York, 1988.

Hogg, Ian V., *The Story of the Gun; From Matchlock to M16*, A&E Books/St. Martin's Press, New York, 1996.

———, *The World's Sniping Rifles with Sighting Systems and Ammunition*, Greenhill Military Manuals, Greenhill Books, London, 1998.

Huebner, Seigfried, *Silencers for Hand Firearms*, Paladin Press, Boulder, Colorado, 1976.

Huntingtton, R. T., *Hall's Breechloaders*, George Shumway Publisher, York, Pennsylvania, 1972.

Katcher, Peter, illustrated by Stephen Walsh, *Sharpshooters of the American Civil War 1861–65 Warrior 60*, Osprey Publishing, Oxford, 2002.

Kelly, Jack, *Gunpowder*, Basic Books, New York, 2004.

Kirkland, Turner E., *Dixie Gun Works Inc., 2002 Catalog No. 151*, Union City, Tennessee, 2002.

Lanning, Michael Lee, *Inside the Crosshairs: Snipers in Vietnam*, Ivy Books, New York, 1998.

Lau, Mike R., *The Military and Police Sniper*, Precision Shooting Inc., Manchester, Connecticut, 1998.

Markham, George, *Guns of the Elite: Special Force Firearms, 1940 to the Present*, Arms and Armour Press, London, 1987.

Mayne, Lt. C. C. B., *The Infantry Weapon and Its Use In War*, London, 1903.

———, *Guns of the Reich: Firearms of the German Forces, 1939–1945*, Arms and Armour Press, London, 1989.

McAleese, Peter, and John Avery, *McAleese's Fighting Manual: The Definitive Soldier's Handbook*, Orion Books Ltd., London, 1998.

McBride, H. W., *A Rifleman Went to War*, Lancer Militaria, Mt. Ida, Arkansas, 1987 (reprint).

McNab, Chris, *The SAS Training Manual*, MBI Publishing Company, St. Paul, Minnesota, 2002.

Minnery, John A., *Firearm Silencers Volume Two*, Desert Publications, El Dorado, Arkansas, 1981.

Mullin, T. J., *Special Operations: Weapons and Tactics*, Greenhill Books, London, 2003.

Myatt, Maj. Frederick, *Modern Small Arms*, Salamander Books, London, 1978.

Naylor, Sean, *Not a Good Day to Die; The Untold Story of Operation Anaconda*, Berkley Books, New York, 2005.

Newark, Tim, *Medieval Warlords*, Blandford Press, New York, 1987.

Nicolle, David, *Arms and Armor of the Crusading Era 1050–1350; Western Europe and the Crusader States*, Greenhill Books, London, 1988, 1999.

Paulson, Alan C., *Silencer History and Performance; Sporting and Tactical Silencers, Volume One*, Paladin Press, Boulder, Colorado, 1996.

Payne-Gallwey, Sir Ralph, *The Crossbow: Mediaeval and Modern, Military and Sporting*, Bramhall House, New York, 1978 (reprint).

Pegler, Martin, *The Military Sniper Since 1914 Elite 68*, Osprey Publishing, Oxford, 2001.

———, *Out of Nowhere: A History of the Military Sniper*, Osprey Publishing, Oxford, 2004.

Peterson, Harold L., ed., *Encyclopedia of Firearms*, E.P. Dutton and Company Inc., New York, 1964.

———, *The Treasury of the Gun*, Golden Press, New York, 1962.

Peterson, Harold L., and Robert Elman, *The Great Guns*, Grosset & Dunlap, Inc., New York, 1971.

Peterson, Roger Tory, *A Field Guide to the Birds; Eastern Land and Water Birds*, Houghton Mifflin Company, Boston, 1947.

Plaster, Major John L., USAR (Ret.), *The Ultimate Sniper: An advanced training manual for military & police snipers*, Paladin Press, Boulder, Colorado, 1993.

Pritchard, Russ A., Jr., *Civil War Weapons and Equipment*, The Lyons Press, Guilford, Connecticut, 2003.

Prokosch, Eric, *The Technology of Killing: A Military and Political History of Antipersonnel Weapons*, Zed Books, London, 1995.

Pushies, Fred J., *Weapons of the Delta Force*, MBI Publishing Company, St. Paul, Minnesota, 2002.

———, *Weapons of the Navy SEALs*, MBI Publishing Company, St. Paul, Minnesota, 2004.

Ried, William, *Arms Through the Ages*, Harper & Row, New York, 1976.

Ripley, Lt. C. Wm. Y. W., *A History of Company F: First United States Sharps Shooter*, Grand Army Press, Rochester, Michigan, 1981 (reprint 1883).

Roberts, Craig, and Charles W. Sasser, *Crosshairs on the Kill Zone: American Combat Snipers, Vietnam Through Operation Iraqi Freedom*, Pocket Books/ Simon & Schuster, New York, 2004.

Sakaida, Henry, illustrated by Christa Hook, *Heroines of the Soviet Union 1941–45 Elite 90*, Osprey Publishing, Oxford, 2003.

SEAL Sniper Training Program, Paladin Press, Boulder, Colorado, 1992.

Senich, Peter R., and Howard Kyle, *The German Sniper: The Man—His Weapons, Volume One*, Normount Technical Publications, Wickenburg, Arizona, 1975.

———, *The Complete Book of U.S. Sniping*, Paladin Press, Boulder, Colorado, 1988.

———, *The German Assault Rifle 1935–1945*, Paladin Press, Boulder, Colorado, 1987.

———, *The German Sniper 1914–1945*, Paladin Press, Boulder, Colorado, 1982.

———, *The Long-Range War: Sniping in Vietnam*, Paladin Press, Boulder, Colorado, 1994.

———, *The One-Round War: USMC Scout-Snipers in Vietnam*, Paladin Press, Boulder, Colorado, 1996.

Serven, James E., *Colt Firearms; From 1836*, Stackpole Books, Harrisburg, Pennsylvania, 1954, 1979.

Sharpe, Philip B., *The Rifle in America*, Special Edition, National Rifle Association, 1995.

Shore, Captain C., *With British Snipers to the Reich*, Greenhill Books, London, 1948, 1997.

SIPRI, *Anti-personnel Weapons*, Crane, Russak & Company, Inc., New York, 1978.

Smith, Graham, ed., *Military Small Arms; 300 Years of Soldiers' Firearms*, Salamander Books, London, 1994.

Spicer, Mark., *Sniper: The Techniques and Equipment of the Deadly Marksman*, Lewis International, Inc., Miami, Florida, 2001.

Swenson, G. W. P., *Pictorial History of the Rifle*, Bonanza Books, New York, 1972.

Tantum, William H. IV, *Sniper Rifles of Two World Wars*, Historical Arms Series No. 8, Museum Restoration Service, Alexandria Bay, New York, 1967.

Tarassuk, Leonid, and Claude Blair, eds., *The Complete Encyclopedia of Arms & Weapons*, Simon & Schuster, New York, 1979.

Textbook of Small Arms 1929, His Majesty's Stationery Office, London, 1929.

Truby, J. David, *Quiet Killers I*, Paladin Press, Boulder, Colorado, 1972.

———, *Quiet Killers II: Silencer Update*, Paladin Press, Boulder, Colorado, 1979.

———, *Silencers in the 1980s: Great Designs, Great Designers*, Paladin Press, Boulder, Colorado, 1983.

———, *Silencers, Snipers, & Assassins: An Overview of Whispering Death*, Paladin Press, Boulder, Colorado, 1972.

White, Terry, *Fighting Techniques of the Special Forces*, Random House, London, 1993.

Zurick, Captain Tim, USAR, *Army Dictionary and Desk Reference*, Stackpole Books, Harrisburg, Pennsylvania, 1992.

JOURNALS, MAGAZINES, NEWSPAPERS, AND OTHER PERIODICALS

Autry, Peyton, "The Slug Rifle" in *The American Rifleman*, pages 40–41, December 1963.

Aveni, Thomas, "The 'Must Shoot vs May Shoot' Controversy" in *Law and Order*, pages 38–39, 41–42, 44, January 2005.

Bailey, DeWitt II, "The Whitworth Rifle" in *Gun Digest*, pages 94–105, 25th Edition, 1971.

Bartlett, Derrick D., "Tactical Disarmament: Most Dangerous Option" in *Tactical Shooter*, pages 60–64, Volume 2, No. 10, November 1999.

Bierman, Harris, "State of the Art Sniper Rifles" in *1988 Guns & Ammo Annual*, pages 138–145.

Blackmore, Howard L., "The British Rifle in America: Official Rifles Used by the British Forces in America" in *The American Rifleman*, pages 28–32, June 1963.

Brown, John W., "The New Bred of Sniper Rifles" in *New Breed*, pages 16–19, 48, September 1986.

Brown, LTC Robert K., "Silent Death in Vietnam" in *Soldier of Fortune*, pages 37–41, 70, January 1978.

Bruce, Robert, "Soviet Sniper Rifle" in *International Combat Arms: Journal of Defense Technology*, pages 24–26, January 1986.

Childs, Gunnery Sgt. Jack, USMC "Sniping in Viet Nam" in *The American Rifleman*, pages 36–38, June 1966.

Coleman, John, "One Shot, One Kill: Army Sniper School is Dead On Target" in *Soldier of Fortune*, pages 44–51, 82, 84, 86, 88, December 1986.

Cooper, Dale B., "Above and Beyond the Call . . . : Delta Snipers Earn America's Highest Award in Somalia's Fiercest Fight" in *Soldier of Fortune*, pages 42–45, 68–71, September 1994.

Cutshaw, Charlie, "Remington M24 Sniper .308" in *Special Weapons for Military & Police*, pages 14–16, 18–19, 61, 2003.

———, ".338 Lapua Dynamic Duo" in *Special Weapons for Military & Police*, pages 6–8, 10–13, 2004.

Dee Haas, Sgt. Douglas Mark, USMC, "Letter from a Marine Sniper" in *The American Rifleman*, page 45, April 1968.

Dunnigan, James A., "The American Sniper" in *1981 Guns & Ammo Annual*, pages 48–53.

Editorial Staff, "DCM Carbine 'First to Reply'" in *The American Rifleman*, pages 20–21, November 1966.

———, "Austen's No. 1 Armed Citizen" in *The American Rifleman*, pages 22–23, November 1966.

———, "Austen's Overlooked Riflemen" in *The American Rifleman*, pages 20–21, November 1966.

Erickson, Leif, "Tools of the Sniper's Trade" in *Special Weapons*, pages 70–73, Volume 3, No. 1, Fall 1984.

Flores, Dan, "Sharpshooters of the Civil War" in *Gun Digest*, pages 6–14, 31st Edition, 1977.

Frigiola, Jim, ".50 Caliber Multi-Purpose Ammunition" in *Very High Power*, pages 7, 9, 1998 #2.

Gardner, Jim, "A History of American Sniping" in *Guns Magazine 2002 Combat Annual*, pages 98–109.

———, "Precision Instruments" in *Police*, pages 24–26, 29, September 2002.

Germani, Justin, "Urban Sniper" in *Guns Magazine 2000 Combat Annual*, pages 70–72, 108, 2000.

Gleason, Foster F. W., "Josiah Gorgas: Armorer of the Confederacy—and His Weapons" in *Gun Digest*, pages 116–121, 19th Edition, 1965.

Goodwin, Dan, "Barrett 99-1 .50BMG" in *Special Weapons for Military & Police*, pages 74–79, 2004.

Hacker, Rick, "Modern Rifles for Sharp Shooters" in *The American Rifleman*, pages 38–41, 64, August 1983.

Hafemeister, Rod, and Rolf Perins, "The Craft of the Sniper" in *Special Weapons*, pages 60–63, Volume 3, No. 1, Fall 1984.

———, "The Making of a Sniper" in *Special Weapons*, pages 64–65, Volume 3, No. 1, Fall 1984.

Halsey, Ashley, Jr., "The Truth About Ferguson and His Rifle" in *The American Rifleman*, page 33, August 1971.

Hargreaves, Major Reginald, M.C. (Ret.), "The Fabulous Ferguson Rifle and Its Brief Combat Career" in *The American Rifleman*, pages 34–37, August 1971.

———, "The Green Jacket and His Rifle" in *The American Rifleman*, pages 38–39, March 1977.

Harrison, E. H., "Dope Bag Questions and Answers: Army Sniper Rifle" in *The American Rifleman*, page 82, December 1969.

———, "Dope Bag Questions and Answers: Whitworth Rifle Performance" in *The American Rifleman*, pages 75–76, July 1970.

Hartov, Steven, "The Enemy's Worst Nightmare: The United States Marine Corps Scout-Snipers" in *Special Operations Report Volume #5*, pages 30–35, Fall 2005.

Held, Robert, "Kentucky Rifle: Fact and Fiction" in *Gun Digest*, pages 193–199, 16th Edition, 1962.

Helsley, Steve, "A Tale of Two Sniper Rifles" in *Tactical Shooter*, pages 54–56, 58–60, Volume 1, No. 8, September 1998.

Heter, E. W., Jr., "Dope Bag Questions and Answers: M1903A4 Rifle" in *The American Rifleman*, page 63, April 1965.

Howell, Cleves, Jr., "The Pursuit of Accuracy" in *The American Rifleman*, pages 34–36, September 1953.

Johnson, Sgt. Gary Paul, "Sniping: An American Marksman's Art" in *Special Weapons*, pages 66–67, Volume 3, No. 1, Fall 1984.

———, "The Guns of Afghanistan: The small arms we're using to fight the war on terrorism" in *Special Weapons for Military and Police*, pages 60–65, Fall 2002, #19.

———, "Saga of the SR25" in *Special Weapons for Military and Police*, pages 38–44, 1998.

Karwan, Chuck, "The Great Rifle Debate—Semiautomatic Sniper Rifle: More kills make a better weapon" in *Soldier of Fortune's Fighting Firearms*, pages 20, 22, 24, Fall 1994.

———, "Sniper!" in *Guns Magazine 1997 Combat Annual*, pages 8–9.

———, "Sniper Rifles" in *Guns Magazine 1997 Combat Annual*, pages 28–35.

———, "Sniper Rifle Technology" in *Guns Magazine 1998 Combat Annual*, pages 42–43.

———, "The M14" in *Guns Magazine 2001 Combat Annual*, pages 102–113.

Kokalis, Peter G., "20-Mike-Mike Mayhem; Aeroteck's Shoulder-Fired Sniping Cannon" in *Soldier of Fortune*, pages 32–36, April 1997.

———, "Combloc Sniper Rifles: Crosshairs of the Warsaw Pact" in *Soldier of Fortune*, pages 54–57, 122–125, April 1998.

Lau, Mike R., and Jason Ferren, "De Oppresso Liber: Sniper Weapons of the Green Berets" in *Tactical Shooter*, pages 39–45, Volume 1, No. 11, December 1998.

Lee, Byron and Darryl Bolke, "ABC's of SWAT" in *Guns Magazine 2000 Combat Annual*, pages 50–55.

Lewis, Jack, "USMC M14 7.62x51" in *Special Weapons for Military and Police*, pages 4–14, Fall 2002, #19.

Liwanag, Lt. Col. David J., "Today's Search for Tomorrow's Heavy Sniper Rifle" in *Tactical Shooter*, pages 27–31, Volume 2, No. 5, June 1999.

Lord, Francis A., "With a Few More Misses, the Yanks Might Have Lost: Few Union lads knew how to shoot" in *The American Rifleman*, page 25, April 1975.

Lynn, Captain Steven, "Battlefield Boogie Men: The Marines Corps Snipers" in *Gung-Ho*, pages 34–37, 42–45, April 1981.

McGuire, Frank G., "Snipers—Specialists in Warfare: Since the early 1500s, snipers have been changing the course of history" in *The American Rifleman*, pages 28–32, July 1967.

Mudgett, Larry II, "The Precision Rifle: Death at Long Range" in *Special Weapons*, pages 64–69, Volume 1, No. 1, Fall 1981.

Muir, Bluford W., "The Father of the Kentucky Rifle" in *The American Rifleman*, pages 30, 75–79, January 1971.

Neumann, George C., "Firearms of the American Revolution; Part II of IV" in *The American Rifleman*, pages 29–33, August 1967.

———, "Revolutionary War Rifles: Accurate but of limited use" in *The American Rifleman*, pages 26–29, October 1973.

Peterson, Harold J., "The Kentucky Rifle" in *The American Rifleman*, pages 28–32, November 1964.

Plaster, Major John L., "Sniping 2000: New Rules of Engagement for Precision Shooting" in *Soldier of Fortune*, pages 52–55, 70, April 1998.

Poyer, Joe, "Big Bore Sniper Rifles" in *International Combat Arms: Journal of Defense Technology*, pages 84–89, September 1986.

———, "Sniper Rifles for Pinpoint Precision" in *International Combat Arms: Journal of Defense Technology*, pages 78–83, July 1986.

Reynolds, Maj. E. G. B., "The No. 4 Rifle Mark 1 (T): Development and modifications of a British Service arm for snipers" in *The American Rifleman*, pages 42–45, November 1964.

Rocketto, Hap, "A Short History of Sniping" in *Tactical Shooter*, pages 30–34, 36–37, 39–45, Volume 1, No. 5, June 1998.

Rosenbarger, Matt, "FBI-Spec FN A3G Special Police Rifle" in *Tactical Response*, pages 48–50, 52, Volume 3, No. 5, Winter 2004.

Ryan, Rodney D., and Hugo Teufel, "The M82A1 Special Application Scoped Rifle" in *Tactical Shooter*, pages 79–81, Volume 2, No. 1, February 1999.

Saepaw, Ron, "The Marine Corps Own: The M40-A1 Sniper Rifle" in *Gung-Ho*, pages 10–11, 14–15, 52, 54, October 1981.

Sanow, Ed, "Ammunition for Tactical Operations" in *Tactical Response*, pages 53–54, 56–57, Volume 3, Number 5, Winter 2004.

Savino, Richard, "AMAC/Iver Johnson .50 Browning" in *Special Weapons for Military and Police*, pages 68–72, Volume 7, No. 1, 1990.

———, "The Barrett 'Light .50'" in *Special Weapons for Military and Police*, pages 34–38, Volume 7, No. 1, 1990.

Scheer, LTC Joseph P., "New Sniper Rifle From H&K" in *Gung-Ho*, pages 62–63, September 1982.

———, "Walther's Big Boomer .300 Magnum Semi Auto Sniper Rifle" in *Gung-Ho*, pages 16–19, 72, 74, July 1988.

Schreier, Konrad F., Jr., "Mauser's WWI Antitank Rifle" in *Gun Digest*, pages 65–68, 52nd Edition, 1998.

Senich, Peter R., "Scharfschutzen—The Master Mechanic" in *Gung-Ho*, pages 22–25, 60–61 December 1982.

Senovich, Peter R., ".50 Caliber Sniping: 2000 Yard Kills" in *Soldier of Fortune*, pages 46–51, 67–71, September 1978.

Serven, James E., "40-Rod Guns Praised for Precision" in *The American Rifleman*, pages 36–38, April 1971.

———, "The Revolving Cylinder Rifle" in *The American Rifleman*, pages 38–39, 73–74, 76, September 1984.

Shceglov, Alexandra, "Women Snipers in Russia" in *Gung-Ho*, pages 22–25, August 1981.

Shults, Jim, editor-in-chief, "The U.S. Army Sniper Program" in *Gung-Ho*, pages 22–31, July 1988.

———, "Big Brass Busters: Two New Super-Caliber Special Purpose Rifles That Can Do" in *Gung-Ho Special Official Weapons Handbook*, pages 32–43, Special #3.

———, "Soviet Mystery Rifle: Shooting the Dragunov (SVD) Sniper Rifle" in *Gung-Ho*, pages 34–37, 58, June 1981.

Simpson, Ross, "Cross Hairs on Baghdad: USMC Scout Snipers Wait to Reach Out and Touch Someone" in *Soldier of Fortune*, pages 44–47, March 1991.

Staff, "Sniper Support of Special Light-Infantry" in *Gung-Ho*, pages 14–18, 20–22, 24–25, 64–66, April 1989.

Stanford, Andy, "SEAL Sniper: Fusion of Man and Machine" in *Guns Magazine 2002 Combat Annual*, pages 52–61.

———, "Modern Marine Marksmanship" in *Guns Magazine 1999 Combat Annual*, pages 92–99.

Stevens, Richard, "Role of the Sniper in Modern Combat" in *New Breed*, pages 48–51, 56–57, February 1985.

Switlik, Matthew C., "Shooting the Famous Ferguson Rifle" in *The American Rifleman*, pages 38–41, August 1971.

Sword, Wiley, "Selecting a Sniper Arm: The Union's Hard Quest" in *The American Rifleman*, pages 22–24, April 1975.

Taylor, John D., "The .338 Lapua Magnum Cartridge: Origin, Development and Future, Part I" in *Tactical Shooter*, pages 52–60, 62–66, Volume 1, No. 12, January 1999.

———, "The .338 Lapua Magnum Cartridge: Origin, Development and Future, Part II—The Men Behind the Cartridge" in *Tactical Shooter*, pages 22–25, Volume 2, No.1, February 1999.

Teufel, Hugo, "Military Match Cartridges and Their Use in Combat: A Brief History, Part I" in *Tactical Shooter*, pages 65–70, Volume 1, No. 7, August 1998.

———, "Military Match Cartridges and Their Use in Combat: A Brief History, Part II" in *Tactical Shooter*, pages 69, 72–75, 77–80, Volume 1, No. 10, November 1998.

———, "The M21 and M25 Semi-Automatic Sniper Weapons Systems" in *Tactical Shooter*, pages 51–55, Volume 1, No. 6, July 1998.

———, "The M24 Sniper Weapon System" in *Tactical Shooter*, pages 8–10, 12–16, Volume 2, No. 5, June 1999.

———, "The SVD: The Sniper Rifle of the Soviet Union and Russia" in *Tactical Shooter*, pages 66–71, Volume 2, No. 4, May 1999.

Tirador, Franco, "U.S. Army Sniping Into the 1990s" in *Gung-Ho*, pages 14–21, July 1988.

Truby, J. David, "Vietnam's Quiet Killers . . . Silencers in Combat" in *Gung-Ho*, pages 56–61, December 1984.

Venturino, Mike, "The Civil War's Rifled Musket: A turning point in battle tactics" in *International Combat Arms: Journal of defense technology*, pages 74–77, 93, January 1989.

Waite, M. D., "U. S. Sniping Rifles: A Survey of Arms and Telescopic Sighting Systems in the U. S. Services" in *The American Rifleman*, pages 44–49, June 1965.

Walker, Greg, "U.S. Army's New M24 Sniper Rifle" in *International Combat Arms: Journal of Defense Technology*, pages 64–69, March 1989.

Walsh, Don, Jr., "Silenced and Subsonic Sniper Systems" in *Gung-Ho Special Official Weapons Handbook*, pages 62–67, Special #7.

Walsh, Steve, "The Great Rifle Debate—Bolt Action Sniper Rifle: Then and now, the professional's choice" in *Soldier of Fortune's Fighting Firearms*, pages 21, 23, 25, Fall 1994.

Weller, Jack, "Civil War Minié Rifles Prove Quite Accurate" in *The American Rifleman*, pages 36–40, July 1971.

Wilson, Mike, "The SR 25" in *Tactical Shooter*, pages 8–14, 16–17, Volume 1, No. 11, December 1998.

Woodman, Major Lyman L., USAF, "Hiram Berdan: Chief of the Sharpshooters" in *The American Rifleman*, pages 11–13, 37–38, March 1951.

MANUALS, PAMPHLETS, AND OTHER OFFICIAL DOCUMENTS

The Confederate Field Manual for the Use of Officers on Ordnance Duty with photographic supplement, USA, 1862/1984.

"Germany—Discovered in Combat II: Sniping at Leading Elements" in *Intelligence Bulletin*, page 65, Volume III, No. 8, War Department, April 1945.

"Germany—In Brief: Sniper Score" in *Intelligence Bulletin*, page 63, Volume III, No. 10, War Department, June 1945.

"Japan—Breakneck Ridge: A Lesson in Jap Defensive Tactics, Snipers" in *Intelligence Bulletin*, page 5, Volume III, No. 8, War Department, April 1945.

"Japan—Combined Attu Reports on Japanese Warfare: Defensive Positions, Sniper and Observation Posts" in *Intelligence Bulletin*, Section I, 3, c., page 37, Volume II, No. 2, War Department, October 1943.

"Japan—Notes By U.S. Observers on Japanese Warfare: Tactics, Snipers" in *Intelligence Bulletin*, Section III, 3, g., page 65, Volume II, No. 2, War Department, October 1943.

Manucy, Albert, *Artillery Through the Ages*, U.S. Government Printing Office, 1949.

Naval Special Warfare Group-2, *SNIPER COI 1–93*, Camp Atterbury, Indiana, 12 April to 14 June 1993.

Operational Feasibility Test of Sniper Rifle Systems, Final Report, U.S. Army Infantry Board, Fort Benning, Georgia, February 1978.

Report R-1896, *Silencers: Principles and Evaluations*, Frankford Arsenal, Department of the Army, August 1968.

Sniper Operations and Equipment, Final Report, ACTIV Project no. ACG-87/67I, U.S. Army Concept Team in Vietnam, 1968.

SW010-AD-GTP-010, *Technical Manual for Small Arms and Special Warfare Ammunition*, Naval Sea Systems Command, April 1987.

SW010-AD-GTP-010, *Technical Manual: Small Arms and Special Warfare Ammunition*, Naval Sea Systems Command, May 1995.

TC 23-14, *Sniper Training and Employment*, U.S. Army, June 1989.

USN Small Arms (General Overview), Naval Sea Systems Command (NAVSEA), 2003.

FILMS AND VIDEOS

ADVANCED SNIPERCRAFT, L.O.T.I. Group Productions, Brentwood Home Video, West Village, California, 1997.

PRO SNIPER, L.O.T.I. Group Productions, Brentwood Home Video, West Village, California, 1994.

U.S.M.C. Scout/Sniper, Video Free America, Brentwood Home Video, West Village, California, 1997.

OTHER UNPUBLISHED SOURCES

Crum, Allen, Travis County, Texas, Witness Report, Offense No. M-968150, 2 August, 1966.

Martinez, Officer R., Supplementary Offense Report, Offense M968150, 1 August, 1966.

McCoy, Officer H., Supplementary Offense Report, Offense M968150, 1 August, 1966.

Paulson, Alan C., *1997 Silencer Trials*, Report BL-71, Bioengineering Laboratory, Conway, Arkansas, 1997.

INDEX

Kevin Dockery has been the armorer in the President's Guard under presidents Nixon and Ford, a radio broadcaster, a gunsmith, and an historian. He spent time in Iraq and Kuwait during Desert Storm as what he refers to as a "corporate mercenary." A noted military historian, he has written a number of books detailing the history of the Navy SEALs and the lives of the men who lived that history, including *The Weapons of Navy SEALs.* He has also written a number of firearms reference books. He currently lives in southeastern Michigan.